The Superfamily of
ras-Related Genes

NATO ASI Series

Advanced Science Institutes Series

A series presenting the results of activities sponsored by the NATO Science Committee, which aims at the dissemination of advanced scientific and technological knowledge, with a view to strengthening links between scientific communities.

The series is published by an international board of publishers in conjunction with the NATO Scientific Affairs Division

A	**Life Sciences**	Plenum Publishing Corporation
B	**Physics**	New York and London
C	**Mathematical and Physical Sciences**	Kluwer Academic Publishers
D	**Behavioral and Social Sciences**	Dordrecht, Boston, and London
E	**Applied Sciences**	
F	**Computer and Systems Sciences**	Springer-Verlag
G	**Ecological Sciences**	Berlin, Heidelberg, New York, London,
H	**Cell Biology**	Paris, Tokyo, Hong Kong, and Barcelona
I	**Global Environmental Change**	

Recent Volumes in this Series

Series A: Life Sciences

The Superfamily of
ras-Related Genes

Edited by
Demetrios A. Spandidos

National Hellenic Research Foundation
Athens, Greece

Plenum Press
New York and London
Published in cooperation with NATO Scientific Affairs Division

Proceedings of a NATO Advanced Research Workshop on
The Super-Family of *ras*-Related Genes,
held May 17–21, 1991,
in Aghia Pelagia, Crete, Greece

Library of Congress Cataloging-in-Publication Data

NATO Advanced Research Workshop on the Super-Family of Ras-Related
Genes (1991 : Hagia Pelagia, Greece)
    The superfamily of ras-related genes / edited by Demetrios A.
Spandidos.
        p.   cm. -- (NATO ASI series. Series A, Life sciences ; v.
220)
    "Published in cooperation with NATO Scientific Affairs Division."
    "Proceedings of a NATO Advanced Research Workshop on the Super
-Family of Ras-Related Genes, held May 17-21, 1991, in Aghia
Pelagia, Crete, Greece"--T.p. verso.
    Includes bibliographical references and index.
    ISBN 0-306-44085-7
    1. Ras oncogenes--Congresses.   I. Spandidos, Demetrios.
II. North Atlantic Treaty Organization. Scientific Affairs
Division. III. Title. IV. Series.
    [DNLM: 1. Genes, ras--congresses. 2. Neoplasms--etiology-
-congresses.   QZ 202 N27774s 1991]
RC268.44.R37N38 1991
616.99'4--dc20
DNLM/DLC
for Library of Congress   91-45146
                                                           CIP

ISBN 0-306-44085-7

© 1991 Plenum Press, New York
A Division of Plenum Publishing Corporation
233 Spring Street, New York, N.Y. 10013

Printed in the United States of America

PREFACE

Since the earliest results on the linkage between <u>ras</u> activation and cell transformation appeared a vast amount of additional information has been generated which emphasizes the importance of <u>ras</u>-related genes in membrane trafficking, cell proliferation, differentiation and cancer. These advances led to the development of new strategies for the diagnosis, prognosis and prevention of cancer. <u>Ras</u>-related genes appear to be central to the mechanism of transformation by other oncogenes and therefore constitute a focal point upon which attempts to intervene in the process of carcinogenesis will be concentrated.

The present volume contains the contributions to the NATO Advanced Research Workshop on the "The Superfamily of <u>ras</u>-Related Genes" held in Agia Pelagia, Heraklion, Crete, Greece, May 1991. At this meeting the leading researchers in this field came together to discuss the most recent progress in analysis of biological function of the <u>ras</u>-related genes. As an organizer I feel that the Workshop was a success, and would like to thank all the participants and the contributors to this volume. I thank the NATO Scientific Affairs Division, Brussels, the Greek Ministry of Industry, Energy and Technology, the International Center for Cancer Research, the National Hellenic Research Foundation and the Medical School, University of Crete for sponsoring the Meeting.

DEMETRIOS A. SPANDIDOS

CONTENTS

TIME-RESOLVED BIOCHEMICAL STUDIES OF RAS PROTEINS BY FLUORESCENCE MEASUREMENTS ON TRYPTOPHAN MUTANTS

Bruno Antonny, Michel Roux, Marc Chabre
and Pierre Chardin

CNRS - Institut de Pharmacologie Moléculaire et Cellulaire
660 route des Lucioles, Sophia Antipolis, 06560 Valbonne, France

ABSTRACT

We have replaced leucine 56 of ras by a tryptophan. The intrinsic fluorescence of this tryptophan was used as an internal conformational probe for time-resolved biochemical studies of the ras protein. Tryptophan fluorescence of mutated ras is very sensitive to magnesium binding, GDP/GTP exchange and GTP hydrolysis. Nucleotide affinities, exchange kinetics and intrinsic GTPase rates of the substituted ras are very close to those of wild-type ras. The SDC 25 gene product enhances GDP/GTP exchange. GAP accelerates GTP hydrolysis by a factor of at least 10^4. A slow fluorescence change follows the binding of GTPγS, its kinetics are close to those of the intrinsic GTPase, suggesting that a "pre-transition" preceeds the GTPase and is the rate limiting step, as proposed by Neal et al. (1990). However, GAP does not accelerate this slow conformational change suggesting that the fast GAP-induced catalysis of GTP hydrolysis bypasses this step and might proceed of a different mechanism. We have also studied another mutant where tyrosine 64 was replaced by tryptophan. The Y64W substitution has very little effects on intrinsic GTP hydrolysis and Y64W ras has a the same affinity for GAP than wild-type ras, however GAP is not able to increase GTP hydrolysis on this mutant, suggesting a role for tyrosine 64 in GAP-induced GTP hydrolysis. The implications of these observations on the mechanism of ras action, and a new model, are discussed.

The Superfamily of ras-Related Genes
Edited by D.A. Spandidos, Plenum Press, New York, 1991

INTRODUCTION

Biochemical studies of ras proteins mostly rely on the binding of radio-labelled guanine nucleotides and filter assays to determine nucleotide exchange and GTP hydrolysis. Other techniques such as the charcoal method to measure Pi release or thin layer chromatography to analyze the bound nucleotide are also used. However, the filter assays have wide standard deviations (usually around 20%) and the experiments need to be repeated at least three times to get decent curves. Furthermore these techniques are hardly applicable for fast measurements in the time scale of a few seconds. For time-resolved studies, one needs internal probes of conformational changes. Very nice studies have been performed with caged GTP and X-ray diffraction studies of the structural changes occuring upon GTP hydrolysis (Schlichting et al., 1990)). However this technique can only be applied for structural studies on crystallized proteins. To study conformational changes occuring on proteins in solution, one way is to introduce a fluorescent probe, since the fluorescence emission of several aromatic groups is very sensitive to the environment. A fluorescent group may be introduced on the guanine nucleotide to study conformational changes occuring on ras (Neal et al., 1990). One problem is that the GDP or GTP analogs with an introduced fluorescent group might not behave exactly as authentic GDP or GTP. We choose a slightly different approach by introducing the fluorescent probe in the ras protein itself. The fluorescence emissions of tyrosine and tryptophan are sensitive to their environment and may provide probes of the local conformation of proteins. Wild-type ras has no tryptophan, but contains 9 tyrosines, and their weak fluorescence changes slightly between the the GDP-bound state and the GTP-bound state. But, given the large number of tyrosines spread over the full length of the polypeptide chain, this change is hard to interpret as one can not identify unambiguously the sensitive tyrosine. Our approach was then to introduce a single tryptophan at a selected position in ras, by site directed mutagenesis, and to monitor the variations of its fluorescence emission upon nucleotide exchange or nucleotide hydrolysis. Tryptophan should be introduced in regions where major conformational changes occur, but this substitution should not modify the biochemical properties of the protein, and should not modify the interaction with other proteins involved in the study (GAP, Exchange Factor, others?)

There is no tryptophan in p21 ras, but in the ras superfamily, the rho and rab proteins all have one conserved tryptophan (Zahraoui et al., 1989), at position 56, that is close to the β and γ phosphates (Valencia et al., 1991). Preliminary fluorescence measurements on rab1 demonstrated that the fluorescence of this tryptophan decreases by 12% upon the GDP/ GTP exchange. This suggested to introduce in ras a tryptophan at position 56, by mutating a leucine.

More sensitive loci for a sensor of conformation could be expected in segments 30-38 and 60-76 where the largest differences between the GDP and GTP structures are seen (Milburn *et al.*, 1990, Schlichting *et al.*, 1990). However these two regions are very sensitive to amino acid substitutions, and many positions can not be changed without altering the biochemical properties of the protein, and/or the interaction with target proteins (GAP, Exchange Factors, etc...). Therefore we choose to introduce only conservative substitutions, i.e. substitute a tryptophan only at positions where an aromatic residue was already found, at tyrosines 32, 64 or 71. We knew that tyrosine 32 was very sensitive to substitutions, since a phenylalanine substitution altered the transforming potential of v-H-ras (Stone *et al.*, 1988). None of the residues in segment 62-76 interact directly with the phosphate oxygens of the bound GTP, their mutation might therefore not interfer with the mechanism of GTP hydrolysis. Tyrosine 71 side chain is buried inside, tyrosine 64 (Y64) appeared as the best candidate to be mutated into a tryptophan.

Both mutations: leucine 56 to tryptophan (L56W) and tyrosine 64 to tryptophan (Y64W) were introduced by site directed mutagenesis and mutant ras proteins produced in E. Coli.

RESULTS

In all our experiments, the ras proteins are initially in the "basal state", that is with a GDP and a magnesium ion bound.

Intrinsic fluorescence emission spectra of wild-type ras and mutants

Wild-type ras fluorescence exhibits the characteristics of pure tyrosine fluorescence which is maximal for an excitation around 275 nm and has an emission maximum around 305 nm. In (L56W)ras or (Y64W)ras, the fluorescence of the single tryptophan is much more intense than that of the 9 or 8 tyrosines. The fluorescence is maximal for an excitation at 280 nm, where both tryptophan and tyrosine absorb, but the emission maximum at 345 nm is characteristic of tryptophan. At 295 nm excitation the tyrosines do not absorb anymore and pure tryptophan fluorescence can be monitored on both ras mutants, with however lower emission yields than for the 280 nm excitation.

Tyrosine fluorescence of wild-type ras is sensitive to magnesium and to the bound nucleotide

Tyrosine fluorescence emission on wild type ras was monitored at 305 nm, (30 nm bandwidth), with excitation at 270 nm (1.5 nm bandwidth). Fluorescence emission of tyrosine is usually much less sensitive to conformation than that of tryptophan. The presence of 9 tyrosines in wild-type ras, only a few of them being in, or near, the "switch" domains, was expected to dilute the effect of a localized conformational change. However tyrosine fluorescence emission was sensitive to the type of nucleotide bound into the site and to the presence of magnesium. Starting from ras-GDP in the presence of an excess free GDP and 1mM MgCl$_2$, the addition of 2 mM EDTA, which reduces the free magnesium concentration below 10^{-6} M, induces an instant 5% increase of the fluorescence, that fully reverses when magnesium is added back. If instead of GDP, an excess GTP is present in the medium, the exchange of the initially bound GDP for GTP during the magnesium free period induces only a very small fluorescence increase after the initial step, but the fluorescence does not decrease back to the initial level when the magnesium is added back after completion of the exchange. Surprisingly, the exchange of GDP for GTPγS induces a much larger change of fluorescence than does GTP. The high fluorescence level with GTPγS bound is unsensitive to magnesium, as is the fluorescence of the GTP-bound state. This indicates that a tyrosine, probably Y32, can distinguish between GTP and GTPγS and also indicates that the conformation of ras-GTPγS is not strictly identical to that of ras-GTP. The fluorescence of the GDP-bound form is much more sensitive to magnesium than that of the GTP-bound form.

Tryptophan fluorescence of the mutants is sensitive to magnesium and to the bound nucleotide

The two ras mutants with a tryptophan at position 56 or 64 provide much more sensitive fluorescence probes of the conformational changes related to the exchange or hydrolysis of the bound nucleotide. In (L56W)ras and (Y64W)ras proteins too, the presence of magnesium influences the tryptophan fluorescence: on excitation at 295 nm, in both mutants the removal of magnesium induces an instant 10% change of the fluorescence emission at 345 nm: a decrease for (Y64W)ras-GDP and an increase for (L56W)ras-GDP. Then, the fluorescence remains stable at this new level if there is only GDP in the medium, and will come back to the initial level when magnesium is added back. In the presence of GTP or GTPγS, after the initial fluorescence jump due to the removal of the bound magnesium, the tryptophan fluorescence in low magnesium evolves within about a minute: it decreases by about 20% for (Y64W)ras and increases by about 10% for (L56W)ras. This evolution reflects the rapid exchange of GDP for GTP or GTPγS in low magnesium. The time course of this exchange is not accelerated by higher concentrations of GTP or GTPγS, it is probably rate limited by the off-rate of the GDP. After the exchange is completed in low magnesium, the fluorescence level of ras remains stable, both with a bound GTP, which does not hydrolyse in ras in the absence of magnesium, and with a bound GTPγS. Adding magnesium back induces only very small fluorescence changes that are not the reversal of the changes that had been initially induced by the removal of magnesium from the corresponding ras-GDP complex. In the presence of magnesium, with excitation at 280 nm, the fluorescence emission levels are almost identical for the GTP and GTPγS-bound forms and differ from that of the GDP-bound form by a 12% increase for the (L56W)ras and a 32% decrease for (Y64W)ras. These results show that fluorescent changes can monitor GDP/GTP exchange and that the fluorescence of the GDP-bound form is much more sensitive to magnesium than that of the GTP-bound form.

Large changes of tryptophan fluorescence monitor the GTPase

With both mutants, after GDP/GTP exchange has been completed in low magnesium, when magnesium is added back the fluorescence level of ras-GTP slowly evolves towards the level initially measured for ras-GDP in the presence of magnesium: the fluorescence of (L56W)ras-GTP decreases, and that of (Y64W)ras-GTP increases. The time course of this fluorescence change reflects the time course of the GTPase. In both mutants the evolution is complete within 2 hours at 37°C. This evolution results from the hydrolysis of GTP in ras, which recovers its GDP-bound conformation.

Nucleotides affinities, exchange kinetics and intrinsic GTP hydrolysis rate of the mutants are very similar to wild-type

The large fluorescence changes that correlate with nucleotide exchange in low magnesium, or with the GTPase in high magnesium allowed us to follow the kinetics of these processes and to estimate the relative affinities and kinetic constants of dissociation or hydrolysis of the nucleotides in the two mutants.

The characteristics of (L56W)ras are very close to that of wild-type ras (John *et al.*, 1988, 1989). Slightly lower intrinsic GTPase rate and nucleotide k_{off} rates are found for (Y64W)ras whose characteristics are however closer to that of a normal ras than to that of any of the known oncogenic mutants.

The mutation of L_{56} that is located rather internally in the protein structure, and does not interact directly with the bound nucleotide, does not modify the binding constants, exchange and hydrolysis of the nucleotide in ras. It also preserves the functional interactions of ras with the exchange protein derived from the SDC 25 gene product and with the GAP protein (see below). The mutation of Y_{64} is not totally harmless, it modifies a little the nucleotide binding, exchange and hydrolysis constants, although Y_{64} is not interacting directly with the nucleotide

(Pai *et al.*, 1990). But Y_{64} does not seem really critical for the binding of the nucleotides in ras and for the mechanism of catalysis of intrinsic GTP hydrolysis. The mutation suppresses the action of GAP (see below), suggesting that the switch II domain also takes part in the interaction with GAP and is essential for GAP induced GTP hydrolysis, as already suggested by Srivastava *et al.* (1989).

A slow conformational change occurs in ras-GTPγS

If GTPγS rather than GTP has replaced GDP into the site, when magnesium is added back a slow and partial reversal of the fluorescence level towards that of the GDP state is still observed for both mutants. The time course of this fluorescence change is the same as that of the GTPase. But for (L56W)ras the amplitude of the fluorescence change following GTPγS binding is about 40% of that observed with GTP, and for (Y64W)ras only 15%. As for the GTPase, magnesium is required for this fluorescence change. Without magnesium, the ras-GTPγS fluorescence is as stable as that of ras-GDP. We checked that the fluorescence evolution was not due to the small contamination of 1% GTP in the GTPγS: doubling this proportion of GTP in the GTPγS did not increase significantly the amplitude of the fluorescence change. The evolution of fluorescence is not due either to the hydrolysis of the GTPγS bound to ras, which is very slow, does not saturate after three hours and was detectable only over a much longer time scale.

The fluorescence evolution must then reflect a slow conformational change that follows the binding of a non-hydrolysable analog of GTP in ras, and occurs independently of GTP hydrolysis. It is probably the conformational change which had been already detected in wild-type ras with fluorescent-labelled GPP-NH-P (mant-GPP-NH-P) and which was suggested to set a rate limiting step for the intrinsic GTPase (Neal *et al.*, 1990).

As magnesium was required into the site to allow the evolution to proceed after the binding of GTPγS, we checked the effect of removing the magnesium after the evolution was complete: would it suppress the effect of the evolution and regenerate the initial state? The addition of EDTA after 90 minutes of evolution of (L56W)ras-GTPγS induces an instant fluorescence change, revelative of the rapid release of magnesium. If magnesium is added back after only a few seconds in EDTA, the fluorescence reverts to the level observed just before the addition of EDTA, not to the inital fluorescence level of rasGTPγS. But if one remains longer with EDTA, a rather fast change is observed which is characteristic of a nucleotide exchange, although here the bound GTPγS can only be exchanged for excess free GTPγS present in the solution. The kinetic constant of this exchange is the same as that observed for a GTPγS/GDP exchange rather than that of a GDP/GTPγS exchange: the rate limiting step must be the k_{off} of the initially bound

GTPγS. After this fast evolution in EDTA is completed, upon the re-addition of magnesium the fluorescence recovers the level it had in the initial ras-GTPγS conformation. This shows that the conformational change is not reversed by the transient release of magnesium, but is fully reversed by the transient release of GTPγS.

GAP does not accelerate the slow conformational change

No effect of GAP was detectable on the fluorescence evolution of the GTPγS-bound form of either of the two ras mutants. This could be expected for (Y64W)ras, since GAP does not accelerate its GTPase (see below). But even on (L56W)ras the slow evolution of fluorescence of the GTPγS-bound form was not modified by concentrations of GAP that accelerated by several orders of magnitudes the GTPase rate in the GTP-bound form of the same ras mutant. The only detectable effect of GAP on ras-GTPγS was to accelerate the hydrolysis of GTPγS, but this was detected only at the time scale of hours. The slow change of conformation of the ras-nucleoside triphosphate complex that had been recently detected in wild type ras complexed with a fluorescent analog of GPP-NH-P (Neal *et al.*, 1990), has been observed here directly through an evolution of the fluorescence of the tryptophan in both (L56W)ras and (Y64W)ras, following the binding of GTPγS. The structural change must therefore concern the switch II domain, that is loop L4 and Helix H2. The corresponding variations of tyrosine fluorescence could not be assessed in wild-type ras, since tyrosine fluorescence is too weak, so we cannot determine whether the conformation of the switch I domain (loop L2) is also concerned. The conformational change of the ras-GTPγS complex having approximately the same time course as GTP hydrolysis, it has been suggested (Neal *et al.*, 1990) that this transition is an obligatory conformational step in the catalytic mechanism of GTP hydrolysis, and that its slow time course rate-limits the intrinsic GTPase of ras. But our observation that concentrations of GAP that accelerate considerably the GTPase rate of (L56W)ras-GTP do not modify the "pre-transition" observed with the GTPγS complex of the same mutant, suggests that the action of GAP is not to accelerate the conformational step presumed to rate-limit the intrinsic GTPase. The fast hydrolysis of GTP in the ras/GAP complex must then bypass this slow conformational change. Either this conformational change is not really a step of the intrinsic GTPase mechanism, or the mechanism of the fast GTP hydrolysis in the ras/GAP complex proceeds through a catalytic pathway different from that responsible for the slow GTP hydrolysis in isolated ras.

Stimulation of GDP/GTP exchange by the carboxyl terminal domain of the yeast SDC25 gene product

The carboxyl-terminal domain of the yeast SDC25 gene product accelerates the exchange of GDP for GTP in the RAS 2 protein of *S. Cerevisiae* and promotes the same effect on human ras (Crechet *et al.*, 1990). The effect of a partially purified preparation of this exchange factor (kindly provided by A. Parmeggiani) was tested on both mutants by monitoring the fluorescence change correlated with the exchange of GDP for GTPγS in the presence of magnesium. Both mutants are sensitive to the exchange factor which could lead to half maximum exchange in about 30 minutes. Similar results were obtained for GDP/GTP exchange, but the data evaluation are complicated by the hydrolysis of the bound GTP which, in the presence of millimolar magnesium, is not neglegible on this time scale. Neither the mutation L56W, nor Y64W, suppressed the nucleotide exchange enhancement induced by the yeast SDC25 gene product. It will be of interest to test the effect of the mammalian ras exchange factors that have been recently described (Huang *et al.*, 1990; Downward *et al.*, 1990; Wolfman and Macara, 1990).

GAP accelerates the GTPase of (L56W)ras but not that of (Y64W)ras

GAP is a 115 kD protein that increases dramatically the rate of GTP hydrolysis in wild-type ras (Trahey and Mc Cormick, 1987). The GTPase activating capacity is localised in the carboxyl terminal domain of the protein (Marshall *et al.*, 1989), and is conserved in a 45 kD fragment preparation (GAPette) that was kindly provided to us by M. Frech and A. Wittinghofer. The hydrolysis rate of GTP in a micromolar solution of (L56W)ras (as monitored by the decay of the tryptophan fluorescence) was considerably increased by nanomolar concentrations of GAP or of "GAPette". In a buffer devoid of monovalent ions, at 37°C, with 1 µM (L56W)ras-GTP and 10 nM GAP, the rate was of 0.6 nM GTP hydrolyzed/sec/nM GAP, that is comparable to the rate observed under similar conditions for the catalysis of GTP hydrolysis in wild-type ras by this GAP preparation (M. Frech, personal communication). The catalytic activity of GAP (or GAPette) is reduced by a factor of about 2.5 when 120 mM KCl (that approximates the cytoplasmic concentration) are added to the buffer. This effect is not specific to K^+ cations or Cl^- anions, similar effects are observed with equivalent concentrations of NaCl or KNO_3.

No effect of GAP or of GAPette on the GTPase rate of the mutant (Y64W)ras protein was detectable by fluorescence. However (Y64W)ras-GTPγS interacts with GAP and is able to compete with wild-type ras-GTP for GAP interaction: using exactly the assay described by Frech *et al.* (1990) we found that the affinity of (Y64W)ras-GTPγS for GAP is similar or slightly higher (2-3 µM) than the affinity of wild-type ras-GTPγS. The affinity of (L56W)ras-GTPγS for GAP (5 µM) is the same as that of wild-type ras-GTPγS. These results indicate that the region around position 64 is also involved in GAP interaction and is essential for GAP-induced GTP hydrolysis.

ras/GAP interaction and catalytic constant of GTP hydrolysis

The rate of GTP hydrolysis induced by 10 nM GAP in (L56W)ras is strictly linear with concentrations of ras at least up to 10 μM. This agrees with previous measurements on wild-type p21 ras by the classical methods using ^{32}P labelled GTP (Vogel *et al.*, 1988). But competition experiments performed with excess cold ras-GTP or cold ras-GPP-NH-P to inhibit the rate of GAP-induced GTP hydrolysis in labelled ras GTP have given $IC_{50\%}$ values of 110 μM for ras-GTP (Vogel *et al.*, 1988) and of only 5-10 μM for ras-GPP-NH-P (Frech *et al.*, 1990; Krengel *et al.*, 1990). Both experiments had been interpreted as measuring the K_d of ras/GAP interaction, and the apparent discrepancy was attributed to the different origins of the GAP preparations. The sensitivity of mutated (L56W)ras to GAP opened the use of the tryptophan fluorescence method for a time-resolved analysis of ras/GAP interaction. The specific monitoring of GTP hydrolysis in the tryptophan-tagged mutant allowed competition studies with a large excess of wild-type ras, taking advantage of the fact that wild-type ras will not be detected in our tryptophan fluorescence assay. This helped to point-out the origin of discrepancies between the apparent affinities of GAP for ras-GTP (Vogel *et al.*, 1988) and for ras-GPP-NH-P (Frech *et al.*, 1990; Krengel *et al.*, 1990).

We find that the competition by 10 μM wild-type ras-GTPγS inhibits by more than 50% the GAP-induced GTPase in (L56W)ras-GTP, but the same amount of wild type ras-GTP or ras-GDP inhibits by about 10% only .

The k_{cat} for GTP hydrolysis in the ras-GTP/GAP complex is very high, much higher than the rate of dissociation in ras-GTP + GAP. The fast GTP hydrolysis and the quick subsequent dissociation of the ras-GDP/GAP complex in ras-GDP + GAP reduce the apparent affinity of GAP for ras-GTP. The true affinity of GAP for ras-GTP is probably in the order of 10 μM, as measured with ras-GTPγS, but is never observable. The high values of k_{cat} and low affinity of GAP for ras-GDP insure the unidirectionality of the switch.

DISCUSSION

We have developed a new ras mutant, L56W, that behaves as normal ras in every respect and allows time-resolved studies of conformational changes occuring in p21 ras. This enabled us to study guanine nucleotide exchange and to confirm results obtained by other methods. GDP release *in-vitro*, in a physiological buffer, is extremely slow (> 1 hour) and not compatible with physiological functions, indicating that, *in-vivo*, GDP release must be promoted by an "exchange factor". There's no need for a GDP Dissociation Inhibitor (GDI), but a protein that specifically recognizes GDP-bound ras might have GDI properties, as a side-effect. We also found that GAP accelerates GTP hydrolysis by a factor of at least 10^4. However, GAP does not accelerate the slow conformational change that seems to be a rate-limiting step for GTP hydrolysis on isolated ras-GTP. Furthermore there is no strict correlation between an impairment of GAP-induced GTP hydrolysis and a low intrinsic GTP hydrolysis rate. For instance, the Y64W substitution has very little effects on intrinsic GTP hydrolysis and (Y64W)ras has the same affinity for GAP than wild-type ras, but GAP is not able to increase GTP hydrolysis on (Y64W)ras-GTP. These results indicate that the mechanisms of GTP hydrolysis in isolated ras or in the ras-GAP complex might be different.

Taken together these results also strongly suggest that GTP hydrolysis on ras, *in-vivo* and on a physiological time scale, occurs only after the interaction with GAP. By analogy with elongation factors, we hypothesize that GTP hydrolysis monitors the proper interaction of ras-GTP with GAP or the proper formation of a complex between ras-GTP, GAP and a third component, the identity of which is still unknown. We made several other hypothesis to build a model for ras functional mechanism.

We postulate that i) like all other G-proteins, ras-GDP and ras-GTP interact with different targets, the specific target of the GDP-bound form would be a GDI-like protein and the specific target of the GTP-bound form a GAP-like protein, ii) an exchange factor is needed to promote GDP release on ras-GDP iii) there is a membrane protein interacting with ras-GDP to stabilize its membrane association iv) ras-GTP promotes the association of GAP with another protein the nature of which is still unknown.

For reasons of simplicity, and because ras is a small protein, with a single domain that can not interact with many proteins at the same time, we postulate that the protein stabilizing membrane association and the exchange factor are one and the same. We also postulate that this protein does not dissociate from ras after GDP/GTP exchange has been completed, and that ras promotes the subsequent association of this protein with GAP. We have drawn a highly speculative but precise model for ras function that accounts for all these assumptions.

We suggest that ras-GDP has a specific target: A, a GDI-like protein that has not yet be discovered but might be seen as an element of the sub-membrane cytoskeleton. The ras-GDP/A complex would be recognized by a membrane protein: B, stabilizing membrane association. The ras-GDP/A/B complex would be the major resting state of ras in the cell. A signal: S, the nature of which is still unknown but might be the binding of a small ligand or the interaction with another protein, induces a conformational change of B, and S-B* is able to promote GDP release, GTP rapidly occupies the empty site and A is released. The ras-GTP/S-B* complex would then rapidly recognize C, a GAP-like protein. When this ras-GTP/S-B*/C complex is formed GTP hydrolysis rapidly takes place, ras-GDP is released, C changes conformation to C* and a stable S-B*/C* complex is formed.

In this model, the normal function of ras is to control lateral segregation of

1°) 1st complex formation, Exchange, 1st complex dissociation

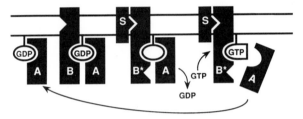

2°) 2nd complex formation, GTP hydrolysis

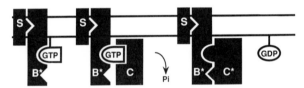

3°) 2nd complex dissociation

Figure 1.

multi-protein complexes at the plasma membrane, via the formation of the ras-GDP/A/B, and S-B*/C* complexes. Mutant ras with impaired GTP hydrolysis upon GAP interaction would impair the dissociation of the ras-GTP/S-B*/C complex and block the cycle at this step that is normally very transient; then all the other steps of the cycle would be completely disturbed. Some of the proteins in these multi-protein complexes are expected to be essential components of already characterized signal transduction pathways, for instance S-B* (or an associated protein) might be a component of the "seven transmembrane domain receptor"/ heterotrimeric G-protein signal transduction system, and C (GAP or a GAP- like protein) might provide a link with the growth factor / tyrosine-kinase signalling system. A large body of evidence indicates that lateral segregation of the receptors and associated proteins in different membrane compartments (also termed receptor patching) controls signal transduction. We suggest that ras plays a major role in this process (Chardin, 1991). This model provides ideas on the mechanism of ras function, it is of course highly speculative. The specific target of the GDP-bound form, the membrane "receptors" and/or "exchange factors" and their regulatory signals, remain to be characterized.

Acknowledgments

We thank Ahmed Zahraoui for a sample of rab1 protein used in preliminary test experiments, Matthias Frech and Alfred Wittinghofer for the gift of GAP and GAPette preparations, J. B. Crechet and A. Parmeggiani for the gift of SCD 25 protein. We also thank Alfred Wittinghofer for very stimulating discussions.
This work was supported by a grant from the Association pour la Recherche sur le Cancer. P. C. is supported by INSERM.

REFERENCES

Barbacid, M. (1987) *Ann. Rev. Biochem.*, **56**, 779-827.

Bourne, H.R., Master, S.B., Miller, R.T., Sullivan K.A. and Heideman, V.Y. (1988) *Cold Spring Harbor Symp. Quant. Biol.*, **53**, 221-227.

Chardin, P. (1991) *Cancer cells*, **3**, 117-126.

Crechet, J.-B., Poullet, P., Mistou, M.-Y., Parmeggiani, A., Camonis, J., Boy-Marcotte, E., Damak, F. and Jacquet, M. (1990) *Science*, **248**, 866-868.

Downward, J., Riehl, R., Wu, L. and Weinberg, R.A. (1990) *Proc.Natl.Acad.Sci. USA*, **87**, 5998-6002.

Frech, M., John, J., Pizon, V., Chardin, P., Tavitian, A., Clark, R., McCormick, F. and Wittinghofer, A. (1990) *Science*, **249**, 169-171.

Halenbeck, R., Crosier, W.J., Clark, R., McCormick, F. and Koths, K. (1990) *J. Biol. Chem.*, **265**, 21922-21928.

Hall, A. (1990a) *Science*, **249**, 635-640.

Hall, A. (1990b) *Cell*, **61**, 921-923.

Higashijima, T., Ferguson, K. M., Sternweis, P. C., Ross, T. M. and Gilman, A. G. (1987) *J. Biol. Chem.* **262**, 752-756.

Huang, Y.K., Kung, H.-F. and Kamata, T. (1990) *Proc.Natl.Acad.Sci. USA*, **87**, 8008-8012.

John, J., Frech, M. and Wittinghofer, A. (1988) *J. Biol. Chem.*, **263**, 11792-11799.

John, J., Schlichting, I., Schiltz, E., Rösch, P. and Wittinghofer, A. (1989) *J. Biol. Chem.*, **264**, 13086-13092.

Krengel, U., Schlichting, I., Scherer, A., Schumann, R., Frech, M., John, J., Kabsch, W., Pai, E. and Wittinghofer, A. (1990) *Cell*, **62**, 539-548.

Marshall, M., Hill, W., Ng, A., Vogel, U., Schaber, M., Scolnick, E., Dixon, R., Sigal, I. and Gibbs, J. (1989) *EMBO J.*, **8**, 1105-1110.

Mc Cormick, F. (1989) *Cell*, **56**, 5-8.

Milburn, M.V., Tong, L., deVos, A.M., Brünger, A., Yamaizumi, Z., Nishimura, S. and Kim, S.-H. (1990) *Science*, **247**, 939-945.

Neal, S.E., Eccleston, J.F. and Webb, M.R. (1990) *Proc.Natl.Acad.Sci. USA*, **87**, 3562-3565.

Pai, E.F., Kabsch, W., Krengel, U., Holmes, K., John, J. and Wittinghofer, A. (1989) *Nature*, **341**, 209-214.

Pai, E.F., Krengel, U., Petsko, G.A., Goody, R.S., Kabsch, W. and Wittinghofer, A. (1990) *EMBO J.*, **9**, 2351-2359.

Schlichting, I., Almo, S.C., Rapp, G., Wilson, K., Petratos, K., Lentfer, A., Wittinghofer, A., Kabsch, W., Pai, E.F., Petsko, G.A. and Goody, R.S. (1990) *Nature*, **345**, 309-315.

Srivastava, S., Di-Donato, A. and Lacal, J.-C. (1989) *Mol. Cell. Biol.*, **9**, 1779-1783.

Stone, J., Vass, W., Wilumsen, B. and Lowy, D. (1988) *Mol. Cell. Biol.*, **8**, 3565-3569.

Trahey, M. and Mc Cormick, F. (1987) *Science*, **238**, 542-545.

Tucker, J., Sczakiel, G., Feuerstein, J., John, J., Goody, R.S. and Wittinghofer, A. (1986) *EMBO J.*, **5**, 1351-1358.

Valencia, A., Chardin, P., Wittinghofer, A. and Sander, C. (1991) *Biochemistry*, in press.

Vogel, U., Dixon, R., Schaber, M., Diehl, R., Marshall, M., Scolnick, M., Sigal, I. and Gibbs, J. (1988) *Nature*, **335**, 90-93.

Wolfman, A. and Macara, I.G. (1990) *Science*, **248**, 67-69.

Zahraoui, A., Touchot, N., Chardin, P. and Tavitian, A. (1989) *J. Biol. Chem.*, **264**, 12394-12401.

PRENYLATION OF RAS AND RAS-RELATED PROTEINS

J. F. Hancock, K. Cadwallader, [+]A. I. Magee,
[+]C. Newman, [+]T. Giannakouros, [‡]E. Fawell,
[‡]J. Armstrong, [#]H. F. Paterson and [#]C. J. Marshall

Dept Haematology, Royal Free Hospital School of
Medicine, London NW3
[+]NIMR, London NW7. [‡]ICRF, London W1
[#]Chester Beatty Laboratories, London SW3

INTRODUCTION

In common with certain other cellular proteins, a CAAX motif (C=cysteine, A=Aliphatic, X=any amino acid) is found at the C-terminus of all *ras* proteins. This motif undergoes a triplet of closely coupled post-translational modifications. First, a prenoid derivative is linked as a thioether to the cysteine residue (Hancock *et al*,1989; Casey *et al*,1989), second, the -AAX amino acids are removed by proteolysis (Gutierriez *et al*,1989) and third, the α-carboxyl group of the now C-terminal cysteine residue is methyl-esterified (Clarke *et al*,1988; Gutierrez *et al*,1989). Although the p21[ras] proteins, the nuclear lamins A and B and the γ-subunit of transducin (Fukada *et al*,1990: Lai *et al*,1990) are all prenylated with C_{15} farnesyl, prenylation of proteins with C_{20} geranylgeranyl is some 10x more common than farnesylation (Epstein *et al*,1990). Recently, certain CAAX containing proteins have been identified which are geranylgeranylated, these include the γ-subunits of brain G-proteins (Yamane *et al*,1990; Mumby *et al*,1990) and the *ras*-related proteins Krev1/rap1A (Kawata *et al*,1990; Buss *et al*,1991) and G25K (Maltese and Sheridan,1990). A common feature of the CAAX motifs of these C_{20} modified proteins is the presence of a leucine residue in the X position. This raises the possibility that the X amino acid determines which isoprenoid residue is used to prenylate a CAAX motif. In addition to the prenylation of CAAX boxes, it is likely that the *ras*-related *rab* proteins, which terminate in CC or CXC motifs, might also undergo prenylation since it is known that for these proteins the C-terminal cysteines are essential for function.

We shall discuss in this review recent results on the prenylation of *ras* and *ras*-related proteins from our own and other laboratories.

PLASMA MEMBRANE TARGETING BY POLYBASIC DOMAINS

One function of the post-translationally modified CAAX motif is to provide a signal for the subcellular localisation of proteins to cellular membranes. For example, the CAAX motif of p21^{ras} combines with a second signal contained within the C-terminal hypervariable domain to target plasma membrane localisation. This second signal comprises either a cysteine palmitoylation site in the case of p21^{H-ras}, p21^{N-ras} and p21$^{K-ras(A)}$, or a polybasic domain comprising six consecutive lysine residues (amino acids 175-180) in the case of p21$^{K-ras(B)}$. Mutant K-*ras(B)* proteins in which the six lysines are mutated to glutamine (K6Q), or mutant H-*ras* proteins which lack palmitoylation sites, are localised to the cytosol rather than the plasma membrane (Hancock *et al*,1990). Similarly the nuclear lamins A and B require both a CAAX motif and a nuclear localisation signal for correct targeting to the nuclear membrane (Holtz *et al*,1989). It is not clear from these observations whether a CAAX motif and the polybasic domain are sufficient for plasma membrane localisation. However, we have now shown that a polybasic domain can act in concert with a CAAX motif to target a heterologous protein, protein A, to the plasma membrane (JFHancock and CJMarshall, unpubl. data).

The polylysine domain could function in two ways. First, since the domain is positively charged at physiological pH, an electrostatic interaction with negatively charged phospholipid head groups may be important. Second, the domain may interact with a specific docking protein for p21$^{K-ras(B)}$ in the plasma membrane. If the latter model is true it is likely that replacing the lysine residues with other positively charged amino acids would compromise the function of the domain. We therefore tested a mutant p21$^{K-ras(B)}$ protein, K6R, which has 6 Lys → Arg substitutions at amino acids 175-180. The K6R protein localises normally to the plasma membrane in NIH3T3 and MDCK cells. Moreover the biological activity of an oncogenic mutant K6R (Gly12

→ Val) protein in focus assays on NIH3T3 cells is equal to, or perhaps marginally greater than, an oncogenic mutant p21$^{K-ras(B)}$ (Gly12 → Val) protein with a wild type polylysine domain (JFHancock and CJMarshall, in prep). In addition, when the C-terminal sequences of K6R are cloned onto the C-terminus of protein A the chimeric protein localises to the plasma membrane. We have noted previously that progressive replacement of lysine residues with glutamine results in a progressive loss of the targeting function of the polybasic domain (Hancock et al,1990). These two sets of data, taken together, strongly suggest that the polybasic domain functions predominantly through a charge interaction with the plasma membrane.

METHYLATION AND PROTEOLYSIS ARE REQUIRED FOR EFFICIENT MEMBRANE BINDING OF FARNESYLATED p21$^{K-ras(B)}$

It is not possible to determine the individual contributions of the CAAX modifications to membrane binding in vivo since mutations which block farnesylation (eg. Cys186 → Ser) necessarily block all subsequent processing events (Hancock et al,1989). However, we have recently described an in vitro system that can fully process p21$^{K-ras(B)}$ and which can be manipulated to produce partially processed p21$^{K-ras(B)}$ proteins (Hancock et al,1991).

A nuclease treated (message dependent) rabbit reticulocyte lysate is able to farnesylate in vitro translated p21$^{K-ras(B)}$. The prenylated protein does not however undergo -AAX proteolysis or methyl-esterification. Addition of a microsomal membrane fraction to the soluble lysate results in full processing of the translated protein, thus demonstrating that the -AAX protease, like the methyltransferase (Hrycyna and Clarke,1991), is membrane associated. Translation of p21$^{K-ras(B)}$ in the presence of microsomes and 3mM methylthioadenosine blocks methylation of the ras protein but not -AAX proteolysis. It is therefore possible to generate, in vitro, p21$^{K-ras(B)}$ proteins which are arrested after farnesylation, arrested after farnesylation and "-AAXing" or which are fully processed. These variously processed forms of p21$^{K-ras(B)}$ can then be assayed for binding to acceptor membranes in vitro. Such experiments clearly demonstrate (Hancock et al,1991) that whilst prenylated non-AAXed p21$^{K-ras(B)}$ binds to membranes, the extent of binding is doubled following -AAXing and doubled again following methylation. Therefore both proteolysis and methylation are required for the efficient membrane binding of p21$^{K-ras(B)}$.

We speculate that retention of the -AAX amino acids adjacent to the prenylated cysteine residue may sterically interfere with the insertion of the farnesyl group into the lipid bylayer. In addition, methylation of the α-carboxyl group on the C-terminal cysteine neutralises a negative charge which may otherwise cause electrostatic repulsion from negatively charged phospholipid headgroups, weakening the avidity of membrane binding.

SUBCELLULAR LOCALISATION OF GERANYLGERANYLATED P21[K-ras(B)]

The observation that *ras* proteins are farnesylated while some *ras*-related proteins are geranylgeranylated suggests that the length of of prenoid group may affect the association of proteins with membranes. We therefore investigated the effects of replacing the CVIM motif of p21[K-ras(B)] with the CAIL sequence of a brain Gγ-subunit, and the CCIL sequence of *ral*. The CAIL sequence has been shown to direct geranylgeranylation (Yamane *et al*,1990; Mumby *et al*,1990) and the CCIL motif would also be expected to, since leucine is the X amino acid. The same CAAX box mutations were also made in a K-*ras(B)* protein without a polybasic domain, ie. with the K6Q mutation (Lys[175-180] → Gln) (Hancock *et al*,1990).

An analysis of the post-translational processing of these CAIL and CCIL mutant p21[K-ras(B)] proteins confirmed that they were geranylgeranylated and methylesterified (JFHancock and CJMarshall, in prep). These data are therefore consistent with the hypothesis that CAA(X=L) motifs direct geranylgeranylation rather than farnesylation. Since these motifs are methylated, then, by implication, they must also be AA(X=L) proteolysed.

Subcellular fractionation of COS and NIH3T3 cells expressing the CAIL and CCIL proteins showed that they were localised to the P100 fraction. Immunofluoresence of NIH3T3 cells expressing the proteins showed strong plasma membrane staining. Also the biological activity of the CCIL and CAIL proteins was little different from farnesylated p21[K-ras(B)] in focus assays on NIH3T3 cells. However, subcellular fractionation of COS cells expressing the K6QCAIL and K6QCCIL proteins, which lack the polybasic domain showed that they were >90% localised to the P100 fraction. In contrast, farnesylated K6Q protein is >90% localised to the S100 fraction. However, although these geranylgeranylated K6Q proteins were found in the P100 fraction, examination of NIH3T3 cells expressing K6QCAIL and K6QCCIL by immunofluoresence revealed that the proteins were not localised to the plasma membrane (JFHancock and CJMarshall, in prep). We are currently investigating further to which intracellular membrane compartment these geranylgeranylated, polybasic mutant proteins are being targeted. However, this result clearly demonstrates that the presence of a C_{20} alkyl chain results in a much stronger affinity for membranes than a C_{15} chain.

POST-TRANSLATIONAL PROCESSING OF *S.POMBE* YPT PROTEINS

The *rab* proteins are a group of *ras*-related proteins that are involved in regulating intracellular membrane trafficking (Balch,1990). The C-termini of *rab* proteins consist either of two consecutive cysteine residues (CC) or two cysteines separated by another amino acid (CXC). These sequences suggest that the *rab* proteins may be modified in an analogous fashion to the *ras* proteins, with the exception that AAX proteolysis would not required as a prelude to α-carboxyl-methylation of the C-terminal cysteine residue. Such post-translational modifications may then be expected to have a functional role in targeting the *rab* proteins to specific intracellular membranes. We have now investigated the post-translational processing of the *S.pombe rab*-like proteins YPT1 and YPT3, which have C-terminal CC motifs, and YPT5 which has a CXC C-terminal motif.

When translated in a rabbit reticulocyte lysate all three YPT proteins incorporated label from $[^3H]$-mevalonic acid. Analysis of the labelled isoprenoid attached to each YPT protein revealed C_{20} geranylgeraniol. In contrast to *ras* proteins, however, we could show that none of the YPT proteins were labelled by S-adenosyl-$[^3H$-methyl]-methionine in lysates supplemented with microsomal membranes (Fawell *et al*,1991).

All three YPT proteins incorporated label from $[^3H]$-mevalonic acid when expressed *in vivo*, in COS cells and in *S.pombe*. We also find that in *S.pombe*, as *in vitro*, YPT1 and YPT3 do not incorporate labelled methyl groups from $[^3H$-methyl]-methionine. However, YPT5 clearly is methylated *in vivo*, both in *S.pombe* and when expressed in COS cells (Fawell *et al*,1991). It is possible therefore, that the YPT5 protein and the *ras* proteins may be methylated by different methyltransferases. Alternatively in the reticulocyte lysate the only access to a single methyltransferase may be via the -AAX protease, such that proteins without a prenylated CAAX motif cannot be methylated. In contrast with a previous study (Molenaar *et al*,1988), we did not detect palmitoylation of any of these YPT proteins *in vivo* (Fawell *et al*,1991).

These data show that a C-terminal CC motif directs geranylgeranylation, but not methylation, whereas a C-terminal CXC motif directs both geranylgeranylation and methylation. The biological differences between methylated and nonmethylated C_{20} modified proteins remain to be addressed. For example, given that farnesylated p21[K-ras(B)] requires methylation for efficient membrane binding, non-methylated geranylgeranylated proteins may be less avidly attached to cellular membranes than methylated forms.

SUMMARY

Prenylation is an essential component of the association with membranes of *ras* and *ras*-related proteins. The presence of C_{15} and C_{20} side chains results in different affinity for cellular membranes. However, prenylation is only one component of the post-translational modifications at the CAAX box and both AAX proteolysis and carboxylmethylation are also required for efficient membrane binding. CAAX post-translational modifications are insufficient on their own to target plasma membrane association for which they need to combine with other signals. At least for p21[K-ras(B)] the presence of the polybasic domain is sufficient to cooperate with the CAAX motif to give plasma membrane localisation.

REFERENCES

Casey, P.J., Solski,P.A., Der,C.J. and Buss,J. (1989) p21[ras] is modified by a farnesyl isoprenoid. *Proc.Natl.Acad.Sci.USA,* **86**, 1167-1177.

Balch,W.E. (1990) Small GTP binding proteins in vesicular transport. *Trends in Biochem. Sci.,* **15**,473-477.

Buss,J.E., Quilliam,L.A., Kato,K., Casey,P.J., Solski,P.A., Wong,G., Clark,R., McCormick,F., Bokoch,G.M. and Der,C.J. (1991) The COOH-terminal domain of the rap1A (Krev-1) protein is isoprenylated and supports transformation by an H-ras:Rap1A chimeric protein. *Mol.Cell.Biol.,* **11**,1523-1530.

Clarke,S., Vogel,J.P., Deschenes,R.J. and Stock,J. (1988) Post-translational modification of the H-ras oncogene protein: evidence for a third class of protein carboxyl methyl transferase. *Proc.Natl.Acad.Sci.USA,* **85**, 4643-4647.

Epstein, W.W., Lever, D.C. and Rilling,H.C. (1990) Prenylated proteins: synthesis of geranylgeranyl cysteine and identification of this thioether as a component of proteins in CHO cells. *Proc.Natl.Acad.Sci.USA,* **87**,7352-7354.

Fawell,E., Hancock,J.F., Giannakouros,T., Newman,C., Armstrong,J. and Magee,A.I. (1991) Post-translational processing of *S.pombe* YPT proteins. *MSS submitted.*

Fukada,Y., Takao,T., Ohuguro,H., Yoshizawa,T., Akino,T. and Shimonishi,Y. (1990) Farnesylated γ-subunit of photoreceptor G-protein indispensable for GTP binding. *Nature,* **346**, 658-660.

Gutierrez, L., Magee,A.I., Marshall,C.J. and Hancock,J.F. (1989) Post-translational processing of p21^{ras} is two step and involves carboxyl-methylation and carboxy-terminal proteolysis. *EMBO J.,* **8**,1093-1098.

Hancock,J.F., Magee,A.I., Childs,J. and Marshall,C.J. (1989) All *ras* proteins are polyisoprenylated but only some are palmitoylated. **57**, 1167-1177.

Hancock,J.F., Patterson,H. and Marshall,C.J. (1990) A polybasic domain or palmitoylation is required in addition to a CAAX motif to localise p21^{ras} to the plasma membrane. *Cell,* **63**, 133-139.

Hancock,J.F., Cadwallader,K. and Marshall,C.J. (1991) Methylation and proteolysis are essential for efficient membrane binding of prenylated p21$^{K-ras(B)}$. *EMBO J.,* **10**, 641-646.

Holtz,D., Tanaka,R.A., Hartwig,J. and McKeown,F. (1989) The CAAX motif of lamin A functions in conjunction with the nuclear localisation signal to target assembly to the nuclear envelope. *Cell,* **59**,969-977.

Hyrcyna,C.A. and Clarke,S. (1990) Farnesyl cysteine C-terminal methyltransfertase activity is dependent upon the STE14 gene product in *Saccharomyces cerevisiae. Mol.Cell.Biol.,* **10**, 5071-5076.

Kawata,M., Farnsworth,C.C., Yoshida,Y., Gelb, M.H., Glomset,J. and Takai,Y. (1990) Posttranslationally processed structure of the human platelet protein smg p21B: Evidence for geranylgeranylation and carboxyl methylation of the C-terminal cysteine. *Proc.Natl.Acad.Sci.USA,* **87**, 8960-8964.

Lai,R.K., Perez-Sala, D., Canada,F.J. and Rando R.R. (1990) The γ subunit of transducin is farnesylated. *Proc.Natl.Acad.Sci.USA,* **87**,7673-7677.

Maltese, W.A. and Sheridan, K.M. (1990) Isoprenoid modification of G25K (G$_p$), a low molecular mass GTP-binding protein distinct from p21^{ras}. *J.Biol.Chem.,* **265**, 17883-17890.

Molenaar,G.M.T., Pranger,R. and Gallwitz,D. (1988) A carboxyl-terminal cysteine residue is required for palmitic acid binding and biological activity of the *ras*-related yeast YPT1 protein. *EMBO J.,* **7**, 971-976.

Mumby, S.M., Casey.P.J., Gilman,A.G., Gutowski,S. and Sternweis, P.C. (1990) G protein γ subunits contain a 20-carbon isoprenoid. *Proc.Natl.Acad.Sci.USA,* **87**, 5873-5877.

Yamane, H.K., Farnsworth,C.C., Hongying, X., Howald,W., Fung, B.K., Clarke,S., Gelb, M.H. and Glomset,J. (1990) Brain G protein γ subunits contain an all-*trans*-geranylgeranyl-cysteine methyl ester at their carboxyl termini. *Proc.Natl.Acad.Sci.USA,* **87**,5868-5872.

THE YPT GENE FAMILY OF

SCHIZOSACCHAROMYCES POMBE

John Armstrong, Erica Fawell[1], Sally Hook, Alison Pidoux
and Mark Craighead

Membrane Molecular Biology Laboratory
Imperial Cancer Research Fund
P.O.Box 123
Lincoln's Inn Fields
London WC2A 3PX

[1]present address:
Dept. of Cellular and Molecular Physiology
Harvard Medical School
25 Shattuck St.
Boston MA 02115
USA

INTRODUCTION

The ypt/rab family is a group of related proteins within the ras superfamily of small GTP-binding proteins. The original members of the group, Ypt1 and Sec4, were identified in the budding yeast *Saccharomyces cerevisiae*, where they appear to function at different stages of the secretory pathway, namely traffic from the endoplasmic reticulum to the Golgi complex and from the Golgi complex to the plasma membrane respectively [1,2]. From these observations arose the hypothesis that the protein family may play a general role in eukaryotic membrane traffic, with a different member of the family determining the specificity of fusion of each class of vesicle mediating transport between the different compartments in the cell[2]. Consistent with this hypothesis, members of the family (known as rab proteins) have been identified in mammalian cells[3,4,5] and in several cases the proteins have been shown to localise to different intracellular compartments[6,7,8]. In two cases the proteins localise to distinct populations of endosomes[6], suggesting that the rab family is involved in endocytic as well as exocytic processes. Indeed, convincing evidence has been provided that the rab5 protein is directly involved in endosome fusion[9]. Except where such cell-free assays are available, however, it is difficult to determine the function of the mammalian proteins, since the genetic methods used in budding yeast are not generally applicable. In contrast, budding yeast and mammalian cells, while clearly resembling each other in many aspects of their membrane traffic, also differ significantly in other respects: most noticeably, budding yeast normally lacks a morphologically defined Golgi complex comparable to the organelle seen in higher cells.

The fission yeast *Schizosaccharomyces pombe* has proven to be a useful model eukaryotic cell in combining genetic tractability with greater similarity to higher cells in several aspects of its biology than is the case with budding yeast. This particularly applies to the regulation of its cell cycle[10]. Unlike budding yeast, it has an easily visible and well-defined Golgi complex[11] and adds galactose to glycoproteins in the secretory pathway[12]. For these reasons we have chosen this organism to investigate the function of the Ypt/rab proteins, with the aim of testing directly the hypothesis that they function in different stages of membrane traffic. Our preliminary results show that the number of genes so far discovered (five reported here and a sixth described by others[13]) appears to be consistent with the hypothesis, and that all but one of the proteins are strikingly close in sequence to mammalian homologues. In addition, one member of the family is a clear homologue of mammalian rab5, raising the possibility that fission yeast may be amenable to a genetic analysis of endocytosis.

MATERIALS AND METHODS

Following the approach described by Touchot *et al.*[3], two oligonucleotides were designed that encode six conserved amino-acids, DTAGQE, involved in GTP binding in the *ras*-related proteins. By hybridising in conditions which allow G-T base pairing [14], the number of permutations required for each strand was reduced to eight. To favour identification of the ypt-related members of the *ras* family, the first two bases of an arginine codon were added. Thus the following 20-mers were synthesised: GATACKGCKGGKCAGGAGCG, and CGTTCTTGKCCKGCKGTGTC. The oligonucleotides were end-labelled with T4 polynucleotide kinase and [γ-^{32}P]ATP in excess, then used as an equimolar mixture.

Nylon filters for hybridisation were incubated for 5 minutes at room temperature in 0.5M sodium phosphate, pH7.2, 7% SDS, 1mM EDTA, 1% bovine serum albumin[15], then in the same solution containing 10^6 cpm/ml of probe overnight at 37°. Filters were rinsed 3 times in 8xSSC, 0.1% SDS then once in 'TMAC' solution (3M tetramethyl ammonium chloride, 50mM Tris-HCL, pH7.6, 2mM EDTA, 0.1% SDS)[14], all at room temperature. Stringent washes were then performed with 'TMAC' solution at 59° twice for 2 minutes each. Libraries used were a partial Mbo1 digest of *S.pombe* genomic DNA in lambda-DASH (Stratagene), and a *S.pombe* cDNA library in lambda-gt11, kindly provided by Drs.A.Craig and V.Simanis respectively. Standard procedures[16] were used for plaque purification, Southern analysis, subcloning into plasmids and sequence determination with modified T7 polymerase [17].

cDNA's for YPT1 and YPT5 were obtained by polymerase chain reaction as follows. Total RNA was isolated from *S. pombe* [18]. First-strand cDNA was prepared from 10µg RNA using 20 units of AMV reverse transcriptase (Gibco-BRL) and oligo-dT (Pharmacia). Oligonucleotides corresponding to 20 nucleotides of the 5' and 3'-complementary coding regions of each gene, with additional BamH1 or BglII sites at the 5' end were prepared, and a polymerase chain reaction performed using 100pmol of each oligonucleotide, one quarter of the cDNA preparation, and 2.5 units Taq polymerase (Cetus-Perkin Elmer) according to the manufacturer's instructions, for 25 cycles of 92^0 (1 min), 50^0 (1 min) and 70^0 (2 min). The products were digested with BamH1 or BglII, cloned into plasmids and sequenced entirely to confirm the absence of mutations.

A peptide corresponding to the predicted amino-acid sequence from residues 33 to 15 from the C-terminus, plus an additional cysteine for coupling, was prepared. The peptide was coupled to keyhole limpet haemocyanin (Calbiochem) using m-maleimidobenzyl-N-hydroxy-sulphosuccinimide (Pierce)[19] and used to immunise rabbits.

For overexpression of ypt5 in *S. pombe*, the cDNA was inserted into the vector pREP[20] and transformed into *S. pombe* strain 556[21] by the lithium method[22]. Transformants were selected by growth in minimal medium lacking leucine[21]. Cells were grown in the presence or absence of 4µM thiamine[20] and examined by immunofluorescence[23].

For expression of ypt5 in monkey CV1 cells, the cDNA was inserted into the vector pCMUIV[24] and transfected into CV1 cells by the calcium phosphate or DEAE-dextran methods. Cells were fixed with paraformaldehyde, permeabilised with 0.2% Triton-X100 and labelled with primary and secondary antibodies[25].

RESULTS AND DISCUSSION

Previously we reported the isolation and sequence analysis of the *S. pombe* ypt1 gene[26]. Subsequently Miyake and Yamamoto reported the same gene, but postulated the presence of an additional intron at the 5' end of the sequence which was confirmed by analysis of cDNA clones[27]. We have repeated this result by PCR of the ypt1 cDNA. The resulting amended protein sequence is even closer to the mouse ypt1 (rab1) homologue than previously supposed (fig.1), emphasising the evolutionary divergence between the two yeasts and the greater similarity of the fission yeast homologue to the mammalian than the budding yeast gene.

```
Mouse YPT1:  MssMNPEYDYLFKLLLIGDSGVGKSCLLLRFADDTYTESYISTIGVDFKI
S.pombe:        MNPEYDYLFKLLLIGDSGVGKSCLLLRFADDTYTESYISTIGVDFKI
S.cerevisiae:   MNsEYDYLFKLLLIGnSGVGKSCLLLRFsDDTYTndYISTIGVDFKI

M        RTiELDGKTiKLQIWDTAGQERFRTITSSYYRGAHGIIvVYDVTDQESFN
P        RTVELeGKTVKLQIWDTAGQERFRTITSSYYRGAHGIIIVYDVTDQdSFN
C        kTVELDGKTVKLQIWDTAGQERFRTITSSYYRGsHGIIIVYDVTDQESFN

M        NVKQWLQEIDRYAsEnVNKLLVGNKCDLttKKVVdYttAKEFADSLgIPF
P        NVKQWLQEIDRYAvEgVNrLLVGNKsDmvDKKVVEYsVAKEFADSLnIPF
C        gVKmWLQEIDRYAtstVdKLLVGNKCDLkDKrVVEYdVAKEFADankmPF

M        LETSAKnaTNVEQsFmTMAaeIKkRMGpgatAggaeKSnVK iQsTpVkQ
P        LETSAKDSTNVEQAFLTMsRQIKERMGnntfAssnaKSsVKvGQgTnVsQ
C        LETSAlDSTNVEDAFLTMARQIKEsMsqqnlnettqKkedK GnvnlkgQ

M        S     GGGCC
P        SssN    CC
C        SltNtGGGCC
```

		Mouse	S.cerevisiae
Differences:	S.pombe	40	53
	S.cerevisiae	55	

Figure 1. Alignment of the predicted protein sequences of ypt1 from mouse[28] (M), budding yeast[29](C) and fission yeast.

We have also described the isolation of clones 23 and 8[30] (now referred to as ypt2 and ypt3 respectively[27,31]). Further analysis of hybridising genomic clones yielded two further members of the ypt family, which we have termed ypt4 and ypt5, as well as homologues of the GTP-binding proteins rho and rac. Gene ypt2 may be the equivalent of the budding yeast sec4 gene[31] although it is notable that the conservation of sequence does not particularly extend throughout its length. The remaining genes ypt3, 4 and 5 do not resemble any budding yeast genes so far reported. In contrast, four of the five gene products showing a striking similarity to mammalian rab proteins (Table 1). Gene products ypt1-4 all have the C-terminal structure -CC usually found on these proteins, while ypt5 has one of the variants of this, -CSC. We have found that both of these structures are subject to modification with a geranylgeranyl group (Hancock *et al.*, this volume).

Table 1. Comparison of the *S. pombe* ypt genes with mammalian rab proteins[4,5,28].

S.pombe gene	C-Term	Mammalian homologue	% Identity
ypt1	-CC	rab1 (ypt1)	70
ypt2	-CC	rab8	71
ypt3	-CC	rab11	82
ypt4	-CC	?	
ypt5	-CSC	rab5	71

Including the ryh1 gene[13], which is a homologue of rab6, the total so far is now six ypt genes from fission yeast. It was of particular interest to discover a homologue of rab5, since this protein is thought to be involved in endocytosis[9], raising the possibility that the ypt5 protein may play a similar role. To begin to investigate this possibility, we have examined the intracellular localisation of ypt5 when expressed both in fission yeast and mammalian cells. To detect the protein, an antibody was raised to a peptide from the C-terminal region of the predicted sequence. When this antibody was used for immunofluorescence of Triton-permeabilised fission yeast, a uniform fluorescence across the cell was seen, perhaps representing staining of the plasma membrane. The specificity of this reaction was confirmed by transforming cells with a multicopy plasmid containing ypt5 cDNA connected to the thiamine-repressible nmt promoter[20]. Cells grown in the presence of thiamine showed immunofluorescence similar to untransformed cells. However, cells grown in the absence of thiamine showed a markedly increased fluorescence.

Given the small size of the fission yeast cell, it was not possible to draw any definitive conclusions concerning the location of the ypt5 protein. Therefore we examined the behaviour of the protein when expressed in mammalian cells. Monkey CV1 cells were transfected with a plasmid containing ypt5 cDNA inserted into the vector pCMUIV[24]. When calcium phosphate was used as transfecting agent, expressing cells showed a very strong immunofluorescence covering the inside of the plasma membrane. It was also noticeable that the expressing cells tended to have a radically altered morphology, forming a contracted cell body from which long processes emerged. Since overexpression of other proteins using this vector has not been reported to result in such changes in morphology, the effect appears to be specific to ypt5. It may reflect an interference by the protein in the cell's membrane balance, perhaps via an endocytic mechanism. When CV1 cells were transfected by DEAE-dextran, a much lower level of expression was observed, with no obvious morphological effects. In this case the immunofluorescence pattern was punctate throughout the cell, suggesting that the protein was localised to a vesicular compartment. Double-labelling with the antibody 1B5, which recognises a membrane protein confined to lysosomes and late endosomes (M.Marsh, Institute of Cancer Research, London), showed no coincidence of labelling. Thus the ypt5 protein localises to a compartment which is clearly not the endoplasmic reticulum, Golgi complex, lysosomes or late endosomes. It is therefore possible that this compartment is the early endosomes, the location of the protein's homologue rab5, and consistent with a role of ypt5 in endocytosis rather than secretion. This possibility is currently being examined by double-labelling with markers for the early endosome.

Therefore analysis of the ypt gene family in fission yeast has shown that in this respect, as in many others, the organism appears to be a closer model of higher eukaryotic cells, and that the likely total of genes is consistent with the proposed general role of these proteins in controlling the different stages of eukaryotic membrane traffic. We are now developing genetic methods to attempt to test this hypothesis directly.

REFERENCES

1. Segev,N., Mulholland,J. and Botstein,D. (1988) *Cell* **52**, 915-924.

2. Salminen,A. and Novick,P.J. (1987) *Cell* **49**, 527.

3. Touchot,N., Chardin,P. and Tavitian,A. (1987) *Proc.Natl.Acad.Sci.USA* **84**, 8210.

4. Zahraoui,A., Touchot,N., Chardin,P. and Tavitian,A. (1989) *J.Biol.Chem.* **264**, 12394.

5. Chavrier,P., Vigneron,M., Simons,K. and Zerial,M. (1990) *Molec. Cell. Biol.* **10**, 6578.

6. Chavrier, P., Parton, R.G., Hauri, H.P., Simons, K. and Zerial, M. (1990) *Cell*, **62**, 317.

7. Goud, B., Zahraoui, A., Tavitian, A. and Saraste, J. (1990) *Nature*, **345**, 553.

8. Fischer v. Mollard, G., Mignery, G.A., Baumert, M., Perin, M.S., Hanson. T.J., Burger, P.M., Jahn, R. and Sudhof, T.C. (1990) *Proc. Natl. Acad. Sci. USA*, **87**, 1988.

9. Gorvel, J.-P., Chavrier, P., Zerial, M. and Gruenberg, J. (1991) *Cell*, **64**, 915.

10. Lee,M. and Nurse,P. (1987) *Nature* **327**, 31.

11. Smith, D.G., and Svoboda, A. (1972). *Microbios*, **5**, 177.

12. Moreno, S., Ruiz, T., Sanchez, Y., Villanueva, J.R., and Rodriguez, L. (1985). *Arch. Microbiol.*, **142**, 370.

13. Hengst, L., Lehmeier, T., and Gallwitz, D. (1990). *EMBO J.*, **9**, 1949.

14. Wood,W.I., Gitschier,J., Lasky,L.A. and Lawn,R.M. (1985) *Proc.Natl.Acad.Sci.USA* **82**, 1585.

15. Church,G.M. and Gilbert,W. (1984) *Proc.Natl.Acad.Sci.USA* **81**, 1991.

16. Sambrook, J., Fritsch, E.F. and Maniatis, T. (1989) Molecular Cloning: A Laboratory Manual. 2nd Ed. Cold Spring Harbor Laboratory Press, Cold Spring Harbor.

17. Tabor, S. and Richardson, C.C. (1987) *Proc.Natl.Acad.Sci.USA* **84**, 4767.

18. Kaufer, N.F., Simanis, V. and Nurse, P. (1985). *Nature (London)*, **318**, 78.

19. Green, N., Alexander, H., Olson, A., Alexander, S., Shinnick, T.M., Sutcliffe, J.G. and Lerner, R.A. (1982) *Cell*, **28**, 477.

20. Maundrell, K. (1990) *J. Biol. Chem.*, **19**, 10857.

21. Moreno, S., Klar, A., and Nurse, P. (1990). In *Methods in Enzymology* (ed. C. Guthrie and G.R. Fink.), Vol.194, pp.795-823. Academic Press, London.

22. Broker, M. (1987). *Biotechniques*, **5**, 516.

23. Hagan, I.M., and Hyams, J.S. (1988). *J. Cell Sci.*, **89**, 343.

24. Nilsson, T., Jackson, M.and Peterson, P.A. (1989) *Cell*, **58**, 707.

25. Ash, J.F., Louvard, D. and Singer, S.J. (1977) *Proc. Natl. Acad. Sci. USA*, **74**, 5584.

26. Fawell,E., Hook,S. and Armstrong,J. (1989) *Nucl. Acids Res.* **17**, 4373.

27. Miyake, S., and Yamamoto, M. (1990). *EMBO J.*, **9**, 1417.

28. Haubruck,H., Disela,C., Wagner,P. and Gallwitz,D. (1987) *EMBO J.* **6**, 4049.

29. Gallwitz, D., Donath, C. and Sander, C. (1983) *Nature*, **306**, 704.
30. Fawell, E., Hook, S., Sweet, D., and Armstrong, J. (1990). *Nucl. Acids Res.* **18**, 4264.
31. Haubruck, H., Engelke, U., Mertins, P., and Gallwitz, D. (1990). *EMBO J.*, **9**, 1957.

PHOSPHORYLATION OF <u>RAP</u> PROTEINS BY THE cAMP-DEPENDENT PROTEIN KINASE

Isabelle Lerosey, Véronique Pizon, Armand Tavitian and
Jean de Gunzburg

INSERM U-248, Faculté de Médecine Lariboisière - Saint-Louis
10 av. de Verdun, 75010 Paris, France

INTRODUCTION

The products of *rap* genes (*rap*1A, *rap*1B, *rap*2A, *rap*2B) are small molecular weight GTP-binding proteins that exhibit striking similarities with *ras* p21s[1-4]. In particular, *ras* and *rap* proteins share a conserved "effector" region spanning residues 32-42 through which *ras*-p21s are thought to exert their biological effects; they also have a C-terminal CAAX sequence (where A is an aliphatic residue and X any amino acid) responsible for posttranslational modification and membrane binding of *ras* proteins[5]. The identity of the "effector" domain between *ras* and *rap* proteins had suggested that *rap* proteins could antagonize the activity of *ras* proteins by competing for a common effector. Independently, M. Noda's group isolated a cDNA, *Krev*-1, whose overexpression could revert the transformed phenotype of Kirsten sarcoma-virus transformed NIH 3T3 cells[6] ; the sequence of the *Krev*-1 protein was identical to that of the *rap*1A protein. Moreover, in vitro, the *rap*1A protein has been shown to be able to compete efficiently with *ras* p21 for interaction with GAP[7](GTPase Activating protein), which may constitute the effector of *ras* p21[8-12].

It had previously been reported that *rap*1 proteins extracted from bovine brain or human platelets could be phosphorylated on serine residues by the cAMP-dependent protein kinase (PKA) as well as in intact platelets stimulated by prostaglandin E_1 [13-17]. In this report, we show that the *rap*1A protein purified from recombinant bacteria is phosphorylated in vitro by the catalytic subunit of PKA and that the deletion of the 17 C-terminal amino-acids leads to the loss of this phosphorylation. This suggests that the serine residue at position 180 constitutes the site of phosphorylation of the *rap*1A protein by PKA. Using specific antibodies directed against the *rap*1 or *rap*2 proteins, it was possible to search for phosphorylation of these products in intact fibroblasts. We could demonstrate that a *rap*1 protein was phosphorylated in vivo when PKA is stimulated by specific agonists. This phenomenon is independent of the proliferative state of the cells. In contrast, protein kinase C (PKC) does not phosphorylate *rap*1 proteins, neither in vitro nor in vivo. Finally, the 60% homologous *rap*2 protein is neither phosphorylated in vitro, nor in vivo by PKA or PKC.

MATERIALS AND CHEMICALS

The catalytic subunit of PKA, its inhibitor (PKI) as well as histones H2B and H1 were from Sigma. PKC was from Calbiochem. $[\gamma\text{-}^{32}P]ATP$ (3000 Ci/mmol) and carrier free $[^{32}P]$orthophosphate (10 mCi/ml) were from Amersham. Human *rap*1A and *rap*2A proteins were purified from recombinant bacteria carrying a ptac-*rap*1A[7] or a ptac-*rap*2A vector[18]. A truncated *rap*1A protein in which lysine 168 was changed to a stop codon was obtained by site-directed mutagenesis using oligonucleotide 5' ATA AAT AGG TAA ACA CCA CTG 3'[18]. The obtention and characterization of antibodies directed against *rap*1 and *rap*2 is described in detail elsewhere[19].

RESULTS

In vitro phosphorylation of human rap1A and rap2A proteins by PKA and PKC

We have investigated if *rap*1A and *rap*2A proteins purified from recombinant bacteria were substrates for phosphorylation by PKA and PKC in vitro. Figure 1 shows that the *rap*1A protein was phosphorylated by the catalytic subunit of PKA (lane 1). This phosphorylation was abolished in the presence of the specific inhibitor of PKA, which shows that the reaction could be attributed to PKA (figure 1, lane 2). The *rap*1A protein contains in its C-terminal domain a consensus site for phosphorylation by PKA[20], which is Lys-Lys-Lys-Ser_{180}. It had previously been shown that the *rap*1A protein was phosphorylated on serine residues by PKA[13,14]. In order to determine if the serine at position 180 actually constitutes the site of phosphorylation, we constructed a truncated rap1A protein in which the 17 C-terminal amino acids were deleted; this truncated protein, which has lost its serine 180, could no longer be phosphorylated by PKA (figure 1, lane 3). It is therefore likely that serine 180 constitutes the site of phosphorylation of the rap1A p21 by PKA. Using the same conditions, we have searched for a phosphorylation of the rap2A protein by PKA. The rap2A protein, that shares 61% identity with the rap1A protein was not a substrate for PKA (figure 1, lane 4). This was not surprising since the rap2A product does not contain a consensus site for phosphorylation by PKA. Under the same conditions, histones H2B which served as a positive control, were efficiently phosphorylated (figure 1, lane 5).

H-ras and K-ras (4B) p21s have been shown to be phosphorylated in vivo as well as in vitro by PKA and also by PKC[21-23]. The site of phosphorylation was serine 177 for H-ras p21 and serine 181 for K-ras (4B)[21-23]. We performed experiments in vitro to investigate if PKC was able to phosphorylate rap1A or rap2A proteins in vitro. No phosphorylation could be detected, whereas histones H1 were efficiently modified under the same conditions (figure 2).

In vivo phosphorylation of rap1 and rap2 proteins

Since we had obtained specific antibodies directed against the rap1 and rap2 proteins, it was possible to search for phosphorylation of these products in intact fibroblasts which contain low level of rap proteins. Rat-1 fibroblasts were labeled with [32P]orthophosphate, in the presence of specific activators of PKA or PKC. Cell lysates were immunoprecipitated with rabbit serum directed against rap1 or rap2 proteins as previously described[24]. Figure 3A shows that the rap1A and/or rap1B is phosphorylated in intact Rat-1 cells when PKA is stimulated directly by the addition of 8-Br-cAMP (lane 4) or in response to agents that activate adenylate cyclase such as forskolin (lane 5) or prostaglandin E_1 (lane 6), in the presence of the phosphodiesterase inhibitor isobutylmethylxanthine (IBMX).

The rap1A and rap1B products could not be discriminated since the antibodies were raised against a peptide common to both proteins and the two genes are expressed in Rat-1 fibroblasts (J. de Gunzburg, unpublished results). Treatment of cells with agents that stimulate PKC, the phorbol ester TPA alone (lane 2) or in combination with the calcium ionophore A23187 (lane 3), did not result in phosphorylation of the rap1 protein.

In the case of the rap2 protein, no phosphorylation could be observed neither when PKA nor PKC were stimulated (figure 3B).

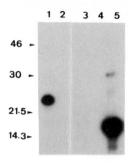

Figure 1. In vitro phosphorylation of human rap1A and rap2A proteins by PKA.

Phosphorylation of purified rap proteins (250 ng) and histones H2B (1 μg) by the catalytic subunit of PKA (10 units) was carried out in a 20 μl reaction mixture containing 50 mM Tris-HCl pH=7.5, 10 mM MgCl$_2$, 1 mM DTT and 20 μM [γ-^{32}P]ATP (10 000 cpm/pmol) for 30 minutes at 30°C. Reactions were stopped by the addition of SDS-gel sample buffer and samples were boiled 3 minutes. They were subjected to SDS-PAGE followed by autoradiography.

Lane 1: rap1A protein; lane 2: rap1A protein in the presence of 10 μg of PKI; lane 3: rap1A protein truncated at residue 168; lane 4: rap2 protein; lane 5: histone H2B.

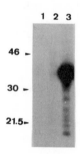

Figure 2. In vitro phosphorylation of human rap1A and rap2A proteins by PKC.

Phosphorylation of purified rap proteins (250 ng) and histones H1 (1 μg) by PKC (50 units) was performed in a 20 μl reaction mixture containing 20 mM Tris-HCl pH=7.5, 5 mM Mg(CH$_3$COO)$_2$, 5 μM [γ-^{32}P]ATP (10 000 cpm/pmol), 20 μg/ml phosphatidylserine, 100 nM TPA and 0.5 mM CaCl$_2$ for 30 minutes at 30°C. Reactions were stopped as described above and analyzed by SDS-PAGE followed by autoradiography.

Lane 1: rap1A protein; lane 2: rap2A protein; lane 3: histones H1.

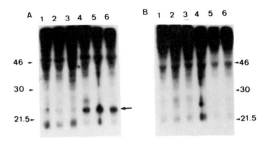

Figure 3. Phosphorylation of *rap*1 and *rap*2 proteins in Rat-1 cells Rat-1 fibroblasts were grown in Dulbecco's modified Eagle's Medium (DMEM) containing 10% calf serum. Prior to labeling for 3 hours with 625 μCi/ml carrier-free [^{32}P]orthophosphate, cells were incubated for 3 hours in phosphate-free DMEM. Phosphate deprivation and labeling were carried out in the presence of 5% calf serum. During the last hour of labeling, cells were treated with specific activators of PKA or PKC. Cell lysates were immunoprecipitated with antisera directed against the *rap*1 (panel A) or *rap*2 (panel B) proteins.
Lane 1: no treatment; lane 2: 100 ng/ml TPA; lane 3: 100 ng/ml TPA and 4 μM Ca^{2+} ionophore A23187; lane 4: 0.5 mM 8-Br-cAMP; lane 5: 100 μM each forskolin and IBMX; lane 6: 50 μM PGE$_1$ in the presence of 100 μM IBMX.

Phosphorylation of the rap1 protein in quiescent and proliferating cells

It had been shown that treatment of growing Rat-1 cells with agents that elevate the intracellular level of cAMP induced a rapid arrest of cellular proliferation. Since it had been suggested that *rap*1 proteins could play an antagonistic role to the *ras* proteins and therefore act in a signalling pathway mediating antiproliferating effects, we were interested in investigating whether the phosphorylation of *rap*1 p21s by PKA in Rat-1 cells was related to their proliferative state. Growth-arrested cells were obtained after 24 or 48 hours of serum deprivation. The level of DNA synthesis in the cells was assayed by monitoring the incorporation of 5'-bromodeoxyuridine. Figure 4A shows that the nuclei of growing cells incorporated 5'-bromodeoxyuridine (panel 1) and that serum starvation was sufficient to cause a significant growth arrest (panels 2 and 3). These cells were labeled with [^{32}P]orthophosphate and lysates were immunoprecipitated with the anti-*rap*1 antiserum. No significant modification of the endogenous level of *rap*1 protein phosphorylation could be detected between dividing or growth-arested cells (figure 4B). *Rap*1 proteins were phosphorylated only when PKA was stimulated by 8-Br-cAMP (figure 4B, lanes 2 and 7); although the extent of phosphorylation was somewhat lower in quiescent cells, it appears that the sites for PKA were available under both conditions.

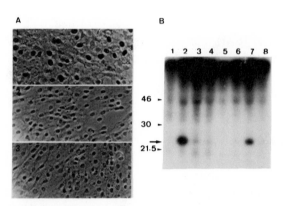

Figure 4. Phosphorylation of the *rap*1 protein in quiescent and proliferating cells.

Panel A: Bromodeoxyuridine incorporation into Rat-1 cells at different states of growth was assayed by immunofluorescence using a commercial kit (Amersham). Labelled nuclei represent cells that are synthesizing DNA. 1: proliferating cells; 2: growth-arrested cells (24h of serum deprivation); 3: growth-arrested cells (48h of serum deprivation).

Panel B: immunoprecipitation of *rap*1 proteins from [^{32}P]orthophosphate-labeled Rat-1 cells. lane 1: proliferating cells; lane 2: proliferating cells treated with 0.5 mM 8-Br-cAMP; lane 3: cells were deprived of serum at the time of labeling; lane 4: 6 hours of serum deprivation; lane 5: 24 hours of serum deprivation; lane 6: 48 hours of serum deprivation; lane 7: cells deprived of serum for 48 hours were treated with 0.5 mM 8-Br-cAMP; lane 8: cells deprived of serum for 48 hours were stimulated with 5% serum at the time of [^{32}P]orthophosphate labeling.

$$
\begin{array}{ccc}
& 170 & 180 \\
\end{array}
$$

Rap1A : N R K T P V E K K K P K K K S̲ C L L L

Rap1B : N R K T P V P G K A R K K S̲ S C Q L L

Rap2A : Y A A Q P D K D D P C C S A C N I Q

Rap2B : Y A A Q S N G D E G C C S A C V I L

Figure 5. Alignment of the C-terminal sequences of the different *rap* proteins.
The *rap*1A and *rap*1B proteins possess a consensus site for phosphorylation by PKA (B-B-X-Ser/Thr-X) and are readily substrates of PKA. In contrast, *rap*2 proteins that do not contain such a sequence are not phosphorylated by PKA. The serine residue modified by phosphorylation in *rap*1 proteins is underlined.

DISCUSSION

In the present study, we have investigated the possible phosphorylation of *rap* proteins in vitro as well as in intact fibroblasts by PKA and PKC. Indeed, phosphorylation of *rap*1 proteins has been described in human platelets by several groups[13-17] but until now has not been investigated in fibroblasts.

We have first shown that the recombinant *rap*1A protein is phosphorylated in vitro by the catalytic subunit of PKA and that a truncated protein missing the 17 C-terminal residues is no more phosphorylatable by PKA. Moreover, the sequence B-B-X-Ser/Thr-X, where B is a basic amino-acid and X any residue, constitutes the consensus site for phosphorylation by PKA on serine or threonine residues[20]. Figure 5 shows the alignment of the C-terminal sequences of the *rap* proteins (*rap*1A, *rap*1B, *rap*2A, *rap*2B). The *rap*1A protein possesses only one consensus site for phosphorylation by PKA which is 177Lys-Lys-Lys-Ser-Cys thereby identifying serine 180 as the most likely residue phosphorylated by PKA. The *rap*1B protein[2], which shares 96% identity with the *rap*1A product has recently been shown to be phosphorylated on serine 179[26]. The *rap*2 proteins are not substrates for PKA, which is not surprising as they do not contain the B-B-X-Ser/Thr-X sequence[1-4]. In contrast with the case of H-*ras* and K-*ras* proteins[21-23], neither *rap*1 nor *rap*2 proteins were substrates for PKC.

As we had developed specific antibodies directed against the *rap*1 or *rap*2 proteins, it was possible to search for phosphorylation of these products in intact fibroblasts. Rat-1 cells were labeled with [^{32}P]orthophosphate and cell lysates were immunoprecipitated with anti-*rap* antisera. These experiments demonstrated that a *rap*1 protein could be phosphorylated in vivo when PKA is stimulated by agonists. However, as our antiserum recognizes both *rap*1A and *rap*1B proteins, we cannot determine whether one or both is a substrate for PKA. Our observation confirm previous reports showing that a *rap*1/smg p21 protein could be phosphorylated by PKA in human platelets[13-17], and extend them to the case of murine fibroblasts which contain a low level of *rap*1 proteins. The experiments of M. Noda's group have shown that the *rap*1A/*Krev*-1 product could exert an antagonistic action to that of *ras* p21s in cellular proliferation[6]. Since a *rap*1 protein could be phosphorylated in intact Rat-1 cells in response to agents that elevate the intracellular concentration of cAMP and also inhibit cellular proliferation[25], we sought to investigate whether this phosphorylation of the *rap*1 proteins in fibroblasts could be related to the proliferative state of the cells. However, our observations do not permit us to establish a relation between these two parameters. The physiological function of the phosphorylation of the *rap*1 proteins remains unclear, nevertheless these results indicate that the *rap*1 p21 may be affected by agents acting through a cAMP-dependent pathway. This does not seem to be the case of the *rap*2 protein that may be regulated by a different set of signals.

Acknowledgments: This work was supported by grants from the INSERM, the Association pour la Recherche contre le Cancer and the Ligue Nationale contre le Cancer. I.L. is the recipient of a fellowship from the Ministère de la Recherche et de la Technologie.

REFERENCES

1- Pizon, V., Chardin, P., Lerosey, I., Olofsson, B. and Tavitian, A. (1988) *Oncogene* **3**, 201-204
2- Pizon, V., Lerosey, I., Chardin, P. and Tavitian, A. (1988) *Nucleic Acids Res.* **16**, 7719
3- Kawata, M., Matsui, Y., Kondo, J., Hishida, T., Teranishi, Y. and Takai, Y. (1988) *J. Biol. Chem.* **263**,18965-18971
4- Ohmstede, C.-A., Farrell, F.X., Reep, B.R., Clemetson, K.J. and Lapetina, E.G. (1990) *Proc. Natl. Acad. Sci. USA* **87**, 6527-6531
5- Barbacid, M. (1897) *Annu. Rev. Biochem.* **56**, 779-827
6- Kitayama, H., Sugimoto, Y., Matsuzaki, T., Ikawa, Y. and Noda, M. (1989) *Cell* **56**, 77-84

7- Frech, M., John, J., Pizon, V., Chardin, P., Tavitian, A., Clark, R., Mc Cormick, F. and Wittinghofer, A. (1990) *Science* **249**, 169-171

8- Trahey, M. and Mc Cormick, F. (1987) *Science* **238**, 542-545

9- Calés, C., Hancock, J.F., Marshall, C.J. and Hall, A. (1988) *Nature* **332**, 548-551

10- Adari, H., Lowy, D.R., Willumsen, B.M., Der, C.J. and Mc Cormick, F. (1988) *Science* **240**, 518-521

11- Vogel, U.S., Dixon, R., Schaber, M.D., Diehl, R.E., Marshall, M.S., Scolnick, E.M., Sigal, I.S. and Gibbs, J.B. (1988) *Nature* **335**, 90-93

12- Mc Cormick, F. (1989) *Cell* **56**, 5-8

13- Hoshijima, M., Kikuchi, A., Kawata, M., Ohmori, T., Hashimoto, E., Yamamura, H. and Takai, Y. (1988) *Biochem. Biophys. Res. Commun.* **157**, 851-860

14- Kawata, M., Kikuchi, A., Hoshijima, M., Yamamoto, K., Hashimoto, E., Yamamura, H. and Takai, Y. (1989) *J. Biol. Chem.* **264**, 15688-15695

15- Lapetina, E.G., Lacal, J.C., Reep, B.R. and y Vedia, L.M. (1989) *Proc. Natl. Acad. Sci. USA* **86**, 3131-3134

16- Siess, W., Winegar, D.A. and Lapetina, E.G. (1990) *Biochem. Biophys. Res. Commun.* **170**, 944-950

17- Fisher, T.H., Gatling, M.N., Lacal, J.C. and White, G.C. (1990) *J. Biol. Chem.* **265**, 19405-19408

18- Lerosey, I., Chardin, P., de Gunzburg, J. and Tavitian A. (1991) *J. Biol. Chem.* **266**, 4315-4321

19- Béranger, F., Goud, B., Tavitian, A. and de Gunzburg, J. (1991) *Proc. Natl. Acad. Sci. USA* **88**, 1606-1610

20- Edelman, A.M., Blumenthal, D.K. and Krebs, E.G. (1987) *Ann. Rev. Biochem.* **56**, 567-613

21- Ballester, R., Furth, M.E. and Rosen, O.M. (1987) *J. Biol. Chem.* **262**, 2688-2695

22- Jeng, A.Y., Srivastava, S.K., Lacal, J.C. and Blumberg, P.M. (1987) *Biochem. Biophys. Res. Commun.* **145**, 782-788

23- Saikumar, P., Ulsh, L.S., Clanton, D.J., Huang, K.-P. and Shih T.Y. (1988) *Oncogene Res.* **3**, 213-222

24- de Gunzburg, J., Rielh, R. and Weinberg, R.A. (1989) *Proc. Natl.Acad. Sci.* **86**, 4007-4011

25- Burgering, B.M.T., Snijders, A.J., Maasen, J.A.,Van der Eb, A.J. and Bos, J.L. (1989) *Mol. Cell. Biol* **9**, 4312-4322

26- Hata, Y., Kaibuchi, K., Kawamura, M., shirataki, H. and Takai, Y. (1991) *J. Biol. Chem.* **266**, 6571-6577

REGULATION OF THE *RAS* PATHWAY IN THE FISSION YEAST

SCHIZOSACCHAROMYCES POMBE

David A. Hughes, Yoshiyuki Imai and Masayuki Yamamoto

Department of Biophysics and Biochemistry, Faculty of Science, University of Tokyo, Tokyo 113, Japan

INTRODUCTION

Homologs of the mammalian *ras* genes have been found in the simple unicellular yeasts *Saccharomyces cerevisiae* (budding yeast) and *Schizosaccharomyces pombe* (fission yeast) (reviewed in Gibbs and Marshall, 1989). Using classical and molecular genetic approaches considerable progress has been made in understanding the function and regulation of *ras* in these organisms. In this paper we will describe our recent results on the regulation of the *ras* pathway in *S. pombe*.

S. pombe appears to have a single *ras* gene, designated *ras1*, that encodes a protein of 219 amino acids with considerable homology, particularly in the N-terminal half, to mammalian and *S. cerevisiae* Ras proteins (Fukui and Kaziro, 1985; Nadin-Davis et al., 1986a). The C-terminus of the *ras1* gene product (CCVIC) is similar to the C-terminal regions of other Ras proteins and is likely to be the site of post-translational modifications required for membrane-association (Hancock et al., 1989). The so-called effector domain (residues 37-45 in Ras1 corresponding to residues 32-40 in H-ras) is strictly conserved in the Ras1 protein and point mutations in this region abolish or strongly inhibit *ras1* function (DAH, unpublished results).

S. pombe cells carrying a null mutation in the *ras1* gene can proliferate normally (Fukui et al., 1986a; Nadin-Davis et al., 1986b), unlike *S. cerevisiae* where the *RAS* genes are essential for cell growth (Tatchell et al., 1984; Kataoka et al., 1984). However, *S. pombe ras1* mutants are defective in sexual differentiation (Fig. 1): haploid *ras1* mutants are unable to mate (or conjugate) with cells of opposite mating-type, i.e. they are sterile, and *ras1* diploids undergo meiosis and sporulation poorly, at a few percent of the wild-type frequency. An unexplained feature of these mutant cells is that they have a rounded or "dumpy" phenotype suggesting some defect in the polarization of cell growth.

The reason for the sterility of *ras1* mutant cells is unclear but it could be due to their inability to respond to mating pheromones. They are, however, able to secrete at least one of the pheromones (the M-factor produced by cells of h^- mating-type) (Fukui et al., 1986b). Further evidence that *ras1* is somehow involved in the response to mating pheromones comes from the phenotype of cells carrying the *ras1*^{Val17} mutation (analogous to the mammalian *ras*^{Val12} oncogene). These cells appear to be supersensitive to the M-factor pheromone, forming excessively long conjugation tubes, and mate rather poorly (Nadin-Davis et al., 1986b). The phenotypes conferred by loss or activation of *ras1* suggest a role for *ras1* in modulating the sensitivity of the mating pheromone-signalling pathway.

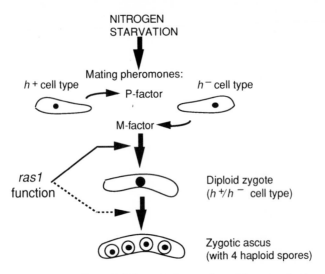

NITROGEN
STARVATION

Mating pheromones:

h + cell type   P-factor

h − cell type

M-factor

ras1
function

Diploid zygote
(*h* +/*h* − cell type)

Zygotic ascus
(with 4 haploid spores)

FIGURE 1. Sexual differentiation and *ras1* function in *S. pombe*

THE *STE6* GENE PRODUCT ENCODES AN ACTIVATOR OF THE *RAS1* PROTEIN

The *ste6* gene of *S. pombe* (Fig. 2A) encodes a protein of 105-kDa that has significant similarity in its C-terminal region to the product of the *CDC25* gene of *S. cerevisiae* (Hughes et al., 1990; Camonis et al., 1986; Broek et al., 1987). Significant similarity is also found with *S. cerevisiae SDC25*, isolated as a multicopy suppressor of *cdc25* mutants (Boy-Marcotte et al., 1989; Damak et al., 1991). In *S. cerevisiae*, *CDC25* is thought to function upstream of *RAS*, possibly as a factor stimulating GDP-GTP exchange on the Ras proteins (Broek et al., 1987; Robinson et al., 1987; Camonis and Jacquet, 1988). Rather indirect evidence in favor of this model of *CDC25* function has come from the finding that the C-terminal region of *SDC25* expressed in bacteria can stimulate GDP-release and GDP-GTP exchange on both *S. cerevisiae* and human Ras proteins (Crechet et al., 1990). More recently it has been found that *S. cerevisiae* cells overexpressing *CDC25* have enhanced Ras GDP-GTP exchange activity (Jones et al., 1991), suggesting that the *CDC25* gene product is indeed a guanine nucleotide-releasing protein (GNRP) for the *S. cerevisiae* Ras proteins.

Does the *ste6* gene product play a similar role in regulating the activity of the *S. pombe* Ras1 protein? Several lines of evidence suggest that this is so (Hughes et al., 1990; Imai et al., 1991). First, both *ste6* and *ras1* mutants are unable to mate or to respond to mating pheromones. Second, activating *ras1* mutations, such as $ras1^{Val17}$ or $ras1^{Leu66}$, restore mating and pheromone-sensitivity to cells carrying a *ste6* null mutation, i.e. activated *ras1* is epistatic to loss of *ste6*. Third, inactivation of the putative *ras* GTPase-activating protein also suppresses the *ste6* null mutation. This genetic analysis is consistent with the *ste6* gene product functioning as an activator of the Ras1 protein. The homology with *S. cerevisiae CDC25* and *SDC25* suggests that the Ste6 protein is a GNRP for the *S. pombe* Ras1 protein.

A problem with this interpretation of *ste6* function is the difference in phenotype between *ras1* and *ste6* mutants, which can undergo meiosis and sporulation normally and have a normal cell morphology. It is possible that there is another *ras* GNRP so that *ste6* mutants are only partially defective in *ras* activation. Alternatively, the intrinsic GDP-GTP exchange rate of the Ras1 protein could provide sufficient activity for meiosis and maintainence of cell shape. An heretical possibility, which is nevertheless consistent with the genetic data, is that some functions of *ras1* are not dependent on the GTP-bound form of the protein. Investigation of GNRP activity in *ste6* mutants should help to settle this question.

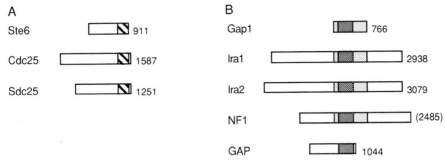

FIGURE 2. A. Guanine nucleotide-releasing proteins (GNRPs). Regions that are similar in all three proteins are indicated by cross-hatching. Cdc25 and Sdc25 are similar throughout their lengths (not shown). The total number of amino-acid residues in each protein is shown. B. GTPase-activating proteins (GAPs). Regions similar to the core "catalytic" region of mammalian GAP are shown by heavy shading. Gap1 of *S. pombe*, Ira1/2 and NF1 have similarity outside the core domain, shown by light shading. Ira1/2 and NF1 are similar throughout most their lengths (not shown).

The similarity between *S. pombe* Ste6 and *S. cerevisiae* Cdc25 and Sdc25 is limited to short regions (~270 amino-acid residues) near the C-terminus of each protein (Fig. 2A). Cdc25 and Sdc25, however, show lower but significant similarity outside this conserved domain, but the C-terminal domains are sufficient for function *in vivo* and can partially rescue *S. pombe ste6* mutants (DAH, M. Jacquet, and MY, unpublished results). We suspect that *ras* GNRPs in other organisms will also have this conserved domain.

THE *GAP1* GENE PRODUCT IS A HOMOLOG OF MAMMALIAN GAP AND NF1

The *gap1* gene of *S. pombe* (Fig. 2B) encodes a 88-kDa protein with similarity to the *ras* GTPase activating proteins GAP and NF1 from mammalian cells and Ira1 and Ira2 from *S. cerevisiae* (Imai et al., 1991; Trahey et al., 1988; Vogel et al., 1988; Xu et al., 1990; Tanaka et al., 1989, 1990). The homology of *S. pombe* Gap1 with mammalian GAP is limited to the catalytic domain but is more extensive with NF1 and Ira1/2. In *S. cerevisiae*, the *IRA1* and *IRA2* gene products function as inhibitors of *RAS* activity, probably by stimulating the conversion of the active GTP-bound form of Ras to the inactive GDP-bound form. The mammalian GAP and NF1 proteins can stimulate *ras* GTPase activity, but it is unclear at present whether they can also function as effectors of *ras*.

Is *S. pombe gap1* an inhibitor or an effector of *ras1*? Cells carrying a null mutation in *gap1* were found to have a phenotype similar to cells carrying *ras1*[Val17], an activating allele of *ras1*, and did not show any of the phenotypes characteristic of loss of *ras1* activity (Imai et al., 1991). The phenotype of *gap1 ras1* double mutants was the same as that of *ras1* mutants showing that the "activated" phenotype conferred by loss of *gap1* is dependent on *ras1* function. This genetic analysis indicates that the Gap1 protein acts as an inhibitor of Ras1 activity and not as an effector.

A comparison of *ras* GTPase activating proteins (Fig. 2B) suggests that *S. pombe* Gap1, *S. cerevisiae* Ira1/2 and human NF1 are most closely related and share homology outside the domain required for GTPase stimulation. Genetic evidence indicates that the yeast proteins are negative regulators of *ras* and their strong structural similarity with human NF1 may suggest that NF1 is also a negative regulator and not an effector of *ras*. Mammalian GAP, however, is not related to the other members of the family outside the catalytic domain suggesting that it could have unique functional properties, perhaps as a *ras* effector (McCormick, 1989).

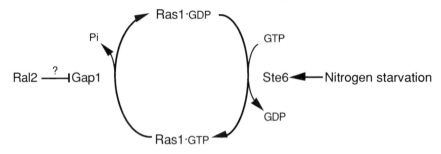

FIGURE 3. Regulation of the *ras* GTPase cycle in *S. pombe*

THE *RAL2* GENE PRODUCT IS A NOVEL REGULATOR OF *RAS* ACTIVITY

The *ral2* gene was identified in a screen for mutants with a similar phenotype to that conferred by loss of *ras1* function (Fukui and Yamamoto, 1988). Strains carrying a *ral2* null mutation are sterile, inhibited in meiosis and sporulation, and have the same dumpy morphology as *ras1* mutants (Fukui et al., 1989). The *ral2* gene appears to function upstream of *ras1* since activating *ras1* alleles restore normal cell shape, mating pheromone sensitivity and a low frequency of mating to cells carrying a *ral2* null mutation. Loss of *gap1* function can also suppress a *ral2* mutation (Imai et al., 1991). Thus *ral2* behaves in a similar fashion to *ste6* in its interaction with *ras1*. However, the 70-kDa *ral2* gene product has no similarity to Ste6 or to *S. cerevisiae* Cdc25 or Sdc25 and so is unlikely to function catalytically as a GNRP, although it is possible that it is a component or regulator of a GNRP complex. An alternative and exciting possibility is that Ral2 acts as an inhibitor of the GAP encoded by the *gap1* gene (Fig. 3).

REGULATION OF THE *RAS* GTPASE CYCLE IN *S. POMBE*

The inability of *ste6* mutants to mate and the inefficient mating caused by activating *ras1* mutations suggests that the amount of active GTP-bound Ras1 protein must be carefully controlled to ensure successful completion of the mating process. This could be achieved by regulating either the GNRP activity of the Ste6 protein or the GAP activity of Gap1, or both (Fig. 3).

When *S. pombe* cells are nitrogen starved, which is the physiological signal that initiates the mating process, the amount of *ste6* mRNA is highly induced (at least ten-fold) (DAH, unpublished results), whereas expression of *gap1*, *ral2* and *ras1* mRNAs is unaffected (YI, S. Miyake and MY, unpublished results; Nadin-Davis and Nasim, 1990). Thus it is likely that part of the Ras1 activation mechanism during mating is through stimulation of GDP-GTP exchange catalyzed by Ste6 protein. It is of course possible that GAP activity is also regulated, perhaps through Ral2 (Fig. 3). To address this question it will be necessary to assay GNRP and GAP activities in cells undergoing the mating process.

INTERACTION OF *RAS* AND HETEROTRIMERIC G PROTEIN-MEDIATED SIGNAL TRANSDUCTION

Genetic analysis of the *S. pombe ras* pathway indicates that it modulates the sensitivity of cells to mating pheromones. The pheromone receptors, encoded by the *mam2* and *map3* genes (K. Kitamura and C. Shimoda, pers. comm.; K. Tanaka, YI and MY, unpublished results), are putative transmembrane proteins of the rhodopsin family, and appear to be coupled to heterotrimeric G protein. The alpha-subunit of the G protein, encoded by the *gpa1* gene, is directly involved in transducing the pheromone signal (Obara et al., 1991). (The putative beta and gamma subunits have not been identified as yet). How and at what level does the Ras1 protein interact with the pheromone-signalling pathway?

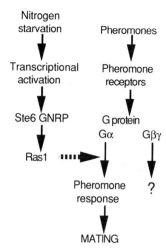

FIGURE 4. Schematic representation of the interaction between *ras1* and the pheromone-signalling pathway in *S. pombe*.

The decision to exit from the cell cycle and initiate mating requires the integration of signals concerning both the availablity of nutrients and the presence of mating partners. Our model is that *ras1* is involved in integrating these two signals by increasing the sensitivity of the pheromone pathway when nutrients are in short supply (Fig. 5). The Ras1 protein itself is activated through GDP-GTP exchange catalyzed by the Ste6 GNRP. Transcription of the *ste6* gene is affected by pheromone as well as by nutritional signals (DAH, unpublished results) suggesting that in the presence of pheromone there is positive feedback to increase the sensitivity of the pheromone-response through activation of Ras1. The molecular mechanism by which the *ras* effector regulates the pheromone pathway is not known. To understand this signal transduction pathway it will be necessary to identify the effectors of both the Ras1 protein and the G protein.

In more complex organisms than yeast, including mammals, heterotrimeric G proteins mediate the effects of many different extracellular signals including some of those regulating cell proliferation (Neer and Clapham, 1988). It will be interesting to find out whether mammalian Ras proteins interact with G protein-mediated pathways to regulate cell proliferation and differentiation. Understanding the interaction between *ras* and the pheromone-signalling pathway in *S. pombe* should provide clues to the more complex interactions between signal transduction pathways in higher cells.

ACKNOWLEDGEMENTS

We thank Dr. C. Shimoda for unpublished results, members of Yamamoto-ken for discussions, and Ms. A. Sugimoto and Mr. H. Kunitomo for help in preparing the manuscript. DAH was supported by a fellowship from the Human Frontier Science Program. This work was supported by the Ministry of Education, Science and Culture of Japan.

REFERENCES

Boy-Marcotte, E., Damak, F., Camonis, J., Garreau, H., and Jacquet, M., 1989, The C-terminal part of a gene partially homologous to the *CDC25* gene suppresses the *cdc25-5* mutation in *Sacharomyces cerevisiae*, *Gene*, 77:21-30.

Broek, D., Toda, T., Michel, T., Levin, L., Birchmeier, C., Zoller, M., Powers, S., and Wigler, M., 1987, The *Saccharomyces cerevisiae CDC25* gene product regulates the *RAS*/adenylate cyclase pathway, *Cell*, 48:789-799.

Camonis, J., and Jacquet, M., 1988, A new *RAS* mutation which suppresses the *CDC25* gene requirement for growth in *Saccharomyces cerevisiae, Mol. Cell. Biol.*, 8:2980-2983.

Camonis, J. H., Kalekine, M., Gondre, B., Garreau, H., Boy-Marcotte, E., and Jacquet, M., 1986, Characterization, cloning and sequence analysis of the *CDC25* gene which controls the cAMP level of *Saccharomyces cerevisiae, EMBO J.*, 5:375-380.

Crechet, J. B., Poullet, P., Mistou, M. Y., Parmeggiani, A., Camonis, J., Boy-Marcotte, E., and Jacquet, M., 1990, Enhancement of the GDP-GTP exchange of *RAS* proteins by the carboxy-terminal domain of *SDC25, Science*, 248:866-868.

Damak, F., Boy-Marcotte, E., Le-Roscouet, D., Guilbaund, R., and Jacquet, M., 1991, *SDC25*, a *CDC25*-like gene which contains a *RAS*-activating domain and is a dispensable gene of *Saccharomyces cerevisiae, Mol. Cell. Biol.*, 11:202-212.

Fukui, Y., and Kaziro, Y., 1985, Molecular cloning and sequence analysis of a *ras* gene from *Schizosaccharomyces pombe, EMBO J.*, 4:687-691.

Fukui, Y., and Yamamoto, M., 1988, Isolation and characterization of *Schizosaccharomyces pombe* mutants phenotypically similar to *ras1⁻, Mol. Gen. Genet.*, 2156:26-31.

Fukui, Y., Kozasa, T., Kaziro, Y., Takeda, T., and Yamamoto, M., 1986a, Role of a *ras* homolog in the life cycle of *Schizosaccharomyces pombe, Cell*, 44:329-336.

Fukui, Y., Kaziro, Y., and Yamamoto, M., 1986b, Mating pheromone-like diffusible factor released by *Schizosaccharomyces pombe, EMBO. J.*, 5:1991-1993.

Fukui, Y., Miyake, S., and Yamamoto, M., 1989, Characterization of the *Schizosaccharomyces pombe ral2* gene implicated in activation of the *ras1* gene product, *Mol. Cell. Biol.*, 9:5617-5622.

Gibbs, J. B., and Marshall, M. S., 1989, The *ras* oncogene - an important regulatory element in lower eukaryotic organisms, *Microbiol. Rev.*, 53:171-185.

Hancock, J. F., Magee, A. I., Childs, J. E., and Marshall, C., 1989, All *ras* proteins are polyisoprenylated but only some are palmitoylated, *Cell*, 57:1167-1177

Hughes, D. A., Fukui, Y., and Yamamoto, M., 1990, Homologous activators of *ras* in fission and budding yeast, *Nature* (London), 344:355-357.

Imai, Y., Miyake, S., Hughes, D. A., and Yamamoto, M., 1991, Identification of a GAP homolog in *Schizosaccharomyces pombe, Mol. Cell. Biol.*, in press.

Jones, S., Vignais, M.-L., and Broach, J. R., 1991, The *CDC25* protein of *Saccharomyces cerevisiae* promotes exchange of guanine nucleotides bound to Ras, *Mol. Cell. Biol.*, 11:2641-2646.

Kataoka, T., Powers, S., McGill, C., Fasano, O., Golfarb, M., Broach, J. R., and Wigler, M., 1984, Genetic analysis of yeast *RAS1* and *RAS2* genes, *Cell*, 37:437-445.

McCormick, F., 1989, *ras* GTPase activating protein: signal transmitter and signal terminator, *Cell*, 56:5-8.

Nadin-Davis, S. A., and Nasim, A., 1990, *Schizosaccharomyces pombe ras1* and *byr1* are functionally related genes of the *ste* family that affect starvation-induced transcription of the mating-type genes, *Mol. Cell. Biol.*, 10:549-560.

Nadin-Davis, S. A., Yang, R. C. A., Narang, S. A., and Nasim, A., 1986a, The cloning and characterization of a *RAS* gene from *Schizosaccharomyces pombe, J. Mol. Evol.*, 23:41-51

Nadin-Davis, S. A., Nasim, A., and Beach, D., 1986b, Involvement of *ras* in sexual differentiation but not in growth control in fission yeast, *EMBO J.*, 5:2963-2971.

Neer, E. J., and Clapham, D. E., 1988, Roles of G protein subunits in transmembrane signalling, *Nature* (London), 333:129-134.

Obara, T., Nakafuku, M., Yamamoto, M., and Kaziro, Y., 1991, Isolation and characterization of the gene encoding G protein α-subunit from *Schizosaccharomyces pombe*: Involvement in mating and sporulation pathways, *Proc. Natl. Acad. Sci. USA.*, in press.

Robinson, L. C., Gibbs, J. B., Marshall, M. S., Sigal, I. S., and Tatchell, K., 1987, *CDC25*: a component of the *RAS*-adenylate cyclase pathway in *Saccharomyces cerevisiae, Science*, 235:1218-1221.

Tanaka, K., Matsumoto, K., and Toh-e, A., 1989, *IRA1*, an inhibitory regulator of the *RAS*-cyclic AMP pathway in *Saccharomyces cerevisiae, Mol. Cell. Biol.*, 9:757-768.

Tanaka, K., Nakafuku, M., Tamanoi, F., Kaziro, Y., Matsumoto, K., and Toh-e, A., 1990, *IRA2*, a second gene of *Saccharomyces cerevisiae* that encodes a protein with a domain homologous to mammalian *ras* GTPase-activating protein, *Mol. Cell. Biol.*, 10:4303-4313.

Tatchell, K., Chaleff, D., DeFoe-Jones, D., and Scolnick, E. M., 1984, Requirement of either of a pair of *ras*-related genes of *Saccharomyces cerevisiae* for spore viability, *Nature* (London), 309:523-527.

Trahey, M., Wong, G., Halenbeck, R., Rubinfield, B., Martin, G., Ladner, M., Long, C., Crosier, W., Watt, K., Koths, K., and McCormick, F., 1988, Molecular cloning of two types of GAP complementary DNA from human placenta, *Science*, 242:1697-1700.

Vogel, U., Dixson, R., Schaber, M., Diehl, R., Marshall, M., Scolnick, E., Sigal, I., and Gibbs, J., 1988, Cloning of bovine GAP and its interaction with oncogenic *ras* p21, *Nature* (London), 335:90-93.

Xu, G., O'Connell, P., Viskochil, D., Cawthon, R., Robertson, M., Culver, M., Dunn, D., Stevens, J., Gesteland, R., White, R., and Weiss, R., 1990, The neurobibromatosis type 1 gene encodes a protein related to GAP, *Cell*, 62:599-608.

CHARACTERIZATION OF NORMAL AND MUTANT HUMAN KIRSTEN-RAS (4B)

p21 AND OF THE CATALYTIC DOMAIN OF GAP

Peter N. Lowe, Susan Rhodes, Susan Bradley and
Richard H. Skinner

Department of Cell Biology, Wellcome Research
Laboratories, Langley Court, Beckenham, Kent BR3
3BS, U.K.

INTRODUCTION

Point mutations in the Harvey-, N and Kirsten-ras genes
are associated with 20-30% of human tumours[1]. Kirsten-ras is
the gene most frequently activated in tumours. Certain tumour
types contain high frequencies of K-ras mutation , for example
colorectal (c. 50%) and pancreatic (>95%) tumours[2-4]. The high
incidence of K-ras mutations in clinically important tumours
makes the K-ras gene product of particular interest as a
target for the discovery of potentially highly selective
anti-tumour agents. Hitherto, biochemical studies on the ras
proteins has focused on the products of the H- and N-ras
genes. The product of the viral Kirsten-ras gene has been
studied but it is the counterpart of a rare transcript of the
cellular K-ras gene (transcript 2A or 4A). The commonly
occuring form of K-ras (2B or 4B) expressed in human cells
differs from the 4A transcript in having a lysine-rich
C-terminus terminus, which is not modified by palmitoylation[5].

We have overcome problems of C-terminal proteolysis in
the lysine-rich region by using an appropriate E.coli
expression strain[6]. We have compared the biochemical
characteristics of the protein with the H-ras gene.

In order to facilitate studies of the interaction
between ras and its GTPase-activating protein (GAP) we have
developed an immunoaffinity purification procedure for the
C-terminal catalytic domain and devised a fluorimetric assay
for assay of GTPase activity.

The Superfamily of ras-Related Genes
Edited by D.A. Spandidos, Plenum Press, New York, 1991

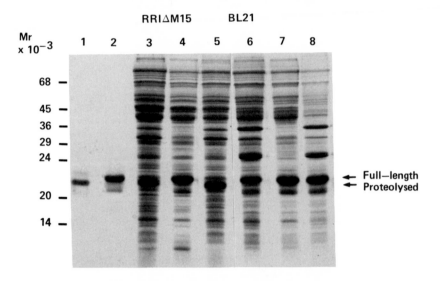

Fig. 1. Broken RR1ΔM15 cells but not BL21 cells degrade
Kirsten-ras (4B) protein. Cultures of E.coli
strains RR1ΔM15 and BL21 were grown overnight and
harvested. The cells were broken by sonication and
centrifuged at 150,000g. The pellets were washed and
resuspended in a volume of buffer equal to the
supernatant. Each of these samples was incubated with
purified full-length K-ras (4B) protein for 10min at
4°C.Samples were then analyzed by SDS polyacrylamide
gel electrophoresis. Lane 1, purified proteolysed
K-ras p21; lane 2, purified full-length K-ras p21;
lanes 3-5, RR1ΔM15 after incubation with full-length
K-ras p21: broken cell suspension (lane 3), 150,000g
supernatant (lane 4), 150,000g pellet (lane 5); lanes
6-8, BL21 after incubation with full-length K-ras p21:
broken cell suspension (lane 6), 150,000g supernatant
(lane 7), 150,000g pellet (lane 8).

RESULTS

Expression of Kirsten-ras(4B) protein

cDNA clones of the human Kirsten ras(4B) gene
transcript were expressed in E.coli under the control of the
lac promoter[6]. Expression in E.coli strain RR1ΔM15 resulted in
high level expression of recombinant protein of the
anticipated Mr. However, when extracts of broken cells were
analysed by SDS-polyacrylamide gel electrophoresis, it was
apparent that the mobility of the band had increased. This
increase in mobility was due to C-terminal proteolysis.

However, when the protein was expressed in BL21 cells, cell breakage did not induce proteolysis. We attribute this difference in behaviour between strains to the lack of the ompT protease in strain BL21[7]. The ompT protease cleaves between adjacent basic residues[7] such as are found in the C-terminal region of K-ras(4B). Consistent with this hypothesis, broken RR1ΔM15 cells, but not BL21 cells, causes the mobility of purified full-length Kirsten-ras p21 to increase. Furthermore, the proteolysis activity is localized to a 150,000g pellet fraction, where the ompT protease would be found (Fig 1). Since several ras-related proteins have C-termini with similar basic regions, this observation might be useful for the expression of these proteins in a full-length form.

The protein expressed in BL21 cells was soluble and could be purified easily in the absence of denaturants.[6] The full-length Kirsten-ras protein, unlike Harvey-ras or N-ras p21, is basic. This facilitates purification since it binds strongly to cation exchangers such as S-Sepharose and Mono S. The protein which has been subjected to the ompT protease ("proteolysed" p21) neither binds to Q-Sepharose nor S-Sepharose, consistent with the loss of most of the lysine residue in the C-terminus. In addition to these two forms we have expressed and purified a truncated protein corresponding to residues 1-166 of the protein, an analogous fragment to that of the H-ras gene which has produced well-ordered crystals of that protein.

Kinetic and biological properties of recombinant Kirsten-ras p21

The rates of dissociation of GDP and of hydrolysis of bound GTP of the c-Kirsten-ras protein are very similar to that of c-Harvey-ras protein[8-10]. Mutation of residue 12 to valine results in similar changes in both K and H-ras proteins. As with H-ras[10], truncation of the protein does not significantly affect these parameters. However, one difference between Kirsten-ras and Harvey ras proteins was in the relative binding constants of GDP and GTP (Table 1).

The full-length Kirsten-ras protein (but not the processed form) was active when microinjected into Swiss 3T3 cells, confirming the quality of the protein produced[6].

Expression, purification and assay of GAP

It has been demonstrated that the C-terminal region of GAP (residues 701-end) can be expressed and is fully functional in binding to ras protein and activating ras-GTPase activity.[11] Levels of expression of this catalytic domain in E.coli (in a soluble form) are not very high. Thus, we have developed an immunoaffinity purification to facilitate its production in amounts sufficient for binding and structural studies. The method used is based on that described by

<u>Table 1 Nucleotide binding and hydrolysis properties of recombinant K- and H-ras proteins</u>

Protein	Length	k_{off}(GDP) (min^{-1})	k_{hyd}	k_d(GTP)/k_d(GDP)
K-Val 12	Full	0.0020	n.d.	4.2
K-Val 12	Truncated	0.0022	0.0011	4.0
H-Val 12	Full	0.0023[a]	0.0027	1.3
K-Gly 12	Proteolyzed	0.0048	0.013	2.7
K-Gly 12	Truncated	0.0056	0.011	2.4
H-Gly 12	Full	0.0063	0.011	0.8

37ºC, 50mM-Tris/HCl, pH7.5, 50mM NaCl, 0.8mM EDTA and 10mmM MgCl$_2$, or without MgCl$_2$ for binding ratios.
[a]Data from John et al[10]

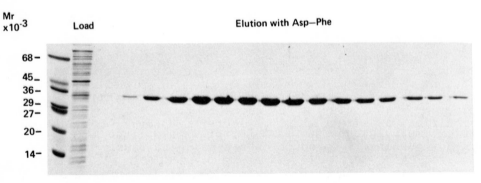

Fig. 2. Immunoaffinity purification of the C-terminal catalytic domain of GAP. E.coli cells containing a plasmid expressing the C-terminal catalytic domain of human GAP engineered so as to add Glu-Glu-Phe as the C-terminal three amino acids were harvested and broken by sonication. A 150,000g supernatant fraction was loaded on to a column of YL1/2 antibody covalently coupled to a column of Sepharose. After extensive washing, the column was eluted with 5mM-Asp-Phe dipeptide, which specifically eluted the recombinant GAP. Samples were analysed by SDS polyacrylamide gel electrophoresis, followed by staining with Coomassie blue R250.

Stammers et al[12] for the purification of HIV proteins. Using PCR, residues 704-1047 of human GAP were cloned and engineered so as to be preceeded by a methionine residue and followed by the three amino acids Glu-Glu-Phe, as the C-terminus of the encoded protein.

Glu-Glu-Phe-COOH is an epitope for the YL1/2 monoclonal anti-tubulin antibody[12-14]. The affinity for the epitope is quite low (ca. 1μM) and thus the interaction can be easily disrupted, for example by 5mM Asp-Phe dipeptide. However, the antibody/epitope affinity is sufficient to allow binding of proteins with this C-terminal epitope to a column of immobilized antibody, and to achieve specific elution of highly purified protein with the dipeptide[12]. The engineered GAP protein was expressed in E.coli and although most of the protein was insoluble, high yields of recombinant GAP protein could be purified from the supernatant by elution from the immobilized antibody column (Fig 2). The eluted protein was very pure and had a similar activity to that of full-length GAP.[15]

GAP increases the rate of conversion of p21.GTP to p21.GDP and P_i. Previously, discontinuous radiometric assays have been used either to measure conversion of GTP to GDP or to monitor production of P_i. To assist kinetic measurements we desired an easy continuous procedure. Since P_i is released from the p21.GDP complex, we utilized a fluorimetric assay for P_i production, previously used for measurement of ATPase activity[16]. In this procedure, purine nucleoside phosphorylase is used to catalyse the following reaction:

$$7\text{-Methylguanosine} \; + \; P_i \longrightarrow 7\text{-Methylguanine} \; + \; P_i$$

Since 7-methylguanosine has a higher fluorescence than 7-methylguanine, P_i production can be continuously monitored as a reduction in the fluorescence (λ_{ex}=300, λ_{em}=410). The reaction can either be performed under single-turnover conditions starting with p21.GTP complex (Fig 3), or by appropriate choice of EDTA and Mg^{2+} concentrations under multiple turnover conditions so that a linear rate of P_i release occurs. This procedure should prove useful for determing the catalytic parameters for GAP activation and for measurement of inhibition constants by proteins such as H-ras leu-61.GTP.

ACKNOWLEDGEMENTS

We would like to thank Dr D.K. Stammers for making unpublished data available and for use of a column of immobilized YL1/2. We are indebted to Dr. F.W. Studier for use of BL21/DE3.

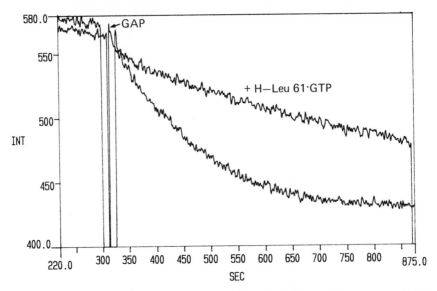

Fig. 3. Fluorimetric assay of GAP activity. Measurements were performed at 37°C in a buffer consisting of 10mM-Tris/HCl, pH7.5, 5mM-MgCl$_2$, 150μM 7-methylguanosine and 0.5 Units/ml of calf-spleen purine nucleoside phosphorylase. Fluorescence was monitored at λ_{ex}=300, λ_{em}=410. Truncated K-ras p21, complexed with GTP, was added to a concentration of 0.5mg/ml. Recombinant GAP (1μg/ml) was then added. In the upper trace, H-ras leu-61.GTP complex (0.2mg/ml) was also present.

REFERENCES

1. J.L. Bos, The ras gene family and human carcinogenesis, <u>Mutat. Res.</u>, 195:255 (1988).
2. S.Kahn, F. Yamamoto, C. Almoguera, E. Winter, K. Forrester, J. Jordano, and M. Perucho, 1987, The c-K-ras gene and human cancer (review), <u>Anticancer Res.</u> 7:639 (1987)
3. C. Almoguera, D. Shibata, K. Forrester, J. Martin, N. Arnheim and M. Perucho, Most human carcinomas of the exocrine pancreas contain mutant c-K-ras genes, <u>Cell</u>, 53:549.
4. V.T.H.B.M. Smit, A.J.M. Boot, A.M.M. Smits, G.J. Fleuren, C.J. Cornelisse and J.L. Bos, Kras codon 12 mutations occur very frequently in pancreatic adenocarcinomas, <u>Nucl. Acids Res.</u> 16:7773 (1988).
5. J.E. Hancock, A.I. Magee, J.E. Childs and C.J. Marshall, All ras proteins are polyisoprenylated but only some are palmitoylated, <u>Cell</u>, 57:1167(1989)

6. P.N. Lowe, M.J. Page, S. Bradley, S. Rhodes, M. Sydenham, H. Paterson and R.H. Skinner, Characterization of recombinant human Kirsten-ras(4B) p21 produced at high levels in Escherichia coli and insect baculovirus expression systems, J. Biol. Chem. 266:1672(1991).

7. J. Grodberg and J.J. Dunn, ompT encodes the Escherichia coli outermembrane protease that cleaves T7 RNA polymerase during purification, J. Bacteriol. 170:1245 (1988).

8. J. Feuerstein, H.R. Kalbitzer, J. John, R.S. Goody and A. Wittinghofer,Characterisation of the metal-ion-GDP complex at the active sites of transforming and nontransforming p21 proteins by observation of the ^{17}O-Mn superhyperfine coupling and by kinetic methods, Eur. J. Biochem. 162:49(1987).

9. J. John, M. Frech, and A. Wittinghofer,Biochemical properties of Ha-ras encoded p21 mutants and mechanism of the autophosphorylation reaction, J. Biol. Chem. 263:11792(1988).

10. J. John, I. Schlichting, E. Schilz, P. Rosch, and A. Wittinghofer, C-terminal truncation of p21^H preserves crucial kinetic and structural properties, J. Biol. Chem. 264:13086 (1989).

11. M.S. Marshall, W.S. Hill, A.S. Ng, U.S. Vogel, M.D. Schaber, E.M. Scolnick, R.A.F. Dixon, I.S. Sigal and J.B. Gibbs, A C-terminal domain of GAP is sufficient to stimulate ras p21 GTPase activity, EMBO J. 8:1105 (1989)

12. D.K. Stammers, M. Tisdale, S. Court, V. Parmar, C. Bradley, and C.K. Ross, Rapid purification and characterisation of HIV-1 reverse transcriptase & RNaseH engineered to incorporate a C-terminal tripeptide α-tubulin epitope, FEBS Lett. in press (1991).

13. J. Wehland, H.C. Schroder, and K. Weber, Amino acid sequence requirements in the epitope recognized by the α-tubulin-specific rat monoclonal antibody YL1/2, EMBO J. 3:1295 (1984)

14. J.V. Kilmartin, B. Wright, and C. Milstein, Rat monoclonal antitubulin antibodies derived by using a new nonsecreting rat cell line, J.Cell Biol. 93:576 (1982)

15. R. Halenbeck, W.J. Crosier, R. Clark, F. McCormick and K. Koth,Purification, characterization , and Western blot analysis of human GTPase-activating protein from native and recombinant sources, J.Biol.Chem. 265:21922 (1990).

16. U. Bank and S. Roy, A continuous fluorimetric assay for ATPase activity, Biochem. J. 266:611 (1990).

THE SIGNAL TRANSDUCTION PATHWAY UPSTREAM OF CDC25 - RAS - ADENYLATE CYCLASE IN THE YEAST *SACCHAROMYCES CEREVISIAE* AND ITS RELATIONSHIP TO NUTRIENT CONTROL OF CELL CYCLE PROGRESSION

Johan M. Thevelein, Linda Van Aelst, Peter Durnez and Stefan Hohmann

Laboratorium voor Cellulaire Biochemie
Katholieke Universiteit te Leuven
Kardinaal Mercierlaan 92
B-3001 Leuven-Heverlee, Flanders, Belgium

INTRODUCTION

In recent years several groups have made great efforts to unravel the function of the *RAS* genes in yeast in the hope of providing a model which could lead to a better understanding of the physiological role of mammalian ras genes and their oncogenic alleles. This research has led to two crucial breakthroughs: 1. The RAS proteins in yeast regulate adenylate cyclase activity in a way similar to the G_S proteins of mammalian adenylate cyclase (Gibbs and Marshall, 1989) and 2. cAMP and hence also the RAS proteins, are involved in the control of progression over the 'start' (or decision) point in the G_1 phase of the yeast cell cycle (Gibbs and Marshall, 1989). Particularly striking findings were that strains with yeast homologues of mammalian ras oncogenes and strains with elevated cAMP levels or elevated activity of cAMP-dependent protein kinase were unable to arrest at the 'start' point of the cell cycle under conditions of nutrient deprivation (Toda *et al.*, 1985; Sass *et al.*, 1986). Under these conditions wild type yeast cells arrest at 'start' and subsequently enter a resting state called G_0 (Pringle and Hartwell, 1981). On the other hand, strains with temperature-sensitive mutations in RAS or adenylate cyclase arrested at 'start' when shifted to the restrictive temperature (Matsumoto *et al.*, 1985; De Vendittis *et al.*, 1986). Subsequently, several other components of the RAS-adenylate cyclase pathway have been identified, including CDC25 which is required for activation of RAS (Camonis *et al.*, 1986), the RAS-GTPase-activating proteins IRA1 and IRA2 (Tanaka *et al.*, 1989, 1990), the SRV2 gene product (or 'CAP'), a subunit of adenyl cyclase (Fedor-Chaiken *et al.*, 1990; Field *et al.*, 1990) and the genes coding for cAMP phosphodiesterase (*PDE1, PDE2*) (Sass *et al.*, 1986; Nikawa *et al.*, 1987b) and the catalytic (*TPK1, TPK2, TPK3*) (Toda *et al.*, 1987a) and regulatory (*BCY1*) (Toda *et al.*, 1987b) subunits of cAMP-dependent protein kinase were identified.

In spite of the continuous progress in identifying new components of the pathway, very little progress has been made in understanding the signaling function of the RAS signal transmission pathway: what is the exact nature of the signal transmitted by the RAS pathway to adenylate cyclase? Because of the crucial importance of nutrients for cell proliferation in micro-organisms in general and for progression over the 'start' point of the yeast cell cycle in particular one could reasonably suppose that the presence of an adequate nutritional status was being signaled to the cyclase by means of this pathway (Gibbs and Marshall, 1989; Dumont *et al.*, 1989).

The Superfamily of ras-Related Genes
Edited by D.A. Spandidos, Plenum Press, New York, 1991

GLUCOSE-INDUCED PROTEIN PHOSPHORYLATION CASCADE

In recent years, several groups have been investigating the nature of the post-translational mechanisms responsible for the rapid metabolic changes which are triggered in derepressed (respiring) yeast cells by addition of fermentable sugar (glucose, fructose, sucrose): inhibition of gluconeogenesis (caused by inactivation of fructose-1,6-bisphosphatase, isocitrate lyase and other gluconeogenic enzymes), inhibition of galactose and maltose uptake (caused by inactivation of the respective carriers), stimulation of glycolysis (caused by activation of phosphofructokinase 2) and mobilization of the storage sugar trehalose (caused by activation of trehalase) etc. (Mazon *et al.*, 1982; Purwin *et al.*, 1982; François *et al.*, 1984, 1988; Thevelein, 1984a; Peinado and Loureiro-Dias, 1986, Lopez-Boado *et al.*, 1987, 1988). These studies converged on the finding that a cAMP-triggered protein phosphorylation cascade was responsible for a large number of the rapid metabolic effects (Holzer, 1984; Thevelein, 1984b, 1988). The relationship of this cascade with the mechanism of glucose repression remains unclear. Glucose repression is one of the major metabolic regulation mechanisms in yeast. It is a long-term effect: it blocks the transcription of many yeast genes resulting in the continuous shutdown of important sections of yeast metabolism (mitochondrial respiration, uptake and metabolism of alternative carbon substrates, gluconeogenesis, etc.) when the cells grow on fermentable sugar (Entian, 1986). Although the main glucose repression mechanism does not seem to involve cAMP as second messenger (Matsumoto *et al.*, 1982), for glucose repression of alcohol dehydrogenase II, encoded by the *ADH2* gene (Cherry *et al.*, 1989), catalase T, encoded by the *CTT1* gene (Bissinger *et al.*, 1989), the heat shock protein encoded by *HSP12* (Praekelt and Meacock, 1990), one of the ubiquitin encoding genes (*UBI4*) (Tanaka *et al.*, 1988) and the SSA3 gene product (Werner-Washburne *et al.*, 1989), evidence was presented suggesting involvement of the RAS-adenylate cyclase pathway in the repression mechanism (see further).

GLUCOSE- AND ACIDIFICATION-INDUCED CAMP SIGNALING

The addition of fermentable sugar to derepressed yeast cells causes a rapid signal-like spike in the cAMP level which, based on studies with temperature-sensitive mutants in cAMP synthesis, appeared to trigger the protein phosphorylation cascade (François *et al.*, 1984; Tortora *et al.*, 1984). Transient membrane depolarization, transient intracellular acidification and increased energy supply were ruled out as triggers for the cAMP signal (Purwin *et al.*, 1986; Caspani *et al.* 1985; Eraso *et al.*, 1987; Thevelein *et al.*, 1987a,b). Later on, evidence was obtained that induction of the cAMP signal by fermentable sugar was mediated by the CDC25-RAS-adenylate cyclase pathway (Mbonyi *et al.*, 1988; Munder and Küntzel, 1989; Van Aelst *et al.*, 1990, 1991a). In this way, fermentable sugar was identified as the first physiological activator of the RAS-adenylate cyclase pathway in yeast. Detection of fermentable sugar in the medium is obviously of crucial importance to yeast cells. Activation of the RAS pathway is only triggered by fermentable sugars and only in concentrations ($K_{1/2} = \pm 15\text{-}20$ mM) which result in fermentation of sugar to ethanol. Lower concentrations of these sugars are metabolized by means of respiration and do not activate the pathway (Beullens *et al.*, 1988; Mbonyi *et al.*, 1990; Thevelein and Beullens, 1985). Therefore, the RAS pathway in yeast appears to serve as a specific sensor for the detection of fermentable sugars in concentrations appropriate for fermentation. Whether the low-affinity glucose carriers play an active role in the sensing mechanism or whether there exists a special receptor for the glucose is unclear. Whatever the nature of the glucose receptor, it clearly has a much lower affinity for glucose than the sugar kinases or the high-affinity sugar transport system. A deletion mutant in the *SNF3* gene, which encodes a component of the high-affinity glucose transport system (Neigeborn *et al.*, 1986; Bisson *et al.*, 1987), is not affected in glucose-induced cAMP signaling (Mbonyi and Thevelein, 1988). A tentative model of the signaling pathway is shown in Fig.1. Most probably the pathway is more complicated. Recently, two mutants (*fdp1* and *byp1*) specifically affected in glucose-induced cAMP signaling and one mutant (*lcr1*) affected both in glucose- and acidification-induced cAMP signaling (but not in basal cAMP synthesis) have been identified (Van Aelst *et al.* 1991b, Hohmann S., Huse K., Valentin E., Mbonyi K., Thevelein J.M. and Zimmermann F.K., manuscript in preparation, M. Vanhalewyn, unpublished results). The *FDP1* gene product appears to be required highly upstream in the signaling pathway and might even function as the actual glucose receptor for most, if not all, glucose-induced signaling pathways, since all glucose-induced regulatory phenomena investigated up to now in this mutant are deficient (van de Poll *et al.* 1974, Van Aelst *et al.* 1991b, Van Aelst L. and Becher dos Passos J., unpublished results). The *byp1* mutant shows less deficiencies in glucose-induced regulatory phenomena than the *fdp1* mutant. The *BYP1* gene product therefore appears to be required in the glucose-induced signaling pathway(s) downstream from the *FDP1* gene product (Hohmann S., Huse K., Valentin E., Mbonyi K., Thevelein J.M. and Zimmermann F.K., manuscript in preparation.) Evidence has also been provided that the transient nature of the cAMP increase is due to feedback-inhibition by cAMP-dependent protein kinase (Mbonyi *et al.*, 1990). The existence of strong feedback-inhibition of cAMP synthesis by cAMP-dependent protein kinase was discovered by the group of Wigler. Yeast strains with very low activity of cAMP-dependent protein kinase display 100 to 1000-fold increased cAMP levels (Nikawa *et al.*, 1987a). Experiments with different carbon sources in these strains have confirmed the specificity of fermentable sugar as trigger for the stimulation of cAMP synthesis (Mbonyi *et al.*, 1990).

Intracellular acidification also stimulates the RAS-adenylate cyclase pathway. As shown in Fig.1, this occurs at a point downstream of the first part of the pathway which is only required for stimulation by fermentable sugar (Mbonyi *et al.*, 1988; Argüelles *et al.*, 1990; Van Aelst *et al.*, 1991a). Hence, intracellular acidification apparently constitutes another physiological trigger of the RAS-adenylate cyclase pathway in yeast. This makes sense in view of the stimulating effect of the pathway on the mobilization of reserve carbohydrates and the generation of energy. Increased energy production is an obvious requirement to overcome stress conditions and under such conditions intracellular acidification might generally be expected to occur in yeast cells (Goffeau and Slayman, 1981; Serrano, 1984; D'Amore and Stewart, 1987; Pascual *et al.*, 1988). Acidification-induced stimulation of the RAS-adenylate cyclase pathway might therefore constitute a rescue mechanism for cells suffering from stress conditions.

PHYSIOLOGICAL ROLE OF GLUCOSE-INDUCED SIGNALING

The discovery of fermentable sugar as trigger for the RAS pathway in yeast appeared to support the generally held (Dumont *et al.*, 1989; Gibbs and Marshall, 1989; Broach, 1991), but poorly substantiated idea that nutrients act as stimulators of this pathway. Recent studies however with glucose repression mutants (Beullens *et al.*, 1988; Mbonyi *et al.*, 1990) and with mutants affected in derepression (Argüelles *et al.*, 1990) showed that the sugar-induced activation pathway of RAS-adenylate cyclase is a glucose-repressible pathway. The glucose-induced cAMP signal is absent in glucose-repressed wild type cells and in cells of mutants (*cat1, cat3*) unable to show derepression (Beullens *et al.*, 1988; Argüelles *et al.*, 1990). In addition, mutants lacking glucose repression such as *hxk2* (Beullens *et al.*, 1988) and *hex2* (J.C. Argüelles, unpublished results) and mutants with strongly reduced activity of cAMP-dependent protein kinase showing deficient glucose repression (Mbonyi *et al.*, 1990), show glucose-induced stimulation of cAMP synthesis when grown on glucose. The latter was also observed in strains with overexpression of *CDC25* which were subsequently found to be deficient in glucose repression (Van Aelst *et al.*, 1991a). All results obtained in this respect are consistent with the presence of a glucose-repressible protein in the pathway

(Fig.1), although this protein has not yet been identified. The putative glucose-repressible protein should be located before the point where intracellular acidification triggers the pathway since the acidification-induced cAMP increase still occurs in glucose-repressed cells or mutants deficient in derepression (on condition that these cells, which lack respiration, are provided with an ATP level which is adequate for adenylate cyclase functioning) (Argüelles *et al.*, 1990). The finding that the glucose-activation pathway of RAS-adenylate cyclase is glucose-repressible has important consequences. First of all, it confines the physiological role of this pathway to the transition period between the respirative-gluconeogenic growth mode and the fermentative growth mode. In this way the action of glucose on derepressed yeast cells by means of this pathway is very short-lived. Its action period might be shortened further by glucose-induced inactivation of one of the components of the signal transmission pathway (e.g. the putative glucose-repressible protein), since glucose-induced activation of cAMP signaling disappears faster than would be expected from glucose repression alone (Thevelein and Beullens, 1985). A second consequence is that the pathway cannot act as a trigger for progression over the 'start' point of the yeast cell cycle since the pathway is no longer operative during exponential growth on glucose. This conclusion was confirmed by the finding that yeast mutants completely lacking glucose-induced cAMP signaling show normal growth on glucose (Van Aelst *et al.*, 1990; J.C. Argüelles and M. Vanhalewyn, unpublished results). If fermentable sugar is involved in some way in pushing yeast cells over the 'start' point of the cell cycle, it is not by means of the RAS-adenylate cyclase pathway.

The idea that activation of the RAS-adenylate cyclase pathway by glucose is shut off during exponential growth on glucose is supported by recent data on the involvement of the pathway in glucose repression of the *CTT1* gene which encodes catalase T. An impressive body of genetic evidence is available showing that increased cAMP levels or protein kinase A activity cause repression of transcription of the *CTT1* gene and also of the *UBI4*, *ADH2*, *SSA3* and *HSP12* genes, while lowered cAMP levels or protein kinase A activity lead to derepression (Tanaka *et al.*, 1988; Bissinger *et al.*, 1989; Cherry *et al.*, 1989; Werner-Washburne *et al.*, 1989; Boorstein and Craig, 1990; Praekelt and Meacock, 1990). However, starvation of wild type yeast cells for nitrogen on a glucose-containing medium also leads to derepression of *CTT1* transcription (Bissinger *et al.*, 1989). Apparently, the glucose in the medium is unable to cause repression of *CTT1* transcription and since glucose-induced stimulation of the RAS-adenylate cyclase signaling pathway does not require the presence of a nitrogen source (it is observed in nitrogen-starved cells: K. Hirimburegama, unpublished results), it demonstrates that glucose activation of the RAS-adenylate cyclase pathway is shut off in the cells under these conditions. The extent of CTT1 derepression observed upon nitrogen starvation in the presence of glucose is even higher than derepression during growth on a nonfermentable carbon source, indicating first that the glucose-activation pathway of RAS adenylate cyclase is completely shut off under these conditions and second that the presence of a nitrogen source is more important for CTT1 repression than the nature of the carbon source. Starvation for phosphate and sulfate on a glucose containing medium also causes derepression of *CTT1* transcription (Bissinger *et al.*, 1989). It is not known however whether phosphate and sulfate starvation affect glucose-activation of the RAS-adenylate cyclase pathway. These data therefore cannot be taken as evidence that glucose-activation of the RAS-adenylate cyclase pathway is shut off in glucose-repressed cells. With respect to this problem, it is important to make a clear distinction between the mechanism of glucose repression of the CTT1, UBI4, ADH2, SSA3 and HSP12 gene products and the main glucose repression mechanism (see higher) which is not affected by nitrogen starvation.

NITROGEN-SOURCE INDUCED SIGNALING

Starvation of yeast cells for nitrogen is well known to cause arrest at the 'start' point in G_1 of the cell cycle (Pringle and Hartwell, 1981). No effect at all however

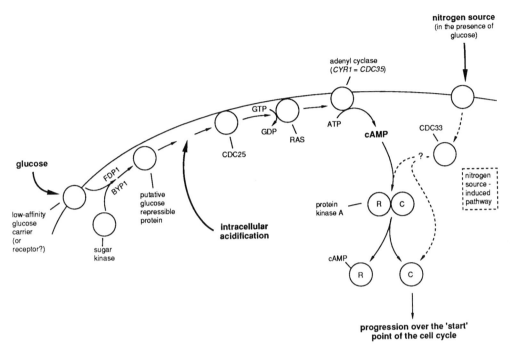

Fig. 1 Tentative model of the glucose- and acidification-induced signal transmission pathway, leading to activation of the RAS-adenylate cyclase system in yeast and its relationship with nitrogen-induced cell cycle progression in cells growing on glucose. CDC25 and RAS activity are required both for glucose- and acidification-induced cAMP signaling. Sugar kinase activity, the putative glucose repressible protein and the *FDP1* and *BYP1* gene products are only required for activation by glucose. Intracellular acidification activates the pathway downstream of the putative glucose-repressible protein. Whether the low-affinity glucose carriers play an active role in the induction mechanism or whether there exists a special glucose receptor is unclear. Because of the presence of the glucose-repressible protein, glucose-induced activation of the RAS-adenylate cyclase system is confined to the transition period during which glucose derepressed cells become glucose repressed. Nitrogen signaling uses a separate pathway which involves the CDC33 gene product and which is postulated to activate protein kinase A at constant cAMP level. This activation is suggested to be responsible for progression over the 'start' point of the cell cycle in glucose-repressed cells. (Only components involved in activation of the pathway are shown. Not shown are the IRA1 and IRA2 gene products which stimulate RAS-GTPase activity, the *PDE1* and *PDE2* encoded low- and high-affinity phosphodiesterases, feedback-inhibition of protein kinase A on cAMP synthesis and the SRV2 or 'CAP' subunit of adenyl cyclase.)

of nitrogen is observed on cAMP levels in vegetative yeast cells (Thevelein, 1984a; Thevelein and Beullens, 1985; François *et al.*, 1988), including nitrogen-starved cells refed with a nitrogen source (K. Hirimburegama and P. Durnez, unpublished results). In spite of this, nitrogen sources are able to cause rapid and drastic activation or inactivation (in the presence of fermentable sugar) of enzymes known to be regulated by cAMP-dependent protein phosphorylation, such as trehalase (Thevelein, 1984a; Thevelein and Beullens, 1985; K. Hirimburegama and P. Durnez, unpublished results), phosphofructokinase 2, glycogen synthase and phosphorylase (François *et al.*, 1988). It is important to note that this nitrogen-source induced activation also requires the presence of glucose or another fermentable carbon source (Thevelein and Beullens, 1985; François *et al.*, 1988). Derepression of *CTT1* transcription upon glucose exhaustion in a nitrogen-containing medium is also consistent with the notion that a glucose-derived factor is required for the nitrogen-source effect. As outlined before however this factor cannot be stimulation of the RAS-adenylate cyclase pathway.

In our view, an explanation for the interaction between cAMP and nitrogen sources for control of progression over the 'start' point of the yeast cell cycle will require a novel mechanism for activation of cAMP-dependent protein kinase, in particular, a mechanism able to activate the kinase at constant cAMP level. We suggest that an unidentified factor generated by the nitrogen source-induced pathway is able to activate protein kinase A synergistically with cAMP (Fig.1). This would allow for an explanation of all cAMP-related effects but at the same time also introduce the possibility for activation of the kinase at constant cAMP level by the nitrogen source in the medium. An alternative model would be that the nitrogen source-induced pathway is able to activate free (and only free) catalytic subunits (Fig.1). Because of the equilibrium between catalytic and regulatory subunits, activation would in this case also indirectly depend on the cAMP level and activation would also be possible at constant cAMP level. The main difference between the two models is that in the first model the maximum protein kinase activity obtainable is not modified by the presence of the nitrogen source whereas in the second model the maximum activity would be higher.

The existence of cell cycle mutants, such as *cdc33* or *cdc60* which arrest at exactly the same point in the cell cycle as nutrient-starved or cAMP-deficient cells, but which are not at all affected in cAMP metabolism (Bedard *et al.*, 1981; Verdier *et al.*, 1989; J. Keleman, unpublished results), appears to support the idea of a separate nitrogen-source induced pathway leading to activation of protein kinase A. Recent results from our laboratory have shown that the CDC33 gene product, which encodes eukaryotic protein synthesis initiation factor 4E (eIF-4E), is required for nitrogen-source induced activation of trehalase (J. Keleman, unpublished results). In addition, the finding by the group of Matsumoto that the *bcy1* mutation suppresses specific 'start' arrest of the *cdc33* mutant (Brenner *et al.*, 1988) supports a localisation of the CDC33 gene product upstream of protein kinase A (Fig.1).

In our view, the true trigger for protein kinase A stimulated progression over the 'start' point of the yeast cell cycle in media containing a fermentable carbon source is the nitrogen source-induced pathway and not the sugar-induced RAS-adenylate cyclase pathway. Functioning of the RAS-adenylate cyclase system in progression over the 'start' point of the cell cycle in glucose-repressed cells is more likely confined to providing a basal level of cAMP rather than providing the second messenger which, in response to an extracellular signal, triggers this process.

ACKNOWLEDGEMENTS

Research carried out in our laboratory was supported by the Belgian National Fund for Scientific Research (FGWO, 'Kom op tegen Kanker'), the National Bank of

Belgium, the Belgian National Lottery, the Research Fund of the K.U.Leuven and the North Atlantic Treaty Organization.

REFERENCES

Argüelles, J. C., Mbonyi, K., Van Aelst, L., Vanhalewyn, M., Jans, A. W. H., and Thevelein, J. M., 1990, Absence of glucose-induced cAMP signaling in the *Saccharomyces cerevisiae* mutants *cat1* and *cat3* which are deficient in derepression of glucose-repressible proteins, *Arch. Microbiol.*, 154: 199.

Bedard, D. P., Johnston, G. C., and Singer, R. A., 1981, New mutations in the yeast *Saccharomyces cerevisiae* affecting completion of 'Start', *Curr. Genet.*, 4: 205.

Beullens, M., Mbonyi, K., Geerts, L., Gladines, D., Detremerie, K., Jans, A. W. H., and Thevelein, J. M., 1988, Studies on the mechanism of the glucose-induced cAMP-signal in glycolysis- and glucose repression-mutants of the yeast *Saccharomyces cerevisiae, Eur. J. Biochem.*, 172: 227.

Bissinger, P. H., Wieser, R., Hamilton, B. and Ruis, H., 1989, Control of *Saccharomyces cerevisiae* catalase T gene *(CTT1)* expression by nutrient supply via the RAS-cyclic AMP pathway, *Mol. Cell. Biol.*, 9: 1309.

Bisson, L. F., Neigeborn, L., Carlson, M., and Fraenkel, D. G., 1987, The *SNF3* gene is required for high-affinity glucose transport in *Saccharomyces cerevisiae, J. Bacteriol.*, 169: 1656.

Boorstein, W. R., and Craig, E. A., 1990, Regulation of a yeast HSP70 gene by a cAMP responsive transcriptional control element, *EMBO J.*, 9: 2543.

Broach, J. R., 1991, *RAS* genes in *Saccharomyces cerevisiae*: signal transduction in search of a pathway, *Trends Genet.*, 7: 28.

Brenner, C., Nakayama N., Goebl, M., Tanaka, K., Toh-e, A., and Matsumoto, K., 1988, *CDC33* encodes mRNA Cap-Binding Protein eIF-4E of *Saccharomyces cerevisiae, Mol. Cell. Biol.*, 8: 3556.

Camonis, J. H., Kalékine, M., Gondré, B., Garreau, H., Boy-Marcotte, E., and Jacquet, M., 1986, Characterization, cloning and sequence analysis of the *CDC25* gene which controls the cyclic AMP level of *Saccharomyces cerevisiae, EMBO J.*, 5: 375.

Caspani, G., Tortora, P., Hanozet, G. M., and Guerritore, A., 1985, Glucose-stimulated cAMP increase may be mediated by intracellular acidification in *Saccharomyces cerevisiae, FEBS Lett.*, 186: 75.

Cherry, J. R., Johnson, T. R., Dollard, C., Shuster, J. R., and Denis, C. L., 1989, Cyclic AMP-dependent protein kinase phosphorylates and inactivates the yeast transcriptional activator ADR1, *Cell*, 56: 409.

D'Amore, T., and Stewart, G. G., 1987, Ethanol tolerance of yeast, *Enzyme Microb. Technol.*, 9: 322.

De Vendittis, E., Vitelli, A., Zahn, R., & Fasano, O., 1986, Suppression of defective RAS1 and RAS2 functions in yeast by an adenylate cyclase activated by a single aminoacid change, *EMBO J.*, 5: 3657.

Dumont, J. E., Jauniaux, J. C., and Roger, P. P., 1989, The cyclic AMP-mediated stimulation of cell proliferation, *Trends Bioch. Sci.*, 14: 67.

Entian, K. D., 1986, Glucose repression: a complex regulatory system in yeast, *Microbiol. Sci.*, 3: 366.

Eraso, P., Mazon, M. J., and Gancedo, J. M., 1987, Internal acidification and cAMP increase are not correlated in *Saccharomyces cerevisiae, Eur. J. Biochem.*, 165: 671.

Fedor-Chaiken, M., Deschenes, R. J., and Broach, J. R., 1990, SRV2, a gene required for RAS activation of adenylate cyclase in yeast, *Cell*, 61: 329.

Field, J., Vojtek, A., Ballester, R., Bolger, G., Colicelli, J., Ferguson, K., Gerst, J., Kataoka, T., Michaeli, T., Powers, S., Riggs, M., Rodgers, L., Wieland, I., Wheland, B., and Wigler, M., 1990, Cloning and characterization of CAP, the *S. cerevisiae* gene encoding the 70 kd adenylyl cyclase-associated protein, *Cell*, 61: 319.

François, J., Van Schaftingen, E., and Hers, H. B., 1984, The mechanism by which glucose increases fructose-2,6-bisphosphate concentration in *Saccharomyces cerevisiae*. A cyclic-AMP-dependent activation of phosphofructokinase 2, *Eur. J. Biochem.*, 145: 187.

François, J., Villanueva, M. E., and Hers, H. G., 1988, The control of glycogen metabolism in yeast. 1. Interconversion in vivo of glycogen synthase and glycogen phosphorylase induced by glucose, a nitrogen source or uncouplers, *Eur. J. Biochem.*, 174: 551.

Gibbs, J. B., and Marshall, M. S., 1989, The ras oncogene - an important regulatory element in lower eucaryotic organisms, *Microbiol. Rev.*, 53: 171.

Goffeau, A., and Slayman, C. W., 1981, The proton-translocating ATPase of the fungal plasma membrane, *Biochim. Biophys. Acta*, 639: 197.

Holzer, H., 1984, Mechanism and function of reversible phosphorylation of fructose 1,6-bisphosphatase in yeast, *in* "Molecular aspects of cellular regulation, Vol. 3", P. Cohen, ed., Elsevier, Amsterdam.

Lopez-Boado, Y. S., Herrero, P., Gascon, S., and Moreno, F., 1987, Catabolite inactivation of isocitrate lyase from *Saccharomyces cerevisiae, Arch. Microbiol.*, 147: 231.

Lopez-Boado, Y. S., Herrero, P., Fernandez, T., Fernandez R., and Moreno, F., 1988, Glucose-stimulated phosphorylation of yeast isocitrate lyase in vivo, *J. Gen. Microbiol.*, 134: 2499.

Matsumoto, K., Uno, I., Toh-e, A., Ishikawa, T., and Oshima, Y., 1982, Cyclic AMP may not be involved in catabolite repression in *Saccharomyces cerevisiae* : evidence from mutants capable of utilizing it as an adenine source, *J. Bacteriol.*, 150: 277.

Matsumoto, K., Uno, I., and Ishikawa, K., 1985, Genetic analysis of the role of cAMP in yeast, *Yeast* , 1: 15.

Mazon, M. J., Gancedo, J. M., and Gancedo, C., 1982, Phosphorylation and inactivation of yeast fructose-bisphosphatase in vivo by glucose and by proton ionophores. A possible role for cAMP, *Eur. J. Biochem.*, 127: 605.

Mbonyi, K., and Thevelein, J. M., 1988, The high-affinity glucose uptake system is not required for induction of the RAS-mediated cAMP signal by glucose in cells of the yeast *Saccharomyces cerevisiae, Biochim. Biophys. Acta* , 971: 223.

Mbonyi, K., Beullens, M., Detremerie, K., Geerts, L., and Thevelein, J. M., 1988, Requirement of one functional RAS gene and inability of an oncogenic ras-variant to mediate the glucose-induced cAMP signal in the yeast *Saccharomyces cerevisiae, Mol. Cell. Biol.*, 8: 3051.

Mbonyi, K., Van Aelst, L., Argüelles, J. C., Jans, A. W. H., and Thevelein, J. M., 1990, Glucose-induced hyperaccumulation of cAMP and absence of glucose repression in yeast strains with reduced activity of cAMP-dependent protein kinase, *Mol. Cell. Biol.*, 10: 4518.

Munder, T., and Küntzel, H., 1989, Glucose-induced cAMP signaling in *Saccharomyces cerevisiae* is mediated by the CDC25 protein, *FEBS Lett.*, 242: 341.

Neigeborn, L., Schwartzberg, P., Reid, R., Carlson, M., 1986, Null mutations in the *SNF3* gene of *Saccharomyces cerevisiae* cause a different phenotype than do previously isolated missense mutations, *Mol. Cell. Biol.*, 6: 3569.

Nikawa, J., Cameron, S., Toda, T., Ferguson, K. W., and Wigler, M., 1987a, Rigorous feedback control of cAMP levels in *Saccharomyces cerevisiae, Gen. Develop.*, 1: 931.

Nikawa, J., Sass, P., and Wigler, M., 1987b, Cloning and characterization of the low-affinity cyclic AMP phosphodiesterase gene of *Saccharomyces cerevisiae, Mol. Cell. Biol.*, 7: 3629.

Pascual, C., Alonso, A., Garcia, I., Romay, C., and Kotyk, A., 1988, Effect of ethanol on glucose transport, key glycolytic enzymes and proton extrusion in *Saccharomyces cerevisiae, Biotechnol. Bioeng.*, 32: 374.

Peinado, J. M., and Loureiro-Dias, M. C., 1986, Reversible loss of affinity induced by glucose in the maltose - H+ symport of *Saccharomyces cerevisiae, Biochim. Biophys. Acta*, 856: 189.

Praekelt, U. M., and Meacock, P. A., 1990, HSP12, a new small heat shock gene of *Saccharomyces cerevisiae* : Analysis of structure, regulation and function, *Mol. Gen. Genet.*, 223: 97.

Pringle, J. H., and Hartwell, L. H., 1981, The *Saccharomyces cerevisiae* cell cycle, *in*: "The Molecular biology of the yeast *Saccharomyces*. Life cycle and inheritance.", J. N. Strathern, E. W. Jones, and J. R. Broach, ed., Cold Spring Harbor Laboratory, Cold Spring Harbor.

Purwin, C., Leidig, F., and Holzer, H., 1982, Cyclic AMP-dependent phosphorylation of fructose 1,6-bisphosphatase in yeast, *Biochem. Biophys. Res. Commun.*, 107: 1482.

Purwin, C., Nicolay, K., Scheffers, W. A., and Holzer, H., 1986, Mechanism of control of adenylate cyclase activity in yeast by fermentable sugars and carbonyl cyanide m-chlorophenylhydrazone, *J. Biol. Chem.*, 261: 8744.

Sass, P., Field, J., Nikawa, J., Toda, T., and Wigler, M., 1986, Cloning and characterization of the high-affinity cAMP phosphodiesterase of *S. cerevisiae*, *Proc. Natl. Acad. Sci. (USA)*, 83: 9303.

Serrano, R., 1984, Plasma membrane ATPase of fungi and plants as a novel type of proton pump, *Curr. Top. Cell. Reg.*, 23: 87.

Tanaka, K., Matsumoto, K., and Toh-e, A., 1988, Dual regulation of the expression of the polyubiquitin gene by cyclic AMP and heat shock in yeast, *EMBO J.*, 7: 495.

Tanaka, K., Matsumoto, K., and Toh-e, A., 1989, Ira1, an inhibitory regulator of the RAS - cyclic AMP pathway in *Saccharomyces cerevisiae, Mol. Cell. Biol.*, 9: 757.

Tanaka, K., Nakafuku, M., Tamanoi, F., Kaziro, Y., Matsumoto, K. and Toh-e, A., 1990, *IRA2*, a second gene of *Saccharomyces cerevisiae* that encodes a protein with a domain homologous to mammalian ras GTPase-activating protein, *Mol. Cell. Biol.*, 10: 4303.

Thevelein, J. M., 1984a, Cyclic - AMP content and trehalase activation in vegetative cells and ascospores of yeast, *Arch. Microbiol.*, 138: 64.

Thevelein, J. M., 1984b, Regulation of trehalose mobilization in fungi, *Microbiol. Rev.*, 48: 42.

Thevelein, J. M., 1988, Regulation of trehalase activity by phosphorylation - dephosphorylation during developmental transitions in fungi, *Exp. Mycol.*, 12: 1.

Thevelein, J. M., and Beullens, M., 1985, Cyclic AMP and the stimulation of trehalase activity in the yeast *Saccharomyces cerevisiae* by carbon sources, nitrogen sources and inhibitors of protein synthesis, *J. Gen. Microbiol.*, 131: 3199.

Thevelein, J. M., Beullens, M., Honshoven, F., Hoebeeck, G., Detremerie, K., den Hollander, J. A., and Jans, A. W. H., 1987a, Regulation of the cAMP level in the yeast *Saccharomyces cerevisiae* : intracellular pH and the effect of membrane depolarizing compounds, *J. Gen. Microbiol.*, 133: 2191.

Thevelein, J. M., Beullens, M., Honshoven, F., Hoebeeck, G., Detremerie, K., Griewel, B., den Hollander, J. A., and Jans, A. W. H., 1987b, Regulation of the cAMP level in the yeast *Saccharomyces cerevisiae* : the glucose-induced cAMP signal is not mediated by a transient drop in the intracellular pH, *J. Gen. Microbiol.*, 133: 2197.

Toda, T., Uno, I., Ishikawa, T., Powers, S., Kataoka, T., Broek, D., Cameron, S., Broach, J., Matsumoto, K., and Wigler, M., 1985, In yeast, Ras proteins are controlling elements of adenylate cyclase, *Cell*, 40: 27.

Toda, T., Cameron, S., Sass, P., Zoller, M., and Wigler, M., 1987a, Three different genes in *Saccharomyces cerevisiae* encode the catalytic subunits of the cAMP-dependent protein kinase, *Cell*, 50: 277.

Toda, T., Cameron, S., Sass, P., Zoller, M., Scott, J. D., McBullen, B., Hurwitz, M., Krebs, E. G., and Wigler, M., 1987b, Cloning and characterization of *BCY1*, a locus encoding a regulatory subunit of the cyclic AMP-dependent protein kinase in *Saccharomyces cerevisiae, Mol. Cell. Biol.*, 7: 1371.

Tortora, P., Burlini, N., Caspani, G., and Guerritore, A., 1984, Studies on glucose - induced inactivation of gluconeogenetic enzymes in adenylate cyclase and cAMP-dependent protein kinase yeast mutants, *Eur. J. Biochem.*, 145: 543.

Van Aelst, L., Boy-Marcotte, E., Camonis, J. H., Thevelein, J. M., and Jacquet, M., 1990, The C-terminal part of the *CDC25* gene product plays a key role in signal transduction in the glucose-induced modulation of cAMP level in *Saccharomyces cerevisiae, Eur. J. Biochem.*, 193: 675.

Van Aelst, L., Jans, A. W. H., and Thevelein, J. M., 1991a, Involvement of the CDC25 gene product in the signal transmission pathway of the glucose-induced RAS-mediated cAMP signal in the yeast *Saccharomyces cerevisiae, J. Gen. Microbiol.*, 137: 341.

Van Aelst, L., Hohmann, S., Zimmermann, F. K., Jans, A. W. H., and Thevelein, J. M., 1991b, A yeast homologue of the bovine lens fiber MIP gene family complements the growth defect of a *Saccharomyces cerevisiae* mutant on fermentable sugars but not its defect in glucose-induced RAS-mediated cAMP signaling, *EMBO J.*, 10: in press.

van de Poll, K. W., Kerkenaar, A., and Schamhart, D. H. J., 1974, Isolation of a regulatory mutant of fructose-1,6-diphosphatase in *Saccharomyces carlsbergensis*, J. Bacteriol., 117: 965.

Verdier, J. M. *et al.*, 1989, Cloning of *CDC33* : a gene essential for growth and sporulation which does not interfere with cAMP production of *Saccharomyces cerevisiae, Yeast* , 5: 79-90.

Werner-Washburne, M., Becker, J., Kosic-Smithers, J., and Craig, E.A., 1989, Yeast Hsp70 RNA levels vary in response to the physiological status of the cell, *J. Bacteriol.*, 171: 2680.

DIFFERENTIAL P21 EXPRESSION AND POINT MUTATIONS OF

RAS GENE FAMILY IN HUMAN CARCINOMA TISSUES

Yuji Mizukami[1], Akitaka Nonomura[1], Fujitsugu Matsubara[1],
Masakuni Noguchi[2] and Shinobu Nakamura[3]

[1]Pathology Section, [2]Surgery and [3]Internal Medicine, Kanazawa
University Hospital, 13-1, Takara-machi, Kanazawa, 920, Japan

INTRODUCTION

The mammalian ras oncogene family consisits of three well-characterized members, N-ras, K-ras and H-ras. The amino acid sequence of the 21-Kd proteins encoded by these three ras members is known to be nearly identical except in two short variable regions[1,2]. Concerning the expression of ras p21 in human tissues, many immunohistochemical studies have been reported using monoclonal antibodies against ras protein[3-5]. However, these antibodies were unable to distinguish N-ras, K-ras and H-ras p21, respectively. Recently, monoclonal antibodies against each of N-ras, K-ras and H-ras p21 have been developed and made available commercially. In the present study, we examined the differential p21 expression of the three members of ras family in human carcinoma tissues of various organs by the immunohistochemical method.

The recent introduction of a new and rapid assay system for the identification of mutated ras genes has made it possible to analyze large numbers of tumor samples for the presence of ras genes and preferential mutational activation of particular ras genes was found in human carcinomas of various organs[6-12]. On the other hand, concerning the expression of ras p21 proteins of each of the genes in cancer tissue, substantial variations are reported among different tumor types; either a mutant or normal form. Activating mutations may only contribute to the development of a tumor when the altered ras p21 protein is substantially expressed in the target cells. Therefore, it is now necessary to clarify the correlation between the type of ras gene mutation and its corresponding protein level in human carcinoma tissues.

The purpose of this study was to determine the differential p21 expression of ras genes in human carcinomas and to evaluate the role of the three p21s in cellular proliferation, in connection with the status of ras gene mutations.

Table 1. Differential p21 Expression of Ras Gene Family in Human Carcinoma Tissues

Breast Carcinoma	histology[1]	ras p21 N	K	H
1.	P	±	-	±
2.	M	+	-	±
3.	W	+	±	2+
4.	IS	+	-	-
5.	W	±	±	-
6.	P	±	-	-
7.	W	+	-	±
8.	IS	+	-	+
9.	W	2+	-	+
10.	P	±	-	±
11.	M	+	-	+
12.	M	+	-	±
13.	P	±	-	±
14.	IS	±	+	2+
15.	M	+	±	±
16.	P	2+	±	±
17.	P	+	-	-
18.	P	-	-	+
19.	M	+	-	+
20.	M	±	-	±
21.	P	±	-	±
22.	P	±	+	+

Thyroid Carcinoma	histology[2]	ras p21 N	K	H
1.	P	+	+	+
2.	P	+	+	±
3.	P	±	±	-
4.	P	+	±	+
5.	F	±	+	+
6.	F	+	+	+
7.	P	±	-	-
8.	P	2+	+	+
9.	P	2+	+	+
10.	F	±	+	+
11.	P	+	+	+
12.	P	2+	±	+
13.	P	+	±	+
14.	F	+	+	+
15.	P	+	±	+
16.	P	+	±	±
17.	P	+	-	±

Lung Carcinoma	histology[3]	ras p21 N	K	H
1	WA	±	-	+
2	MS	±	±	-
3	PA	+	±	2+
4	WA	-	-	±
5	WS	+	-	+
6	WS	±	-	+
7	MS	+	-	+
8	WA	±	±	-
9	MS	±	±	+
10	MA	-	-	+

Gastric Carcinoma	histology[5]	ras p21 N	K	H
1	PA	±	-	+
2	WA	+	-	2+
3	MA	-	-	±
4	MA	-	-	+
5	MA	±	±	+
6	MA	+	-	+

Pancreatic Carcinoma	histology[4]	ras p21 N	K	H
1	MA	+	-	+
2	MA	-	-	2+
3	WA	+	±	+
4	MA	±	-	±
5	WA	±	-	+
6	MA	+	-	+
7	MA	-	-	±
8	MA	±	±	+
9	MA	-	-	±
10	WA	+	-	2+

Colon Carcinoma	histology[6]	ras p21 N	K	H
1	MA	±	-	2+
2	MA	-	-	±
3	PA	-	-	-
4	MA	±	-	+
5	MA	-	-	±
6	MA	±	±	+
7	MA	-	-	+
8	WA	-	-	±

1)W: Well differentiated; M: Moderately differentiated; P: Poorly differentiated; IS: In situ.
2)P: Papillary; F: Follicular.
3)-6)WA: Well differentiated adenocarcinoma; MA: Moderately differentiated adenocarcinoma; PA: Poorly differentiated adenocarcinoma;
WS: Well differentiated squamous carcinoma; MS: Moderately differentiated squamous carcinoma.

MATERIALS AND METHODS

I. Immunohistochemical analysis

Seventy-three surgically obtained carcinoma tissues were used, consisting of 22 breast carcinomas, 17 thyroid carcinomas, 10 lung carcinomas, 10 pancreatic carcinomas, 6 gastric carcinomas and 8 colon carcinomas. The histologic variety of each carcinoma was summarized in Table 1. Formalin-fixed, paraffin-embedded sections were used for the immunohistochemical study. The 5-μm sections were deparaffinized and treated with hydrogen-peroxide-methanol to remove endogeneous peroxidase activity. The immunohistochemical procedure was based on the avidin-biotin-peroxidase complex (ABC) method.

The primary antibodies employed were monoclonal antibodies against human cellular N-ras p21 (c-N-ras), human cellular K-ras p21 (c-K ras) and human cellular H-ras p21 (c-H-ras)(all purchased from Oncogene Science Inc. Manhasset, NY.). Each antibody was used at a 1:10 dilution (10μg IgG/1ml PBS). It has been proven that these three monoclonal antibodies do not cross-react with each other by ELISA and immunoblotting assays using purified recombinant proteins (instruction of the manufacturer).

Both the distribution (percentage of positive cells) and the intensity of staining was assessed in a semiquantitative fashion. The following system was employed to score the distribution and intensity of positive cells; (-)=negative, negative staining; (±)=minimal, marginal staining; (+)=mild, weak staining of all cells or stronger staining of a subpopulation group; (2+)=moderate to intense, strong staining of all cells or a major subpopulation of cells. We judged tissues of (±) to (2+) as positive and tissues of (-) as negative.

II. Ras oncogene point mutations

Paraffin-embedded blocks from formalin-fixed tissue of 14 cases of breast carcinoma, which were selected from the 22 carcinomas examined for differential p21 expression, were used as source material. The clinical data of these 14 patients were summarized in Table 2. The point mutations examined were limited to codons 12 and 13 of the N-ras gene and codon 12 of the K-ras gene. DNA was extracted from one to three 50-μm sections by phenol-chloroform method. One microgram of DNA was amplified using the polymerase chain reaction (PCR) as decribed by Saiki et al[13]. Forty cycles of denaturation (94 ℃, 1 min), annealing (60 ℃, 2 min) and extension (72 ℃, 3 min) were carried out with an automated DNA thermal cycler. The amplified DNA samples were subsequently spotted onto nylon filters. We tested a wild-type oligonucleotide in addition to six additional oligonucleotides, each containing a different single base change corresponding to a different encoded amino acid (mutated oligonucleotides). Filters were prehybridized overnight at 56 ℃ in 3 M tetramethylammonium chloride (TMAC), 50 mM Tris pH 7.5, 2 mM ethylenediamine tetraacetic acid (EDTA), 100μg/ml of denatured salmon testis DAN, 0.3% SDS, and 5X Denhardt's solution. Subsequently they were hybridized for 3 hours at 56 ℃ in the same mixture containing oligonucleotide probe labelled with γ -phosphate 32 (^{32}P) ATP. The filters were rinsed well twice with SSPE/0.1 % SDS at room temperature and washed twice SSPE/0.1% SDS at 60 ℃. After rinsing with 3M TMAC buffer removed

Table 2. Clinical Data of 14 Patients with Breast Carcinoma

Case No.	Original* Case No.	Age	Clinical Stage	ER Status	PgR Status	Prognosis
1	1	63	I	+	+	well
2	2	72	II	+	+	well
3	3	69	I	+	ne	well
4	6	74	I	-	ne	well
5	7	28	II	+	+	well
6	8	64	III	ne	ne	well
7	9	56	I	+	ne	well
8	10	40	III	+	-	well
9	11	73	III	+	+	well
10	12	60	III	ne	ne	death by disease
11	14	39	III	+	+	well
12	15	40	II	+	+	well
13	17	51	I	+	+	well
14	18	47	I	-	-	well

*Original Case No. indicates that shown in immunohistochemical results (see Table 1).
Estrogen receptor (ER) and progesterone receptor (PgR) was measured by
dextran-coated charcoal method. More than 5 fmol/mg protein was judged as positive.

0.3% SDS and carrier DNA, the filters were finally washed with the same buffer at 63 ℃ for 30 min. Autoradiography was performed at -80 ℃ for 10 min.

RESULTS

I. Immunohistochemical findings

Table 1 summarizes staining results in 73 carcinoma tissues with N-ras, K-ras and H-ras p21 protein.

a) Breast carcinoma

Twenty-one of the 22 carcinomas (95%) showed a positive staining for N-ras p21 and 18 carcinomas (82%) were positive for H-ras p21. K-ras p21 was positive only in 3 carcinomas (14%)(Figs. 1 and 2). In relation to the histologic differentiation of the carcinomas, poorly differentiated carcinomas showed a tendency to poorer H-ras p21 expression (Fig. 3). Adjacent normal breast tissues showed minimal to mild expression of H-ras p21 in the ductal epithelium.

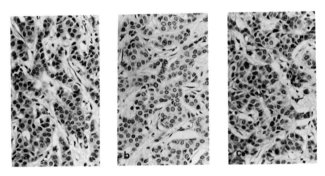

Fig. 1. Breast carcinoma (Case 19). Moderately differentiated type. Mild staining for N-ras p21(a) and H-ras p21 (c), but negative for K-ras p21(b). (ABC, × 200).

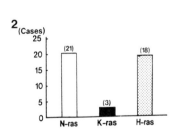

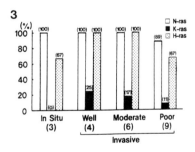

Fig. 2. Positivity of N-, K- and H-ras p21 in 22 breast carcinomas. Parentheses indicate a number of positive cases.

Fig. 3. Positivity of N-, K- and H-ras p21 in 22 breast carcinomas by histologic differentiation. Parentheses indicate a percent of positive cases.

b) Thyroid carcinoma

All 17 carcinomas (100%) were positive for N-ras p21 and 16 carcinomas (94%) were also positive for H-ras p21. K-ras p21 was expressed positively in 11 carcinomas (65%) (Fig. 4). The frequency of K-ras positivity was highest in the thyroid carcinomas among all organs examined in our study. In adjacent normal thyroid tissue, flattened follicular cells were negative for any of the ras p21.

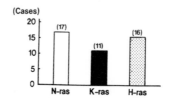

Fig. 4.   Positivity of N-, K- and H-ras p21 in 17 thyroid carcinomas.

c) Lung carcinoma

Eight of the 10 carcinomas (80%) were positive for N-ras p21 and all 10 carcinomas (100%) were positive for H-ras p21. Only one carcinoma (10%) was positive for K-ras p21 (Fig. 5). No difference was observed in ras p21 expression between adenocarcinomas and squamous cell carcinomas. In normal lung tissue, pneumocytes were negative for any of the ras p21, but bronchial surface epithelium was minimally positive for H-ras p21.

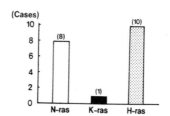

Fig. 5.   Positivity of N-, K- and H-ras p21 in 10 lung carcinomas.

d) Pancreas carcinoma

Eight of the 10 carcinomas (80%) were positive for N-ras p21 and all 10 carcinomas (100%) were positive for H-ras p21. Only 3 carcinomas (30%) were positive for K-ras p21 (Figs. 6 and 7). In the pancreas tissue, ductal epithelium and acinar cells were minimal to mildly positive for N- and H-ras p21 (Fig. 8).

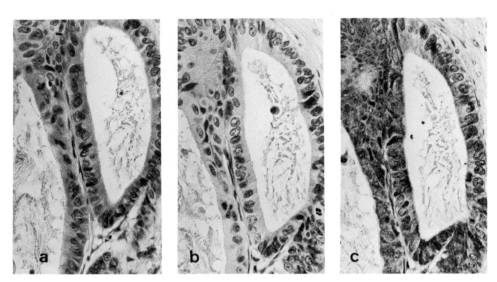

Fig. 6.　Pancreas carcinoma (Case 10).　Well differentiated type.　Mild staining for N-ras p21(a), negative for K-ras p21(b) and intense staining for H-ras p21(c).　(ABC, ✕ 200).

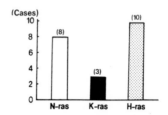

(Cases)

Fig. 7. Positivity of N-, K- and H-ras p21 in 10 pancreatic carcinomas.

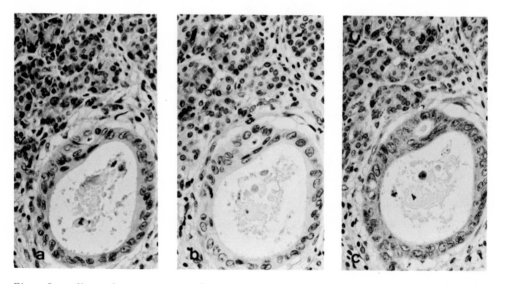

Fig. 8. Normal pancreas. Ductal epithelium and acinar cells. Minimal staining for N-ras p21(a), negative for K-ras p21(b) and mild staining for H-ras p21(c). (ABC, × 200).

e) Gastric carcinomas

Four of the 6 carcinomas (67%) were positive for N-ras p21 and all 6 carcinomas (100%) were positive for H-ras p21. None of the 6 cases were positive for K-ras p21 (Fig. 9). In the adjacent normal gastric mucosa, foveolar epithelium and fundic gland cells showed an occasional positive staining for H-ras p21 (Fig. 10).

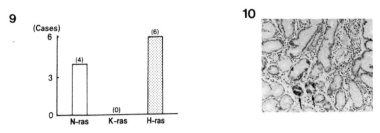

Fig. 9. Positivity of N-, K- and H-ras p21 in 6 gastric carcinomas.

Fig. 10. Normal gastric mucosa. Positive staining for H-ras p21 (arrows) in fundic gland cells. (ABC, ✕ 100).

f) Colon carcinoma

Three of the 8 carcinomas (38%) were positive for N-ras p21 and 7 carcinomas (88%) were positive for H-ras p21. K-ras p21 was not expressed in any of the carcinomas (Fig. 11). One poorly differentiated adeno-carcinoma showed a negative staining for all of ras p21. Adjacent normal foveolar epithelium showed an occasional marginal staining for H-ras p21.

g) Relation between ras p21 positivity and histologic differentiation in 48 adenocarcinomas

In order to examine the relation between differential p21 expression and histologic differentiation of the carcinomas, 48 adenocarcinomas; excluding 3 carcinomas in situ of the breast, 17 thyroid carcinomas and 5 squamous carcinomas of the lung, were divided into three histologic groups (well, moderately and poorly differentiated). The frequency of positive cases for H-ras p21 was decreased in the poor histologic group, but in N- and K-ras p21 expression, such a relation was not evident (Fig. 12).

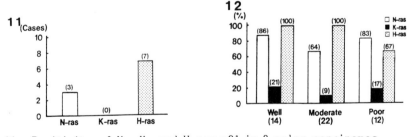

Fig. 11. Positivity of N-, K- and H-ras p21 in 8 colon carcinomas.

Fig. 12. Positivity of N-, K- and H-ras p21 in 48 adenocarcinomas by histologic differentiation. Parentheses indicate a percent of positive cases.

II. Ras gene point mutations

Using oligonucleotide hybridization and polymerase chain reaction technique, we screened for ras oncogene point mutations at codons 12 and 13 for K- and N-ras gene. Hybridization with the wild-type oligonucleotide showed a strong signal for each tumor. In no case was a signal of similar intensity produced using any of the other mutated for oligonucleotides (Figs. 13 and 14). To test whether our analysis with these mutated oligonucleotides can actually detect a known mutation, we used SW 480 cells, a colon carcinoma cell line known to contain a K-ras mutation (G to T mutation, at codon 12), PA-1 cells, an ovarian teratocarcinoma cell line (N-ras at codon 12, G to A mutation) and KS cells, derived from human leukemic cells (N-ras at codon 13, G to A mutation). These positive controls indicated hybridization of both wild-type and mutant-encoding oligo- nucleotides, whereas in DNA from breast carcinomas, only wild-type oligonucleotides hybridized.

Detection of K-ras Codon 12 Mutations

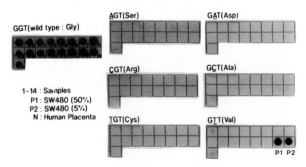

Fig. 13. Dot blot analysis for mutations of K-ras at codon 12 in 14 breast carcinomas. Each dot (1-14) contains amplified DNA from a different tumor. The names of amino acids on each blot represent the encoded product of the point mutation tested for at codon 12. DNA from SW480 cells (P1, P2) is indicated as a positive control.

Detection of N-ras Codon 12 Mutations

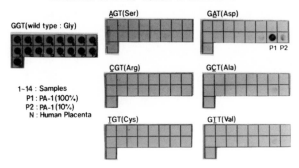

Fig. 14. Dot blot analysis for mutations of N-ras at codon 12. Point mutations are not identified in any of the 14 breast carcinomas. P1 and P2 indicate a positive control.

DISCUSSION

We have shown the differential p21 expression of three ras gene members in human carcinoma tissues of various organs. The most striking finding of this study was that K-ras p21 was most poorly and most infrequently expressed in carcinoma tissues compared with intense expression of N- and H-ras p21 througout the various organs which we examined.

Previous studies suggested that ras p21 expression is correlated with its neoplastic transformation and subsequently many immunohistochemical studies have indicated that ras gene product is increased in carcinoma tissues compared with benign and normal tissues[3-6]. However, these immunohistochemical studies were performed using antibodies against pan-ras p21. The three monoclonal antibodies used in this study are specific for N-, K- and H-ras p21 respectively, therefore, we are able to evaluate the differential expression of N-, K- and H-ras p21 in human tissues using these antibodies. We initially expected that the expressional pattern of p21 of three ras members may be characteristic to organs evidencing carcinoma. However, our results indicated that the expressional pattern of three p21s is similar in all carcinoma tissues in respect to predominant expression of N- and H-ras p21, and poorer and infrequent expression of K-ras p21, except thyroid carcinomas.

It is well established that high levels of normal ras proteins can induce neoplastic transformation of NIH/3T3 murine fibroblasts[14, 15]. On the other hand, activated ras oncogene by mutation also contributes to tumorigenesis[6, 7]. Recently, with advanced technology to identify these mutations, the frequency of ras gene mutation in various human carcinomas has been reported[8-12]. Ras gene mutations were found in approximately 90% of pancreatic carcinomas (K-ras)[8], 50% of thyroid carcinomas (N-, K-, and H-ras)[9], 40% of colon carcinomas (K-ras)[10] and 30% of lung carcinomas (K-ras)[11, 12]. These reports indicated that ras gene mutations in human carcinomas occur preferentially in K-ras gene. In these reports, only thyroid carcinoma seems to lack any specificity. The monoclonal antibody against K-ras p21 used in our study is unable to distinguish a mutant form from a normal form of K-ras p21. However, when we take these contradictory facts into consideration (i.e. poorer and infrequent expression of K-ras p21 and the reported higher incidence of K-ras gene mutation in human carcinoma tissues), we should perhaps doubt the theory which stresses the important participation of ras gene mutation in tumorigenesis, because activating mutations may only contribute to develop a tumor when altered ras p21 protein is subsequently expressed in the target cells. However, our immunohistochemical assay can't detect a very small amount of K-ras p21. The relation between the quantity and quality of mutant form of K-ras p21 should be examined further.

Within the limitations of our examination of point mutations at codons 12 and 13 of the N-ras and codon 12 of the K-ras, none of 14 breast carcinomas showed any point mutations of ras genes. Our data reinforces the reported evidence that breast carcinomas never or very rarely harbor ras gene mutation[7]. This evidence suggests the possibility that, in human breast carcinomas, mutational activation of ras genes might not play an important role in the development of the carcinomas. The question of whether ras gene mutations have a more or less important role in the initiation and progression of human neoplasms should be further investigated in a large scale study.

It must be noted that fully differentiated normal cells also express significant amounts of ras product. Generally, H-ras p21 was most frequently, and N-ras p21 less frequently, expressed in the normal cells in our study. This finding suggests that ras protein, especially H-ras p21, is involved not only in cellular proliferation but also in cellular function[16, 17]. Our data that H-ras p21 expression was increased in accordance with higher histologic differentiation of the carcinomas also supports this hypothesis. Nakagawa et al.[18] reported that induction of v-H-ras oncogene induces differentiation of human medullary carcinoma cells. The role of ras product in cellular differentiation should be examined further in parallel with its role in tumorigenesis.

Acknowledgements: The authors thank Mrs Sanae Itoh for preparing the manuscript and Miss Noriko Tabishima and Mr. Ryoji Aozaki (SRL Co. Ltd, Tokyo) for their technical assistance in analysis of point mutations.

REFERENCES

1. K. Shimizu, D. Birnbaum, M.A. Ruley, O. Fasano, Y. Suard, L. Edlund, E. Taparowsky, M. Goldfarb, and M. Wigler.
    Structure of the Ki-ras gene of the human lung carcinoma cell line Calu-1. Nature 304:497(1983).
2. E. Taparowsky, K Shimizu, M. Goldfarb, and M. Wigler.
    Structure and activation of the human N-ras gene. Cell 34:581 (1983).
3. Y. Mizukami, A. Nonomura, T. Hashimoto, S. Terahata, F. Matsubara, T. Michigishi, and M. Noguchi. Immunohistochemical demonstration of ras p21 oncogene product in normal, benign and malignant human thyroid tissues. Cancer 61:873 (1988).
4. M.V. Viola, F. Fromowitz, M.S. Oraves, S.D. Deb, G. Finkel, J. Lundy, P. Hand, A. Thor, and J. Schlom.
    Expression of ras oncogene p21 in prostate cancer. N Engl J Med 314:133 (1986).
5. A.K. Ghosh, M. Moore, and M. Harris. Immunohistochemical detection of ras oncogene p21 product in benign and mammary tissues in man. J Clin Pathol 39:428 (1986).
6. M. Barbacid. Ras genes. Annu Rev Biochem 56:779 (1987).
7. J.L. Bos. Ras oncogenes in human cancer: a review. Cancer Res 49:4682 (1989).
8. M. Tada, O. Yokosuka, M. Omata, M. Ohta, and K. Isono.
    Analysis of ras gene mutations in biliary and pancreatic tumors by polymerase chain reaction and direct sequencing. Cancer 66:930 (1990).
9. N.R. Lemoine, E.S. Mayall, F.S. Wyllie, D. Williams, M. Goyns, B. Stringer, and D.Y. Thomas. High frequency of ras oncogene activation in all stages of human thyroid tumorigenesis. Oncogene 4:159 (1989).
10. J.L. Bos, E.R. Fearon, S.R. Hamilton, M.V. Vries, J.H. van Boom, A.J. van der Eb, and B. Vogelstein. Prevalence of ras gene mutations in human colorectal cancers. Nature 327:293 (1987).

11. R.J.C. Slebos, R.E. Kibbelaar, O. Dalesio, A. Kooistra, J. Stam, C.J.L.M. Meijer, S.S. Wagenaar, R.G.J.R.A. Vanderschueren, N.V. Zandwijk, W.J. Mooi, J.L. Bos, and S. Rodenhuis. K-ras oncogene activation as a prognostic marker in adenocarcinoma of the lung. N Engl J Med 323:561 (1990).

12. T. Kobayashi, H. Tsuda, M. Noguchi, S. Hirohashi, Y. Shimosato, T. Goya, and Y. Hayata. Association of point mutation in c-K-ras oncogene in lung adenocarcinoma with particular reference to cytologic subtypes. Cancer 66:289 (1990).

13. R.K. Saiki, D.H. Gelfand, S. Stoffel, S.J. Schrf, R. Higuchi, G.T. Horn, K.B. Mullis, H.A. Erlich. Primer-directed enzymatic amplification of DNA with a thermostable DNA polymerase. Science 239:487 (1988).

14. E.G. Chang, M.E. Furth, E.M. Scolnick, and D.R. Lowy. Tumorigenic transformation of mammalian cells induced by a normal human gene homologous to the oncogene of Harvey murine sarcoma virus. Nature 297:479 (1982).

15. D. Defeo, M.A. Gonda, H.A. Young, E.H. Chang, D.R. Lowy, E.M. Scolnick, and R.W. Ellis. Analysis of two divergent rat genomic clones homologous to the transforming gene of Harvey murine sarcoma virus. Proc Natl Acad Sci USA 78:3328 (1981).

16. M.E. Furth, T.H. Aldrich, and C. Cordon-Cardo. Expression of ras protooncogene proteins in normal human tissues. Oncogene 1:47 (1987).

17. J. Leon, I. Guerrero, and A. Pellicer. Differential expression of the ras gene family in mice. Mol Cell Biol 7:1535 (1987).

18. T. Nakagawa, M. Mabry, A. De Bustros, J.N. Ihle, B.D. Nelkin, S.B. Baylin. Introduction of v-Ha-ras oncogene induces differentiation of cultured human medullary thyroid carcinoma cells. Proc Natl Acad Sci USA 84:5923 (1987).

REARRANGEMENT OF THE HUMAN MEL GENE, THE RAB 8 HOMOLOGUE, IN HUMAN MALIGNANT MELANOMAS

Rose Ann Padua[1], David Hughes[1], Elaine Nimmo[2], Peter Schrier[3], and Keith Johnson[2]

[1]LRF Preleukaemia Unit, University of Wales College of Medicine, Cardiff, U.K., [2]Department of Anatomy Charing Cross and Westminster Medical School, London U.K., [3]Laboratorium voor Klinische Oncologie, Leiden Netherlands

ABSTRACT

The human MEL gene was originally isolated following transfection into NIH3T3 mouse fibroblast cells of DNA from a human melanoma cell line NKI4. MEL was mapped to chromosome 19 (p13.2), a region in which a number of translocation breakpoints occur. We have sequenced MEL cDNA clones and shown significant homology to the RAS-related YPT1 proteins and more specifically, identity with the recently described canine RAB 8. The original transfectant from which MEL was isolated appeared to comprise of complex rearrangements involving human sequences mapping on chromosome 8. Furthermore, most of the MEL sequences, apart from the putative first extron, were absent. The presence of the human MEL sequences may be fortuitous in this transfectant and the transforming gene has yet to be isolated. However, 12 melanoma cell lines and 20 fresh melanoma tumours were screened for MEL gene rearrangements and DNA from a single patient with a tumour metastasis from the intestinal tract was found to have rearranged MEL sequences. DNA from this tumour was positive in NIH3T3 focus formation assays, producing anchorage independent foci with human repetitive sequences present. One of these transformants was tested and found to be tumorigenic in nude mice. These transformants showed no evidence of human H, K or N RAS or MEL activation. Six additional metastases from this patient lacked this rearrangement, suggesting that the gene rearrangement observed may be associated with late events in the evolution of the tumour metastasis.

INTRODUCTION

A number of cellular oncogenes have been identified using the NIH3T3 transformation assay. In most cases, transforming activity was found to be due to activation of members of the RAS gene family with point mutations at either codons 12/13 or 61 (Barbacid 1987; Bos 1988). Other genes shown to register in this assay include RAF (Ishikawa et al., 1987; Fukui et al., 1985; Ikawa et al., 1988), TRK (Martin-Zanca et al., 1986; Kozma et al., 1988) and RET (Takahashi & Cooper, 1987; Ishikaza et al., 1988).

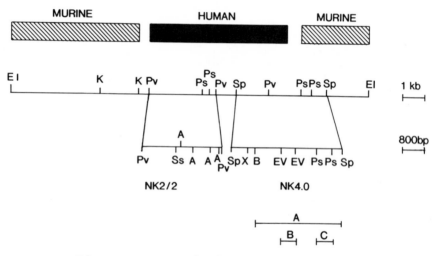

Figure 1. Restriction Map of NK16.

The 5′ and 3′ portions of this clone are mouse in origin. The MEL specific sequences are within the region encompassing the subclone NK2/2. Restriction sites: A - AvaI, B - BamHI, EI - EcoRI, EV - EcoRV, K - KpnI, Ps - PstI, Pv - PvuII, Sp - SphI, Ss - SstI, X - XbaI.

We have previously reported the detection of a novel transforming gene from a human melanoma cell line NKI4, using the NIH3T3 focus formation assay and have cloned a conserved fragment containing human repetitive sequences (Padua et al., 1984). This clone (NK16) did not transform NIH3T3 cells and when hybridised at low stringency showed homology to ki-ras and HRAS.

RESULTS

Characterisation of NK16

Analysis of NK16 defined mouse and human sequences (Figure 1). A subclone from the clone NK2/2 was derived by shotgun cloning and shown to detect a human 5.5 kb Eco RI fragment, a murine 7.5 kb fragment and in addition, a 16 kb fragment fragment in the transfectant (Figure 2). These sequences were mapped by screening somatic cell hybrids to chromosome 19 (19p13.2-19q13.2) (Spurr et al., 1986). A transcript of 3.5 kb was detected in NIH3T3, NKI4 transfectants and human melanoma cell lines using this probe. This gene was designated MEL (for melanoma). Other probes were generated from NK16 which were either human or murine specific (Figure 2). Using probe A (Figure 1), these human sequences were mapped to chromosome 8 by Southern blot analysis of somatic cell hybrid DNA (Spurr and Padua, unpublished observations).

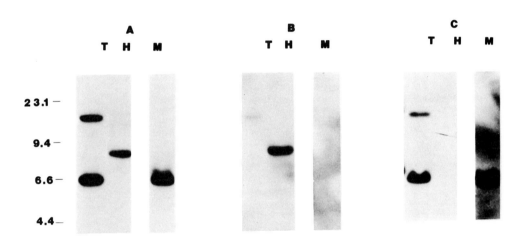

Figure 2. Determination of Human and Mouse specific sequences from NK16.

Genomic DNA (20ug) was digested with EcoRI and analysed by Southern blotting and hybridisation with alpha-^{32}PdCTP labelled probes. M-NIH3T3 DNA, T-NKI4 secondary transfectant, H-Human Leucocyte DNA. Probe A detects an EcoRI fragment in both human and mouse DNA, and the 16 kb transfectant specific fragment. Probe B hybridises to the Human fragment and the transfectant specific fragment, but not the mouse fragment. Probe C hybridises to the mouse fragment and the transfectant fragment but not the human fragment. DNA marker fragment sizes are indicated in kilobases (kb).

```
MEL    MAK-------------TYDYLFKLLLIGDSGVGKTCVLFRFSEDAFNSTFISTIGIDFK
RAB8   MAK-------------TYDYLFKLLLIGDSGVGKTCVLFRFSEDAFNSTFISTIGIDFK
RAB1   MSS----------MNPEYDYLFKLLLIGDSGVGKSCLLLRFADDTYTESYISTIGVDFK
RAB2   MA---------------YAYLFKYIIIGDTGVGKSCLLLQFTDKRFQPVHDLTIGVEFG
RAB3A  MASATDSRYGQKESSDQNFDYMFKILIIGNSSVGKTSFLFRYADDSFTPAFVSTVGIDFK
RAB3B  MASVTDGKHGVKDASDQNFDYMFKLLIGNSSVGKTSFLLRYADDTFTPAFVSTVGIDFK
RAB4   MS-------------ETYDFLFKFLVIGNAGTGKSCLLHQFIEKKFKDDSNHTIGVEFG
RAB5   MASRGATRPNGPNTGNKI--CQFKLVLLGESAVGKSSLVLRFVKGQFHEFQESTIGAAFL
RAB6   MSTGGDFGNPL--------RKFKLVFLGEQSVGKTSLITRFMYDSFDNTYQATIGIDFL
YPT    MNS-------------EYDYLFKLLLIGNSGVGKSCLLLRFSDDTYTNDYISTIGVQFK
SEC4   MSGLRTVSASSGNGKS--YDSIMKILLIGDSGVGKSCLLVRFVEDKFNPSFITTIGIDFK
KRAS   M-----------------TEYKLVVVGAGGVGKSALTIQLIQNHFVDEYDPTIEDSY-
HRAS   M-----------------TEYKLVVVGAGGVGKSALTIQLIQNHFVDEYDPTIEDSY-

                                                              120
MEL    IRTIELDGKRIKLQIWDTAGQERFRTITTAYYRGAMGIMLVYDITNEKSFDNIRNWIRNI
RAB8   IRTIELDGKRIKLQIWDTAGQERFRTITTAYYRGAMGIMLVYDITNEKSFDNIRNWIRNI
RAB1   IRTIELDGKTIKLQIWDTAGQERFRTITSSYYRGAHGIIVVYDVTDQESFNNVKQWLQEI
RAB2   ARMITIDGKQIKLQIWDTAGQESFRSITRSYYRGAAGALLVYDITRRDTFNHLTTWLEDA
RAB3A  VKTIYRNDKRIKLQIWDTAGQERYRTITTAYYRGAMGFILMYDITNEESFNAVQDWSTQI
RAB3B  VKTVYRHEKRVKLQIWDTAGQERYRTITTAYYRGAMGFILMYDITNEESFNAVQDWATQI
RAB4   SKIINVGGKYVKLQIWDTAGQERFRSVTRSYYRGAAGALLVYDITSRETYNALTNWLTDA
RAB5   TQTVCLDDTTVKFEIWDTAGQEGYHSLAPMYYRGAQAAIVVYDITNEESFARAKNWVKEL
RAB6   SKTMYLEDRTVRLQLWDTAGQERFRSLIPSYIRDSTVAVVVYDITNVNSFQQTTKWIDDV
YPT    IKTVELDGKTVKLQLWDTAGQERFRTITSSYYRGSHGIIIVYDVTDQESFNGVKMWLQEI
SEC4   IKTVDINGKKVKLQLWDTAGQERFRTITTAYYTGAMGIILVYDVTDERTFTNIKQWFKTV
KRAS   RKQVVIDGETCLLDILDTAGQEEYSAMRDQYMRTGEGFLCVFAINNTKSFEDIHHYREQI
HRAS   RKQVVIDGETCLLDILDTAGQDDYSAMRDQYMRTGEGFLCVFAINNTKSFEDIHQYREQI

                                                              180
MEL    EEHASA-DVEKMILGNKCDVNDKRQVSKERGEKLALDYGIKFMETSAKANINVENAFFTL
RAB8   EEHASA-DVEKMILGNKCDVNDKRQVSKERGEKLALDYGIKFMETSAKANINVENAFFTL
RAB1   DRYASE-NVNKLLVGNKCDLTTKKVVDYTTAKEFADSLGIPFLETSAKNATNVEQSFMTM
RAB2   RQHSNS-NMVIMLIGNKSDLESRREVKKEEGEAFAREHGLMFMETSAKTASNVEEAFINT
RAB3A  KTYSWD-NAQVLLVGNKCDMEDERVVSSERGRQLADHLGFEFFEASAKDNINVKQTFERL
RAB3B  KTYSWD-NAQVILVGNKCDMEEERVVPTEKGQLLAEQLGFDFFEASAKENISVRQAFERL
RAB4   RMLASQ-NIVIILCGNKKDLDADREVTFLEASRFAQENELMFLETSALTGEDVEEAFVQC
RAB5   QRQASP-NIVIALSGNKADLANKRAVDFQEAQSYADDNSLLFMETSAKTSMNVEI-FMAI
RAB6   RTERGS-DVIIMLVGNKTDLADKRQVSIEEGERKAKELNVMFIETSAKAGYNVKQLFRRV
YPT    DRYATS-TVLKLLVGNKCDLKDKRVVEYDVAKEFADANKMPFLETSALDSTNVEDAFLTM
SEC4   NEHAND-EAQLLLVGNKSDM-ETRVVTADQGEALAKELGIPFIESSAKNDDNVNEIFFTL
KRAS   KRVKDSEDVPMVLVGNKCDLPS-RTVDTKQAQDLARSYGIPFIETSAKTRQGVDDAFYTL
HRAS   KRVKDSDDVPMVLVGNKCDLAA-RTVESRQAQDLARSYGIPYIETSAKTRQVEDAFYTLV

                                                              234
MEL    ARDIKAKMDKKWKATAP-GSNQGVKITPDQ--------QKRSSFFR----CVLL
RAB8   ARDIKAKMDKKLEGNSPQGSNQGVKITPDQ--------QKRSSFFR----CVLL
RAB1   AAEIKKRMGPGATAGGAEKSNVKIQSTPVK--------QS---K-GGGC-C
RAB2   AKEIYEKIQEGVFDINNEANGIKIGPQHAATNATHAGNQGGGQA-GGGC-C
RAB3A  VDVICEKMSESLDTADPAVTGAKQGPQLSDQ-------QVPP---HQDCAC
RAB3B  VDAICDKMSDSLDT-DPSMLGSSKNTRLSDTPPLL---Q-------QNCSC
RAB4   ARKILNKIESGELDPERMGSGIQYGDAALRQLRSPRRTQA---PNAQECGC
RAB5   AKKLPKNEPQNPGANSARGGGVDLTEPTQPTRN-----Q---------C-CSN
RAB6   AAALPGMESTQDRSREDMIDIKLEKPQE----------QP---VSEGGCSC
YPT    ARQIKQSMSQQNLNETTQKKEDKGNVNLKG--------QSLT-NTGGGC-C
SEC4   AKLIQEKIDSNKLVGVGNGQEGNISINSGSG-------NS----SKSNC-C
KRAS   VREIRKHKEKMSKDGKKKKKKSKTK----------------------CVIM
HRAS   EIRQHKLRKLNPPDESGP------------------------MSCKCVLS
```

Figure 3. Alignment of MEL with human RAB, canine RAB8, yeast YPT, SEC4 and human KRAS (exon 4B) and HRAS proteins.

Sets of identical or conservative residues are shown in boldface type. Conservative amino acids substitutions are grouped as follows: cysteine, serine, threonine, proline, alanine and glycine; asparagine, aspartic acid, glutamic acid, and glutamine; histidine, arginine, and lysine; methionine, isoleucine, leucine, and valine; phenylalanine, tyrosine, and tryptophan. The reference numbering is determined by alignment.

MEL cDNA cloning and Sequencing

MEL cDNA clones were isolated from a normal human fibroblast cDNA library constructed in GT10 (Kindly provided by A. Hall). The MEL cDNA clone pc2.2 detected 3.5 kb transcripts in melanoma cell lines as well as in EJ, a bladder carcinoma cell line and in NIH3T3 fibroblasts (Hughes and Padua, unpublished observations). Clone pc2.2 was recloned into M13mp18 and sequenced (Sequenase) using the universal primer and synthetic oligonucleotides. The predicted MEL protein was found to to share approximately 40% homology with members of the RAS family and 87% identity with the murine YPT1 (Nimmo et al., 1991, in press and Figure 3). Sequence comparison with the recently isolated canine RAB 8 (Chavrier et al., 1990) reveals that MEL is the human homologue of RAB 8. MEL/RAB8 differs from the other RAB/YPT proteins in lacking the characteristic CC or CAC C-terminus, but is similar to members of the RAS family in having a CAAX C-terminus motif. The CAAX sequence together with upstream residues is responsible for the membrane localisation of RAS (Willumsen et al., 1984; Gutierrez et al., 1989; Hancock et al., 1990).

MEL Sequences in NK16

Sequencing of the subclone NK2/2 derived from NK16 using an automatic sequencer (Applied Biosystems) revealed identity with the MEL cDNA sequence to the end of the putative first exon. Hybridisation of this clone with synthetic oligonucleotide probes corresponding to 3' sequences of the MEL cDNA reveals that most of the gene is absent in this clone (Hughes and Padua, unpublished observations). Therefore, the MEL sequences in the NKI4 transfectant forms part of a region of complex rearrangement generated by the transfection assay and until this complex is cloned and shown to have transformation properties, MEL cannot be designated the transforming gene in the melanoma cell line NKI4.

MEL Gene Rearrangement

MEL was localised more precisely to chromosome 19p13.2 (Nimmo et al., 1989), a region involved in translocations in a variety of neoplasias including melanoma (Parmiter et al., 1986). DNA from 12 melanoma cell lines and 20 melanoma tumours were analysed for MEL gene rearrangements by Southern blot analysis. One melanoma tumour (HK) which was a metastasis of the intestinal tract, was found to have rearranged MEL sequences (Figure 4). DNA from this tumour transformed NIH3T3 cells to produce anchoragae independent foci with human repetitive sequences (Padua and Schrier, unpublished observations). One of these transformants was shown to be tumorigenic in nude mice. However, human MEL sequences were not detected in these transformants by Southern blot analysis. Further investigations also revealed the absence of activated human RAS genes. Sequential cycles of transfection should enable the identification of the transforming gene.

Analysis of six other tumour metastases from this patient revealed that <u>MEL</u> gene rearrangements were not present. As melanoma metastasis is thought to be a clonal event, this suggests that the rearrangement observed in the single metastatic tumour may represent not only tumour heterogeneity, but possibly, tumour evolution. This is consistent with our findings of <u>RAS</u> mutations in melanomas being both acquired in metastatic tumours and lost from primary melanomas after metastasis (Shukla et al., 1989; Lewis et al., unpublished observations).

CONCLUSION

The human <u>MEL</u> gene which was originally isolated from an NIH3T3 transformant was found by cloning and sequencing to be

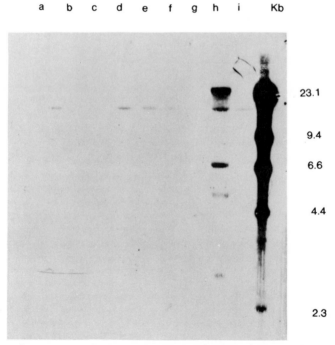

Figure 4. Rearrangement of MEL in a melanoma tumour.

Genomic DNA (20ug) was digested with Eco RI, electrophoresed in a 0.8% agarose gel, transferred to a nitrocellulose membrane and hybridised with alpha-^{32}PdCTP labelled pC2.2. Lanes a-h are melanoma tumour DNA. Lane h is DNA from patient HK. Lane i, NKI4 melanoma cell line. DNA fragment sizes are indicated in kilobases (Kb).

the human homologue of RAB 8.   The original transformant
comprised a complex rearrangement involving MEL sequences and
unrelated human sequences as well as unique murine sequences.
It is unlikely that MEL is the transforming gene in this
transfectant as only the putative first exon was present.   MEL
was localised to chromosome 19 (p13.2), a region which is often
involved in translocations associated with malignancies.   MEL
gene rearrangement was observed in a single melanoma metastatis
but this rearrangement was not observed in six other metastatic
tumours from this patient.   This finding suggests that the MEL·
rearrangement may represent tumour evolution and therefore is a
late event in tumour progression.

Acknowledgements

We thank Lorna Pearn for technical assistance, Peter
Sanders for the computer analyses and Bob Williamson for
helpful discussions.   This work was supported by the Leukaemia
Research Fund of Great Britain.

REFERENCES

Barbacid, M., 1987, RAS genes, Ann. Rev. Biochem., 56:779. Bos,
    J.L., 1988, The RAS family and human carcinogenisis,
    Mutat. Res., 195:255.
Chavrier P.,   Vingron M.,   Sander C.,   Simons K.,   Zerial M.,
    1990, Molecular cloning of YPT1/SEC4-related cDNAs from an
    epithelial cell line, Mol. Cell Biol.,   10:6578.
Fukui, M., Yamamoto, T., Kawai, S., Maruo, K., Toyoshima, K.,
    1985, Detection of a raf-related and two other transforming
    DNA sequences in human tumours maintained in nude mice,
    Proc. Natl. Acad. Sci. (USA), 82:5954.
Gutierrez, L., Magee, A.I., Marshall, C.J., Hancock, J.F.,
    1989,   Postranslational   processing   of p21ras is two-
    step and   involves   carboxy-methylation   and carboxy-
    terminal   proteolysis,   EMBO J, 8:1093.
Hancock, J.F., Magee, A.I., Childs, J.E., Marshall, C.J., 1990,
    All ras proteins are polyisoprenilated but only some are
    palmitoylated, Cell, 57:1167.
Ikawa, S., Fukui, M., Ueyama, Y., Tamaoki, N., Yamamoto, T.,
    Toyoshima, K., 1988, B-raf, a new member of the raf family,
    is activated by DNA rearrangement, Mol. Cell. Biol.,
    8:2651.
Ishikawa, F., Takaku, F., Nagao, M., Sugimura, T., 1987, Rat
    c-raf oncogene activation by a rearrangement that produces a
    fused protein, Mol. Cell. Biol., 7:1226.
Ishizaka, Y., Tahira T., Ochiai, M., Ikeda, I., Sugimura, T.,
    Nagao, M., 1988, Molecular cloning and characterization of
    human ret-II oncogene, Oncogene Res. 3:193.
Kozma, S.C., Redmond, S.M.S., Xiao-Chang, F., Saurer, S.M.,
    Groner, B., Hynes, N.E., 1988, Activation of the receptor
    kinase domain of the trk oncogene by recombination with two
    different cellular sequences, EMBO J., 7:147.

Martin-Zanca, D., Hughes, S., Barbacid, M., 1986, A human oncogene formed by the fusion of truncated tropomyosin and protein tyrosine kinase sequences, <u>Nature</u>, 319:743.

Nimmo, E., Padua, R.A., Hughes, D., Brook, J.D., Williamson, R., Johnson, K.J., 1989, Confirmation and refinement of the localisation of the c-MEL locus on chromosome 19 by physical and genetic mapping, <u>Human</u> <u>Genetics</u>, 81:382.

Nimmo, E.R., Sanders, P.G., Padua, R.A., Hughes D., Williamson, R., Johnson, K.J., 1991, The <u>MEL</u> gene: a new member of the rab/YPT class of ras-related genes. Oncogene, in press.

Padua, R.A., Barass, N., Currie, G., 1984, A novel transforming gene in a human malignant melanoma cell line, <u>Nature</u>, 311:671.

Parmiter, A.H., Balaban, G., Herlyn, M., Clark Jr., W.H., Nowell, P.C., 1986, A t(1;19) chromosome translocation in three cases of human malignant melanoma, <u>Cancer</u> <u>Res.</u>, 46:1526.

Shukla V., Hughes D.C., McCormick F., Padua R.A., 1989, <u>RAS</u> mutations in human melanotic lesions: K-<u>RAS</u> activation is a frequent and early event in melanoma development, <u>Oncogene</u> <u>Res.</u>, 5:121.

Spurr, N.K., Hughes, D.C., Goodfellow, P.J., Brook, J.D., Padua, R.A., 1986, Chromosomal assignment of c-MEL, a human transforming oncogene, to chromosome 19 (p13.2-q13.2), <u>Somat.</u> <u>Cell.</u> <u>Mol.</u> <u>Genet.</u>, 12:637.

Takahashi, M., Cooper, G.M., 1987, <u>ret</u> transforming gene encodes a fusion protein homologous to tyrosine kinases, <u>Mol.</u> <u>Cell.</u> <u>Biol.</u>, 7:1378.

Willumsen, B.M., Norris, K., Papageorge, A.G., Hubbert, N.L., Lowy, D.R., 1984, Harvey murine sarcoma virus p21ras protein: biological and biochemical significance of the cysteine nearest the carboxy terminus, <u>EMBO</u> <u>J</u>, 3:2581.

RAS MUTATIONS IN PRELEUKAEMIA, IN PATIENTS FOLLOWING CYTOTOXIC THERAPY AND IN NORMAL SUBJECTS

G. Carter[1], D. Hughes[1], N. Warren[1], A. Jacobs[1], J. Whittaker[1], E. Thomson[2], and R.A. Padua[1]

[1] LRF Preleukaemia Unit, Department of Haematology University of Wales College of Medicine, Heath Park Cardiff. U.K. [2] Department of Child Health, Llandough Hospital, Penarth, South Glam. U.K.

INTRODUCTION

Point mutations in members of the RAS gene family, NRAS, KRAS and HRAS, are a common molecular lesion in patients with acute myeloid leukaemia (AML). Approximately 30% of such patients show mutations (Bos, 1987; Farr et al., 1988) mostly in NRAS, although KRAS abnormalities are also seen. Mutations in HRAS are a rare abnormality in haemopoietic malignancies (Browett et al., 1989). A similar pattern of mutations is also seen in patients with the preleukaemic Myelodysplastic syndromes (MDS), where mutations have been detected in up to 41% of patients (Padua et al., 1988; Yunis et al., 1989). In MDS RAS mutation appears to be associated with poor prognosis in terms of an increased likelihood of transformation to acute leukaemia, and whilst RAS mutation can be detected in all sub-types of MDS (Padua et al., 1988), the highest incidence is seen those patients a monocytic phenotype (Padua et al., 1988; Yunis et al., 1989).

Patients representing all stages of leukaemic development provide us with a model for the study of leukaemogenesis. We have investigated 3 groups of patients for evidence of RAS mutations; i) Patients with MDS, ii) patients who are disease free but at risk of developing therapy-related MDS/AML, iii) normal subjects in no defined risk group for leukaemic development.

Here, we present data from each of these 3 groups, extending our previous study of MDS (Padua et al., 1988) to include 75 patients. We have also studied patients who are at risk of developing a secondary or therapy-related MDS/AML. We have previously detected RAS mutations in the peripheral blood of 9/70 patients in complete remission from lymphoma (Carter et al., 1990). There was a significant correlation between the presence of a mutant RAS gene and the period of time from the cessation of therapy (> 1000 days). We present preliminary data on patients who are at similar risk of leukaemic development (Levine & Bloomfield, 1986); these include 50 patients in complete remission from childhood acute lymphocytic leukaemia (ALL). All individuals at the time of study were haematologically normal.

In a survey of normal subjects, in no defined risk group for leukaemic development, we describe evidence for the presence of HRAS mutations in 2 cases. DNA from both registered in a Nude mouse tumorigenicity assay, and in 1 female case monoclonality can be demonstrated using an X-linked probe (PGK).

MATERIALS AND METHODS

Bone marrow or peripheral blood DNA samples was studied from 75 patients with MDS. Peripheral blood DNA was analysed from 50 patients in complete remission from childhood ALL, all of these subjects were haematologically normal at the time of study. DNA from haematologically normal subjects in no defined risk-group for leukaemic development, was obtained from peripheral blood or bone marrow obtained at operation from subjects undergoing cardiac or hip-replacement surgery.

Samples were screened for mutations at codons 12, 13 and 61 of the N and K-RAS genes, and codons 12 and 61 of H-RAS. Relevant target codons were amplified by polymerase chain reaction (PCR), and mutations detected by dot-blotting and sequential hybridisation with mutation specific oligonucleotide probes (Padua et al., 1988). NIH3T3 transformation and Nude mouse tumorigenicity assays were carried out on DNA from 46 MDS patients and 2 normal subjects as described previously (Padua et al., 1987).

Clonality analysis was carried out on ficoll separated cell populations from the peripheral blood of a female normal subject (patient DD). Clonality was assessed using the X-linked probe for the Phosphoglycerate kinase (PGK) locus (Vogelstein et al., 1987).

RESULTS

MDS patients

DNA from 75 MDS patients was assessed for evidence of RAS mutations. The incidence data are shown in Table 1, the numbers of patients who have undergone transformation to acute leukaemia (AML) are shown in Table 2. The overall incidence of RAS mutation in our series is 52% (39/75). RAS mutation is found in all sub-types of MDS, but most mutations are seen in patients with chronic myelomonocytic leukaemia (CMML), 84% of such patients show RAS mutation. Most RAS mutations are in NRAS

(42%), but KRAS mutations are also seen (32%). 4 patients had both NRAS and HRAS mutations, and 1 patient showed KRAS and HRAS mutations. We do not know if the co-existent mutations represent separate clones of mutated cells, or single clones with double mutations. Our data suggest that RAS mutation appears to be associated with an increased likelihood of transformation to AML (Table 2). Of the 17/75 patients who transformed to AML during the course of our study, 14 where from the group in whom RAS mutation was detected.

TABLE 1

RAS Mutations in MDS

FAB Subtypes:	RARS	RA	RAEB	CMML	TOTAL
Normal	12	12	7	5	36
Abnormal	4	6	3	26	39
Total	16	18	10	31	75
% Mutated	25	33	30	84	52
Mutations*					
HRAS	3	0	1	10	14
KRAS	1	3	0	13	17
NRAS	2	3	2	15	15

* 5 Patients had co-existent RAS mutations (4 with NRAS).

TABLE 2

Leukaemic Transformation in MDS

	Patients	AML
Mutant RAS	39	14
No mutant RAS	36	4
	75	17

POST CYTOTOXIC THERAPY PATIENTS

DNA from 50 patients in complete remission from childhood acute lymphocytic leukaemia (ALL) was assessed for evidence of RAS gene mutations. To date 4 patients show mutations. 3 patients showed the same G -> T transversion in NRAS leading to a glycine -> valine substitution at position 12 (N12Val). 1 patient showed a KRAS mutation in codon 12, This G -> A transition would give rise to a glycine -> aspartic acid substitution (K12Asp).

Material taken at the time of primary disease presentation was analysed for the presence of RAS mutation. In all 4 cases mutations were not detected, indicating that the present mutations have been somatically acquired.

Three of the four patients remain haematologically normal, however 1 patient with an N12Val mutation has since relapsed with primary disease. Details of these patients is given in Table 3.

TABLE 3

RAS Mutations Following Cytotoxic Therapy for Childhood ALL

Patient	Mutation	Substitution	Time Off Therapy (Days)
GJ	G -> T	N12Val	> 600
KS	G -> T	N12Val	> 4000
MS	G -> T	N12Val	> 1000
GM*	G -> A	K12Asp	> 700

* This patient did not show RAS mutation in a sample taken 6 months previously. 6 month follow-up samples from the other patients are not available.

NORMAL SUBJECTS

DNA from 18 normal subjects was assessed for evidence of RAS gene mutations. 2 subjects showed mutations in HRAS; one female patient (DD) showed evidence of an H12Asp mutation in bone marrow DNA, and one male patient (JR) evidence of an H12Val mutation. Buccal mucosa from DD and JR did not show a detectable RAS mutation, consistent with the somatic origin of these mutations.

Clonality studies of the female patient (DD) using the X-linked probe phosphoglycerate kinase (PGK) showed evidence of monoclonal haemopoiesis in both lymphocytic and granulocytic (myeloid) cell fractions. DNA from patient DD registers in an NIH3T3 transformation and Nude mouse tumorigenicity assay. Human HRAS sequences have been demonstrated in the tumour DNA by Southern blot analysis. To exclude the possibility of contamination from the EJ cell line which has an H12Val substitution, target sequences spanning nucleotide 2719 were amplified from the DNA of patient JR. Hybridisation analysis and direct sequencing showed the products to contain the normal allele at this position and not the EJ-specific form.

CONCLUSIONS

Our studies confirm that RAS mutation is a frequent feature in patients with clinical preleukaemia (MDS). In such patients, the presence of a RAS mutation is associated with a monocytic phenotype and an increased likelihood of transformation to overt AML.

Our studies also show that mutant RAS bearing cells may be detected in the absence of any overt haematological abnormality. This applies to patients who are in a defined risk-group for leukaemic development, and also normal subjects for whom no such risk can be ascribed. This further establishes the principle set by ourselves (Carter et al., 1990) for the detection of RAS mutations in patients following cytotoxic therapy for lymphoma, and for HRAS mutations in normal subjects first described by Nag et al., (Nag et al., 1989). Such observations underscore the need for a large survey of normal subjects to assess the possible incidence of RAS mutations in the population. Continued follow-up of these patients will show whether the presence of a RAS mutation will predict those at risk of developing MDS/AML or a therapy-related MDS/AML.

Acknowledgements

The authors wish to thank Lorna Pearn and Val McTiffin for excellent technical assistance. This work was supported by the Leukaemia Research Fund of Great Britain.

REFERENCES

Bos, J. L. 1989, RAS Oncogenes in Human Cancer, Cancer Research, 49:4682.
Browett, P. J., Yaxley, J. C. & Norton, J. D. 1989, Activation of Harvey RAS Oncogene by Mutation at Codon 12 is Very Rare in Haemopoietic Malignancies, Leukaemia, 3:86.
Carter, G., Hughes, D. C., Clark, R. E., Whittaker, J. A., Jacobs, A. J. & Padua, R. A. 1990 RAS Mutations Following Cytotoxic Therapy for Lymphoma, Oncogene, 5:411.
Farr, C.J., Saiki, R. K., Erlich, H. A., McCormick, F. & Marshall, C. J., 1988, Analysis of RAS Gene Mutations in Acute Myeloid Leukaemia by Polymerase Chain Reaction and Oligonucleotide Probes, Proc. Natl. Acad. Sci. (USA), 85:1629.
Levine, E.G & Bloomfield, C.D. 1986, Secondary Myelodysplastic Syndromes and Leukaemias, Clin. Haematol., 15:1037.
Nag, A., Jones, J. A. & Smith, R. G., 1989, Point Mutation in c- Ha-ras Oncogene in Normal and Abnormal Cells, Lancet, 1:274.

Padua, R. A., Carter, G., Hughes, D. C., Gow, J., Farr, C. J., Oscier, D., McCormick, F., Smith, S., & Jacobs, A. J., 1988   RAS Mutations in Myelodysplasia Detected by Amplification,   Oligonucleotide Hybridisation and Transformation, Leukaemia, 8:503.

Vogelstein, B., Fearon, E. R., Hamilton, S. R., Preisinger, A. C., Willard, H. F., Michelson, A. M., Riggs, A. D. & Orkin, S.   H., 1987. Clonal Analysis Using Recombinant DNA Probes from the   X-Chromosome, Cancer Research, 47:4806.

Yunis, J.J., Boot, A. J. M., Mayer, M. G., & Bos, J. L. 1989, Mechanisms of ras Mutation in Myelodysplastic Syndrome, Oncogene, 4:609.

CHARACTERIZATION OF THE HUMAN N-RAS PROMOTER REGION

J.T. Thorn, A.V. Todd, D. Warrilow, F. Watt[#],
P.L. Molloy[#] and H.J. Iland

The Kanematsu Laboratories, Royal Prince Alfred
Hospital, Missenden Road, Camperdown, NSW, 2050
Australia. [#]CSIRO Division of Biomolecular
Engineering, North Ryde, P.O.Box 184, 2113, NSW
Australia

ABSTRACT

Overexpression of ras protooncogenes has been implicated in
cancer development. We therefore initiated a study of the human
N-ras promoter to determine the regions that control N-ras
expression and their potential for interaction with DNA-binding
proteins. N-ras CAT constructs were stably integrated into K562
cells by electric field-mediated gene transfer. A significant
proportion of promoter activity was found to lie within a 438bp
fragment comprising an untranslated exon (exon -1) with adjacent
5' sequence and a small CpG island. A 107bp fragment at the 5'
end of exon -1 was essential for promoter activity, while a 44bp
deletion from within this region decreased promoter activity by
two thirds. Unlike the human H-ras promoter, the human N-ras
promoter did not exhibit bidirectional activity. DNase
footprinting of the 438bp fragment revealed seven protected
regions, many of which contain sequences homologous to known DNA
binding protein sites (MLTF/myc, CREB/ATF, AP-1, AP-2, myb, and
E4TF1). Using purified MLTF and appropriate competitors in gel
shift and DNase footprinting assays, we demonstrated binding of
MLTF to the MLTF consensus sequence within exon -1.

INTRODUCTION

Overexpression of ras has been reported in 21 to 50% of
human tumors (Tanaka et al, 1986; Barbacid, 1987), and may lead
to deregulated signal transduction in cancer cells representing
an alternative activation mechanism to point mutations. H-ras and
K-ras are overexpressed in pancreatic endocrine tumors (Hofler
et al, 1988), while N-ras is overexpressed in glioblastomas
(Gerosa et al, 1989). Increased ras expression has also been

The Superfamily of ras-Related Genes
Edited by D.A. Spandidos, Plenum Press, New York, 1991

correlated with depth of invasion of colon carcinoma (Thor et al, 1984). Overexpression in conjunction with coding mutations has been observed for N-ras in a guinea pig cell line (Doniger and DiPaolo, 1988) and in human xeroderma pigmentosum tumor cells (Suarez et al, 1989). A similar phenomenon involving H-ras occurs in the human EJ bladder cancer cell line (Cohen and Levinson, 1988). Furthermore, a 20- to 50-fold overexpression of normal N-ras can cause transformation in vitro (McKay et al, 1986). Therefore greater knowledge of the mechanisms controlling ras expression may provide a better understanding of the role ras genes play in carcinogenesis.

Transfection studies have only indicated that the human N-ras promoter lies somewhere within a Pst I/Bgl II 900bp fragment. This region includes the 5' untranslated exon -1, together with 5' and 3' flanking sequences, four putative Spl sites and two start sites (Taparowsky et al, 1983; Brown et al, 1984; Hall and Brown, 1985). Recently, the mouse N-ras promoter has been shown to reside within a 230bp fragment at the 5' end of the gene in which a negative regulatory element is also situated (Paciucci and Pellicer, 1991).

MATERIALS AND METHODS

Assessment of Promoter Function

K562 cells (5×10^6 cells) (Lozzio and Lozzio, 1975) were transfected by electric field-mediated gene transfer according to a protocol optimized by Croaker et al, (1990) in this laboratory. 50μg of the N-ras CAT constructs (in which various portions of the N-ras promoter region had been inserted upstream of a CAT reporter gene) and 5μg of pRSVneo (Gorman et al, 1983) were linearized and cotransfected into K562 cells. Stably transfected cells were selected in 1mg/ml geneticin (Sigma Pharmaceuticals) over a period of four weeks.

To determine transfection efficiency, genomic DNA was isolated and probed for the presence of the CAT gene and the filter reprobed for β-actin as a control for variations in DNA loadings. CAT activity was determined according to Sleigh (1986) except that cells were lysed by freeze-thawing and the final ethyl acetate extraction mixture was extracted through 100μl of Tris HCl (7.8). Total cell number was calculated prior to cell lysis, and results were calculated as [^{14}C]-chloramphenicol acetate/10^7 viable cells with corrections made for transfection efficiency and then expressed as a percentage of the normal full length N-ras promoter (pNRCAT-Fe-1).

DNA Binding Assays

Gel shift assays were performed according to Carthew et al, (1985) with minor modifications and using MLTF-containing fraction (AA) (Watt and Molloy, 1988). DNase Footprinting was set up as for gel shift assays except that 4μl of HeLa nuclear extract (1mg/ml), prepared according to Sharpiro et al, (1988), was used (refer to Thorn et al, 1991).

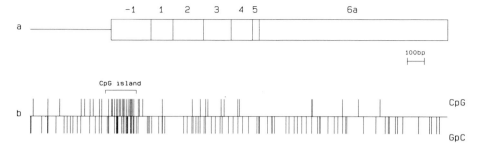

Figure 1. CpG island in the Human N-ras gene. (a) Regions of the human N-ras gene that have been sequenced (Hall and Brown, 1985) which includes the 5' promoter region from the Pst I site (line) and the exons (boxes). (b) Plot of CpG's (above line) and GpC's (below line) corresponding to the regions of the human N-ras gene shown in (a). The position of the CpG island is indicated on the plot.

RESULTS

CpG Island within the Human N-ras Gene

CpG islands are regions of DNA, up to 2 kilobases (Kb) in length, which have a high G+C content and a high frequency of CpG dinucleotides relative to the rest of the genome (reviewed in Gardiner-Garden and Frommer, 1987). CpG islands are predominantly found in 5' regulatory regions and are thought to play a crucial role in transcription control. They can be visualized when the location of CpG's are plotted against the location of GpC's, the latter being indicative of the overall G+C content (Bird, 1986). Using previously reported sequence data (Hall and Brown, 1985), we observed a cluster of CpG's and GpC's localized to 148bp at the 5' end of the first untranslated exon (exon -1) of the N-ras promoter region (Figure 1).

Identification of functionally important regions of the Human N-ras Promoter

N-ras CAT constructs were cotransfected into the hematopoietic cell line K562, together with pRSVneo, enabling geneticin selection of transfectants. Promoter activities are expressed relative to pNRCAT-Fe-1 which contains the 900bp human N-ras promoter region and an additional 200bp of intron sequence (Figure 2). When the N-ras promoter was cloned in the reverse orientation (pNRCAT-Fe-1b), all activity was lost indicating that the human N-ras promoter is not bidirectional. pNRCAT-Fe-2 possesses the central 438bp of the promoter region, which includes exon -1. This fragment retains two thirds of the N-ras promoter activity suggesting that sequences at the 5' and/or 3' extremes of the promoter region do not play a major role in expresssion but are necessary for maximal promoter function.

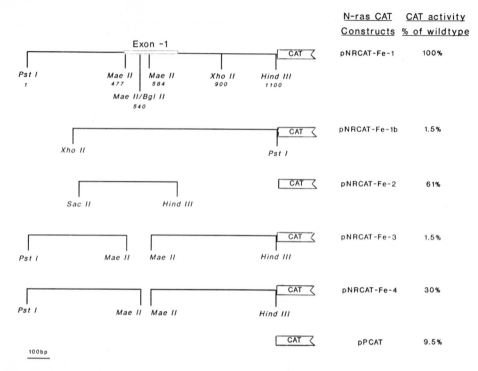

Figure 2. N-ras promoter CAT activities. Deletions of the N-ras promoter region were made and their activities expressed relative to the wildtype promoter, pNRCAT-Fe-1 (100%). The position numbers are relative to the Pst I site at the 5' end of the promoter. The deletion constructs are shown on the left with their corresponding CAT activities shown on the right. The Sac I/Hind III sites in pNRCAT-Fe-2 are PCR introduced cloning sites and not endogenous to the N-ras promoter. The cloning vector, pPCAT, which lacks any N-ras promoter elements, expressed almost 10% CAT activity relative to pNRCAT-Fe-1 and this presumably represents basal expression associated with the CAT gene in this environment.

pNRCAT-Fe-3 contains the full length promoter region with a 107bp Mae II fragment deleted from within exon -1. This deletion abolished all promoter activity indicating that the 107bp fragment is essential for N-ras expression. The major contribution to promoter activity appears to originate from within an even smaller portion of the 107bp Mae II fragment of exon -1, since a full length promoter lacking only 44bp (pNRCAT-Fe-4) retains only one-third activity relative to pNRCAT-Fe-1.

DNase 1 Footprinting of the Human N-ras Promoter

The sense strand of the 438bp Sac I/Hind III N-ras fragment from pNRCAT-Fe-2 was footprinted with HeLa nuclear extract, and 7 regions were protected. Table 1 shows the consensus sequences of known transcription factor binding sites that can be

Table 1. Transcription factor consensus sequences found within the N-ras promoter. The regions protected from DNase digestion are shown along side the consensus sequences of the various transcription factors which show homology (Jones et al, 1988; Hai et al, 1989; Mitchell and Tjian, 1989; Biedenkapp et al, 1988). Sequences are written 5' to 3'. The two mismatches between N-ras protected region I and the E4TF1 consensus sequence are found near the centre of the consensus sequence and may exclude binding; therefore the putative transcription factor in region I has been placed in brackets. Sequencing of the N-ras promoter revealed discrepancies with the published sequence (Hall and Brown, 1985) producing a putative AP-2 site (Thorn et al, in press). AS = Anti-Sense Strand

PROTECTED REGIONS	TRANSCRIPTION FACTORS	CONSENSUS SEQUENCE	N-RAS SEQUENCE
I	(E4TFI)	GGAAGTG	GGAACCG^{AS}
II	E4TFI	GGAAGTG	GGAAGTG
III	CREB/ATF AP-2	TGACGT(A/C)(A/G) CCCCAGGC	TTACGTAG CCCCAGGC^{AS}
IV	MLTF/c-myc	CCACGTGA	CCACGTGG
V	AP-1	TGA(G/C)TCA	TGACTCG
VI	UNKNOWN		
VII	v-myb + UNKNOWN	C(A/C)GTT(A/G)	a. CAGTTT b. CTGTTA

Table 2. Conservation of putative consensus sequences between species. Consensus sequence observed in mouse and guinea pig genes at positions identical to those identified in the human N-ras gene following DNase I footprinting. Base mismatches to consensus sequences are underlined.

BINDING PROTEINS	CONSENSUS SEQUENCE	N-RAS GENE		
		HUMAN	MOUSE	GUINEA-PIG
MLTF/c-myc	CCACGTGA	CCACGTG_G_	CCACGTG_G_	CCACGTG_G_
CRE/ATF	(T/G)(T/A)CGTCA	TACGT_AG_	TACGT_AG_	TACGT_AG_
AP-1	T(T/G)AGTCA	TGACTC_G_	_A_GACTC_T_	TGACTC_G_
E4TF1	GGAAGTG	GGAAGTG	GGAAGTG	GGAAGTG
v-myb	C(A/C)GTT(A/G)	C_T_GTTA	_A_TGTTA	_A_TGTTA

identified in the N-ras DNase protected regions. Many of these consensus sequences are present in the mouse and guinea pig genes at equivalent positions (Table 2).

Binding of MLTF to the Human N-ras Promoter

Within the N-ras gene promoter lies a consensus sequence for Adenovirus major late transcription factor (MLTF) centred at 543 (543bp from the Pst I site). The DNA sequence of the Adenovirus major late promoter (MLP) that binds MLTF is shown in Figure 3a, with the minimal core sequence needed for MLTF binding boxed (Sawadogo and Roeder, 1986). The proposed N-ras MLTF binding site is also shown with mismatches to the Adenovirus MLTF binding site indicated by connecting lines. There is only one base mismatch within the core sequence, and this is at position 547 in N-ras.

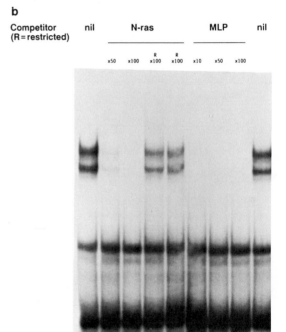

Figure 3. Binding of MLTF to N-ras. (a) Comparison of MLTF binding sites between Adenovirus MLP and N-ras gene. The sequences are aligned with the mismatches indicated by adjoining lines. The core sequence needed for MLTF binding to Adenovirus MLP is boxed (Sawadogo and Roeder, 1986) and is centred at position -58 in Adenovirus MLP and at position 543 in the N-ras promoter. (b) DNA binding assay illustrating N-ras and MLP competing with N-ras for MLTF binding. Labelled N-ras was competed with unlabelled N-ras (lanes 2-5) and MLP (lanes 6-8) in 10-100 molar excess as indicated. Lanes 1 and 9 contain no competitor DNA. MLP DNA competes with N-ras for binding of MLTF. N-ras competes with itself unless the binding site has been disrupted by digestion with Bgl I (lane 4) or Mae II (lane 5).

An end-labelled 103bp Hinp I fragment of N-ras encompassing the proposed MLTF site, and an end-labelled 100bp MLP fragment (Watt and Molloy, 1988) incorporating the Adenovirus MLTF site, were used in gel shift assays to confirm that the N-ras promoter binds MLTF. The AA fraction used contained partially purified MLTF and was incubated with both the MLP and N-ras fragments. Figure 3b shows MLTF binding to the N-ras promoter; the addition of an increasing molar excess of either unlabelled N-ras or MLP fragments successfully competed with N-ras for binding of MLTF, digestion of unlabelled N-ras at the MLTF site prevented this competition. MLP has a greater affinity for MLTF than does N-ras, and we estimate the difference in affinities is in the range of 5- to 10-fold.

DISCUSSION

We have identified a short CpG island in the human N-ras promoter region coinciding with the 5' untranslated exon. We believe its presence is significant with respect to regulation of N-ras expression since it overlaps the 107bp fragment that is essential for N-ras expression. CpG rich islands have been found within housekeeping genes primarily in their 5' regulatory region, and it has been suggested that they represent promoter elements of genes (Bird, 1986).

Footprinting of the 438bp region of the human N-ras promoter has revealed many protected regions which are homologous to consensus sequences for known transcription factors. A significant proportion of these putative binding sites lie within this 107bp N-ras promoter fragment found to be essential for promoter activity, including those for MLTF/myc, AP-1, AP-2, and CRE/ATF, as well as one of the start sites. With the exception of the MLTF site, all other binding sites are putative based on homology between consensus sequences and the footprinted regions.

Our results highlight the potential interactions that may occur between N-ras and several other oncogenes. The AP-1 transcription factor is a heterodimer formed by the c-jun and c-fos gene products (Halazonetis et al, 1988). N-ras can activate the c-fos gene in a transient expression assay whilst it also inhibits nerve growth factor and basic fibroblast growth factor induction of c-fos (Thomson et al, 1990). The finding of an AP-1 consensus binding site in the human N-ras promoter suggests that N-ras may have the potential to modulate its own expression levels in a feedback system by influencing c-fos and c-jun expression.

The N-ras MLTF site only differs from the Adenovirus MLTF core sequence by one base change and this base has been shown to reduce transcriptional activity of the Adenovirus MLP by only 25% (Miyamoto et al, 1985), suggesting that the N-ras MLTF site is likely to be functional. Recent studies have found that the c-myc binding site is identical to that of MLTF (Blackwell et al, 1990) and c-myc can bind to the Adenovirus MLP (Prendergast and Ziff, 1991). Taken together with the well documented cooperation between c-myc and N-ras in inducing cellular transformation, it is possible that c-myc may also bind to the N-ras promoter at the MLTF site identified here.

Sequence comparisons of the 5' regions of the human, mouse and guinea pig N-ras genes reveal extensive sequence homology. In particular, many of the consensus sequences found to be DNase protected in the human N-ras sequence (MLTF/myc, CRE/ATF, AP-1, myb and E4TF1) are also present in these other species of N-ras. Conservation of these potential protein binding sites supports the view by Doniger (1987) that conserved untranslated regions are likely to be important regulatory domains.

The region in the mouse N-ras promoter identified recently as a negative regulatory element (NRE) (Paciucci and Pellicer, 1991) coincides with DNase protected region VII in the human N-ras promoter. We have observed two potential myb binding sites in this area of the human promoter, each matching the myb consensus sequence at five out of six bases. Analysis of the equivalent mouse sense strand sequence, which also exhibited DNase protection, revealed a four out of six match with both of the putative myb binding sites. In addition, the anti-sense strand at this position contains a five out of six match in the reverse orientation (5'-CCATTA-3' compared with the consensus sequence 5'-C(A/C)GTT(A/G)-3'). A double stranded oligonucleotide (3'd-20), which lacks this potential myb binding site (5'-TAATG-3' on the sense strand), failed to exhibit negative regulatory activity in functional assays (Paciucci and Pellicer, 1991). When the data from our human studies are taken together with those of the mouse N-ras promoter, they suggest that myb binding may act as a negative regulator of N-ras transcription.

ACKNOWLEDGEMENTS

A full report of this study is in press in Oncogene. This research was supported by the Australian National Health and Medical Research Council and in part by the Leo and Jenny Leukaemia and Cancer Foundation.

REFERENCES

1. Barbacid, M. 1987, Ras genes, Ann. Rev. Biochem., 56:779.
2. Biedenkapp, H., Borgmeyer, U., Sippel, A. E., & Klempnauer, K.-H. 1988, Viral myb oncogene encodes a sequence-specific DNA-binding activity, Nature, 335:835.
3. Bird, A. P. 1986, CpG-rich islands and the function of DNA methylation, Nature, 321:209.
4. Blackwell, T. K., Kretzner, L., Blackwood, E. M., Eisenman, R. N., & Weintraub, H. 1990, Sequence-specific DNA binding by the c-myc protein, Science, 250:1149.
5. Brown, R., Marshall, C. J., Pennie, S. G., & Hall, A. 1984, Mechanism of activation of an N-ras gene in the human fibrosarcoma cell line HT1080, EMBO J, 3:1321.
6. Carthew, R. W., Chodosh, L. A., & Sharp, P. A. 1985, An RNA polymerase II transcription factor binds to an upstream element in the Adenovirus major late promoter, Cell, 43:439.

7. Cohen, J. B., & Levison, A. D. 1988, A point mutation in the last intron responsible for increased expression and transforming activity of the c-H-ras oncogene, <u>Nature</u>, 334:119.
8. Croaker, G. M., Wass, E. J., & Iland, H. J. 1990, Electric field-mediated gene transfer into K562 cells: optimization of parameters affecting efficiency, <u>Leukemia</u>, 4:502.
9. Doniger, J. 1987, Differential conservation of non-coding regions within human and guinea pig N-ras genes, <u>Oncogene</u>, 1:331.
10. Doniger, J., & DiPaolo, J. A. 1988, Coordinate N-ras mRNA up-regu;ation with mutational activation in tumorigenic guinea pig cells, <u>Nucleic Acids Res.</u>, 16:969.
11. Gardiner-Garden, M., & Frommer, M. 1987, CpG islands in vertebrate genomes, <u>J. Mol. Biol.</u>, 196:261.
12. Gorman, C., Padmanabhan, R., & Howard, B. H. 1983, High efficiency DNA-mediated transformation of primate cells, <u>Nature</u>, 221:551.
13. Gerosa, M. A., Talarico, D., Fognani, C., Raimondi, E., Colombatti, M., Tridente, G., De Carli, L., & Della Valle, G. 1989, Overexpression of N-ras oncogene and epidermal growth factor receptor gene in human glioblastomas, <u>J Natl. Cancer Inst.</u>, 81:63.
14. Hai, T., Liu, F., Coukos, W. J., & Green, M. R. 1989, Transcriptional factor ATF cDNA clones: an extensive family of leucine zipper proteins able to selectively form DNA-binding heterodimers, <u>Genes Dev.</u>, 3:2083.
15. Halazonetis, T. D., Georgopoulos, K., Greenberg, M. E., & Leder, P. 1988, c-jun dimerizes with itself and with c-fos, forming complexes of different DNA binding affinities, <u>Cell</u>, 55:917.
16. Hall, A., & Brown, R. 1985, Human N-ras: cDNA cloning and gene structure, <u>Nucleic Acids Res.</u>, 13:5255.
17. Hofler, H., Ruhri, C., Putz, B., Wirnsberger, G., & Hauser, H. 1988, Oncogene expression in endocrine pancreatic tumors, <u>Virchows. Archiv. B Cell Pathology</u>, 55:355.
18. Jones, N. C., Rigby, P. W. J., & Ziff, E. B. 1988, Trans-acting protein factors and the regulation of eukaryotic transcription: lessons from studies on DNA tumor viruses, <u>Genes Dev.</u>, 2:267.
19. Lozzio, C. B., & Lozzio, B. B. 1975, Human chronic myelogenous leukemia cell line with positive Philadelphia chromosome, <u>Blood</u>, 45:321.
20. Maxam, A. M., & Gilbert, W. 1980, Sequencing end-labelled DNA with base-specific chemical cleavage, <u>Methods Enzymol.</u>, 65:499.
21. McKay, I. A., Marshall, C. J., Cales. C., & Hall, A. 1986, Transformation and stimulation of DNA synthesis in NIH-3T3 cells are a titratable function of normal p21^{N-ras} expression, <u>EMBO J</u>, 5:2617.
22. Mitchell, P. J., & Tjian, R. 1989, Transcriptional regulation in mammalian cells by sequence-specific DNA binding proteins, <u>Science</u>, 245:371.
23. Miyamoto, N. G., Moncollin, V., Egly, J. M., & Chambon, P. 1985, Specific interaction between a transcription factor and the upstream element of the Adenovirus-2 major late promoter, <u>EMBO J</u>, 4:3563.

24. Paciucci, R., & Pellicer, A. 1991, Dissection of the mouse N-ras gene upstream regulatory sequences and identification of the promoter and a negative regulatory element, Mol. Cell. Biol., 11:1334.
25. Prendergast, G. C., & Ziff, E. B. 1991, Methylation-sensitive sequence-specific DNA binding by the c-myc basic region, Science, 251:186.
26. Sawadogo, M., & Roeder, R. G. 1986, DNA-binding specificity of USF, a human gene-specific transcription factor requirred for maximal expression of the major late promoter of Adenovirus, In Cancer cells: DNA tumor viruses 4:147, Cold Spring Laboratories, Cold Spring Habor New York.
27. Sharpiro, D. J., Sharp, P. A., Wahli, W. W., & Keller, M. J. 1988, A high-efficiency HeLa cell nuclear transcription extract, DNA, 7:47.
28. Sleigh, M. J. 1986, A nonchromatogrphic assay for expression of the chloramphenicol acetyltransferase gene in eukaryotic cells, Anal. Biochem., 156:251.
29. Suarez, H. G., Daya-Grosjean, L., Schlaifer, D., Nardeux, P., Renault, G., Bos, J. L., & Sarasin, A. 1989, Activated oncogenes in human skin tumors from a repair-deficient syndrome, xeroderma pigmentosum, Cancer Res., 49:1223.
30. Tanaka, T., Slamon, D. J., Battifora, H., & Cline, M. J. 1986, Expression of p21 ras oncoproteins in human cancers, Cancer Res., 46:1465.
31. Taparowski, E., Shimizu, K., Goldfarb, M., & Wigler, M. 1983, Structure and activation of the human N-ras gene, Cell, 34:581.
32. Thomson, T. M., Green, S. H., Trotta, R. J., Burstein, D. E., & Pellicer, A. 1990, Oncogene N-ras mediates selective inhibition of c-fos induction by nerve growth factor and basic fibroblast growth factor in a PC12 cell line, Mol. Cell. Biol., 10:1556.
33. Thor, A., Horan Hand, P., Wunderlich, D., Caruso, A., Muraro, R., & Schlom, J. 1984, Monoclonal antibodies define differential ras expression in malignant and benign colonic diseases, Nature, 311:562.
34. Thorn, J. T., Todd, A. V., Warrilow, D., Watt, F., Molloy, P. L., & Iland, H. J. 1991, Characterization of the human N-ras promoter region, Oncogene, in press.
35. Watt, F., & Molloy, P. L. 1988, Cytosine methylation prevents binding to DNA of a HeLa cell transcription afctor requirred for optimal expression of the Adenovirus major late promoter, Genes Dev., 2:1136.

RAS AND RAP1 GTPASES MUTATED AT POSITION 64

M.S.A. Nur-E-Kamal and H. Maruta

Melbourne Tumor Biology Branch
Ludwig Institute for Cancer Research
P.O. Royal Melbourne Hospital
Victoria 3050, Australia

ABSTRACT

   Ras and Rap1 GTPases require Gly12, the effector domain (residues 32 to 40) and Ala 59 for activation by GAP1 and GAP3, respectively. The replacement of Gly12 by Val or Ala59 by Thr potentiates the Ras oncogenicity and Rap anti-oncogenicity. However, the mutations in the effector domain, in particular the replacement of Thr35 by Ala, abolishes both Ras oncogenicity and Rap anti-oncogenicity, indicating that the effector domains are involved in the interactions of these signal transducers with their targets as well as the GAPs. Here, we demonstrate that replacement of Tyr64 of the HaRas protein or Phe64 of Rap1A protein by Glu reduces their intrinsic GTPase activities and abolishes their stimulation by GAP1 or GAP3, respectively. Further mutational analysis has revealed that only the Tyr to Phe or Leu (Ras) and Phe to Tyr (Rap1A) substitutions at position 64 still allow these GTPases to be activated by GAPs.

INTRODUCTION

   Mutations in the GTP-binding domains convert the normal Ras proteins (Ha, Ki and N-) into highly oncogenic proteins (1). The Rap proteins (1A, 1B, 2A and 2B) shares 50% sequence identity with the Ras proteins (2-5), and over-expression of the Rap1A gene is able to reverse the malignant transformation caused by the oncogenically mutated Ras genes (3). Mutations in the GTP-binding domains also potentiate the anti-Ras action (anti-oncogenicity) of the normal Rap1A protein (6). Intrinsic GTPase activities of the Ras and Rap1 proteins are stimulated by GAP1 and GAP3, respectively (7,8). Since only the GTP-bound forms of the Ras and Rap proteins serve as signal transducers, GAP1 and GAP3 appear to act as the attenuators of the Ras and Rap signals, respectively.
   Three regions of Ras and Rap GTPases, i.e., Gly12, effector domain (residues 32 to 40) and Ala59, are required for their GAP-dependent activation (7, 9-12). Interestingly, the Ras oncogenicity and Rap anti-oncogenicity are potentiated by the mutations at either position 12 or 59 (1,6), but abolished by the mutations in the effector domains (6, 13, 14), indicating that the residues 32 to 40 are involved not only in the GAP-dependent attenuation, but also in the transduction of the Ras and Rap signals. We have provided an evidence indicating that another

```
        61        63  64  65                      70
Ras :-Gln-Glu-Glu-Tyr-Ser-Ala-Met-Arg-Asp-Gln-
Rap :-Thr-Glu-Gln-Phe-Thr-Ala-Met-Arg-Asp-Leu-
Rsr1:-Ile-Ala-Gln-Phe-Thr-Ala-Met-Arg-Glu-Leu-
Rho :-Gln-Glu-Asp-Tyr-Asp-Arg-Leu-Arg-Pro-Leu-
```

Figure 1. The amino acid residues in positions 61 to 70 of GTP binding proteins.

region (residues 61 to 65) of both Ras and Rap1 proteins plays the key role in their interactions with GAP1 and GAP3, respectively (12).

Gln61 and Ser65 are required for the GAP1-dependent attenuation of the Ras signals but not for the GAP3-dependent attenuation of the Rap1 signals, whereas Thr65 is required for the GAP3-dependent attenuation of the Rap1 signal but not for the GAP1-dependent attenuation of the Ras signals (12). Since residues 61 to 63 are not required for the Ras-dependent signal transduction (13), replacement of Gln61 by Leu or other amino acids significantly potentiates the Ras oncogenicity (1). Conversely, residue 61 of Rap1A appears to be essential for its anti-oncogenicity (6). Replacement of Thr61 by Gln in the Rap1A protein significantly reduces its anti-oncogenicity (6), although it does not affect either the intrinsic GTPase activity or its stimulation by GAP3 (12). Furthermore, replacement of Gln63 by Glu significantly reduces the Rap1 GTPase activity (12) and potentiates the Rap1A anti-oncogenicity (6). However, the Glu63 mutant of the Rap1 GTPase is still highly activated by GAP3 (12). Conversely, replacement of Glu63 by Gln in the Ras GTPase does not significantly affect either the intrinsic GTPase activity or its stimulation by GAP1 (12). Neverthless, Lys63 mutant of the HaRas protein was found to be moderately oncogenic (16). We have been trying to identify the role of Tyr64 in the HaRas protein and Phe64 in the Rap1 protein for the following reasons: (i) The oligopeptide corresponding to residues 56 to 66 of the HaRas protein has been shown previously to be phosphorylated at Tyr64 by EGF receptor in vitro (17). However, the effect of Tyr64 phosphrylation on the biological function of the intact Ras protein still remains to be clarified; (ii) the deletion of the residues 64 to 68 was reported recently to increase the intrinsic GTPase activity of the oncogenically mutated HaRas protein (Arg12/Thr59) and cosequently to reduce its oncogenicity (18), whereas mutations at residues 65 to 68 of the oncogenically mutated HaRas protein (Val12/Thr59) do not significantly affect the oncogenicity (13). These observations suggest that Tyr64, but not residues 65 to 68, is required for the Ras oncogenicity. However, the Leu64 mutant of the normal HaRas protein (Gly12/Ala59) is still able to complement the Ras2 defect in yeast, indicating that Tyr is not absolutely required at position 64 of Ras protein for signal transduction (13). (iii) The non-oncogenic R-Ras GTPase, which shares only 50% sequence identity with three other Ras GTPases (Ha, Ki and N-), contains Phe at positin 64 (19, 20), but is still activated by GAP1 (21). This observation has suggested, but not proved yet, that Tyr and Phe are functionally replaceable at position 64 for the GTPase activation. Interestingly, replacement of Phe64 by Tyr in the Rap1A protein does not affect its anti-oncogenicity (6), indicating that Phe and Tyr are also functionally replaceable at position 64 for the Rap1 signal transduction.

Here we provide an evidence indicating that either Tyr or Phe at position 64 of the Ras and Rap1 GTPases are essential for GAP dependent activation. It still remains to be clarified, however, whether residue 64 is essential for Ras oncogenicity and Rap anti-oncogenicity or not.

MATERIALS AND METHODS

<u>PCR site-directed mutagenesis and subcloning</u> . Replacement of Tyr or Phe
at position 64 of HaRas or Rap1A proteins by other amino acids was
performed by the oligonucleotide-directed mutagenesis through polymerae
chain reaction (PCR) as described previously (12). The PCR products
encoding normal or mutated HaRas were subcloned into the BamHI/EcoRI
sites of the plasmid pGEX-1, whereas those encoding the normal or
mutated Rap1A proteins were subcloned into the BamHI/HindIII sites of
the plasmid pGEX-2TH, to be expressed in <u>E.coli</u> under the control of lac
promoter as glutathione S-transferase (GST) fusion proteins as described
previously (12, 22, 23).

<u>Purification of GST fusion proteins (Ras/Rap/GAP1C) and bovine GAP3</u>.Both
GST/Ras and GST/Rap fusion proteins were overexpressed in <u>E.coli</u> at 25°C
in the presence of isopropyl-beta-D-thiogalactopyranoside (IPTG), and
affinity purified on a glutathione (GSH)-agarose column (12). The
expression and purity of these Ras or Rap fusion proteins were monitored
by SDS-polyacrylamide gel electrophresis. An example is shown in
Fig. 2. The C-terminal domain (residues 720 to 1044) of bovine GAP1
called GAP1C was also affinity-purified as a GST fusion protein as
described above. GAP3 was partially purified from bovine brain cytosol
as described earlier (12). This preparation of GAP3 did not show any
GAP1 activity.

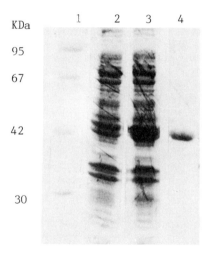

<u>Figure 2</u>. Expression and purification of GST fusion proteins: <u>E.coli</u>
cells were cultured in the presence and absence of IPTG. The cells were
then harvested, lyzed and centrifuged at 11Krpm for 15 min. The GST
fusion proteins were purified on a glutathione-aganose column. Lane 1,
size marker; <u>E.coli</u> grown in the absence (lane 2) and presence (lane 3)
of IPTG. Lane 4, purified GST fusion protein.

<u>Assay for GAP-dependent activation of the Ras or Rap GTPases</u>. The
hydrolysis of (gamma-^{32}P) GTP bound to either Ras or Rap fusion proteins
was determined by incubating at 35°C for various periods in the presence
or absence of either GAP1C or GAP3 preparations, then absorbing the
unhydrolyzed (gamma-^{32}P) phosphate in the supernatant (12, 24). Protein
concentration were determined by Bradford method with bovine serum
albumin as standard (25).

RESULTS AND DISCUSSION

Characterization of HaRas GTPases Mutated at position 64. The three distinct Ras (Ha, Ki and N-) GTPases contain Tyr at position 64, whereas R-Ras GTPase contains Phe at the same position (19). Neverthless, all these four Ras GTPases are activated by GAP1 (7, 21), suggesting that Tyr and Phe at this position are functionally replaceable for GAP-dependent GTPase activation. The deletion of residues 64 to 68 was reported to increase the intrinsic GTPase activity of the oncogenically mutated HaRas protein (Arg12/Thr59) and reduce its oncogenicity (18). Consequently, we have replaced Tyr64 of the normal HaRas protein by several other amino acids to determine more precisely the role that Tyr64 might play in the intrinsic GTPase activity and its stimulation by GAP1. Either Glu64 or Asp64 mutations significantly reduced intrinsic GTPase activity of the Ras protein and completely abolished its stimulation by GAP1C, i.e., the C-terminal domain (residues 720 to 1044) of GAP1, which is sufficient for activation of the normal Ras GTPases.

However, these effects do not appear to be significantly related to the presence of a negative charge. Replacement of Tyr 64 by another hydrophilic amino acid (Gln) also showed the similar effects. Futhermore, replacement of Tyr64 by Larger (Trp) or smaller (Gly) amino acids also significantly reduced intrinsic GTPase activity and also abolished its stimulation by GAP1C, although Gly64 mutation appears to be more effective than the Trp64 mutation. Finally, as expected, Phe64 mutation showed the minimum effects. Compared with the normal Ras protein, the intrinsic GTPase activity of the Phe64 mutant was slightly lower and was less sensitive to GAP1C- dependent activation. Interestingly, Leu64 mutant of the Ras protein still could respond to GAP1C, although its sensitivity was much lower than that of the Phe64 mutant. These observations clearly show that the hydrophobic amino acids (Tyr, Phe or Leu) at position 64 are absolutely essential for activation of the Ras GTPases by GAP1. None of the mutations at position 64 affected the binding of GTP to the HaRas protein.

Replacement of Phe64 by Glu or Tyr in Rap1A GTPase. Like the R-Ras GTPase, the normal Rap contain Phe, instead of Tyr, at position 64 (2,4). However, both Rap1A and 1B GTPases are activated by GAP3, but neither by GAP1 (8, 12). Interestingly, replacement of Gln63 by Glu potentiates the anti-oncogenicity of Rap1A protein (6) and significantly reduces the intrinsic GTPase activity (12). The same mutation does not significantly affect GAP3-dependent activation (12). To determine further role of Phe64 in GAP3-dependent activation of the Rap1A GTPase, we have replaced Phe64 by Glu. The Glu64 mutation significantly reduced the intrinsic GTPase activity of the Rap1A protein and completely abolished its stimulation by GAP3. We then tested the possibility of functional replacement of Phe64 by Tyr. Unlike the Glu64 mutant, Tyr64 mutant of Rap1 GTPase was still activated by GAP3. These observations clearly indicate that the hydrophobic residues (Phe or Tyr) at position 64 are also essential for GAP3-dependent activation of the normal Rap1 GTPases as GAP1-dependent activation of the normal Ras GTPases. Interestingly, the Phe64 mutant of the normal HaRas was not significantly activated by GAP3 whereas the Tyr64 mutant of the Rap1 GTPase was not activated by GAP1C (data not shown). These observations confirm that neither Tyr64 nor Phe64 of the Ras and Rap1 GTPase, respectively, are responsible for distinguishing between GAP1 and GAP3 for their activation.

Ras GTPase mutated at positions 62 or 66. In order to examine if mutations at other neighboring positions affect either the intrinsic GTPase activity or the GAP-dependent stimulation, we have substituted the amino acids at positions 62 and 66 of the Ras protein. The replacement of Ala66 by either Gly, Trp, Glu or Arg, and Glu62 by Ala had no significant effect on the enzymatic property of the Ras protein. These observations confirm that Tyr or Phe at position 64 play a unique role in the signal attenuation of the Ras and Rap proteins by the corresponding GAPs.

We have shown previously that Gln61 and Ser65 of the Ras GTPases are sufficient for distinguishing GAP1 from GAP3 (12), whereas Thr65 of the Rap GTPases is primarily responsible for distinguishing GAP3 from GAP1, in GAP dependent activation of these GTPases (12, 27). Here, we have shown that, like Val12, Thr59 or Leu61 mutations, replacement of Tyr64 by many other amino acids, especially Glu, in the normal HaRas protein significantly reduces the intrinsic GTPase activity and abolishes its stimulation by GAP1. However, neither Phe64 nor Leu64 mutations of the HaRas GTPase significantly change its enzymatic properties. Interestingly, the replacement of Phe64 by Glu, but not by Tyr, in the Rap1 protein also significantly reduces intrinsic GTPase activity and abolishes its stimulation by GAP3. Thus, it appears that both Ras and Rap1 GTPases require the hydrophobic amino acids (Tyr, Phe or Leu) at position 64 for GAP-dependent activation.

If Tyr64 or Phe64 of the Ras and Rap1 proteins are dispensable for their signal transducing activities, it is most likely that, like Val12 or Thr59 mutations, the Glu64 mutation would potentiate both Ras oncogenicity and Rap1 anti-oncogenicity. Interestingly, it has been shown that residues 61 to 68 of the normal HaRas protein are not required for the signal transducing activity as mutations in this region including Leu64 mutation do not abolish the ability of complementing the Ras2 defect in yeast (13). Furthermore, Leu61 or Lys63 mutations of the HaRas protein and Glu63 mutation of the Rap1A protein also potentiate the oncogenicity and anti-oncogenicity, respectively (6, 16). However, it has recently been reported that deletion of residues 64 to 68 of the oncogenically mutated HaRas protein (Arg12/Thr59) sinificantly reduces oncogenicity and increases the intrinsic GTPase activity (18). Since we have shown here that the Glu64 mutation of the normal HaRas protein reduces the intrinsic GTPase activity, it is likely that the elevation of the intrinsic GTPase activity is not due to the deletion of Tyr64, but due to other residues (65 to 68). Furthermore, replacement of Ala66 by either Thr or Val in an N-Ras protein of C.elegans , called let60, reduces the signal transducing activity that is essential for vulva formation (26). Thus, it is conceivable that either replacement or deletion of Ala66 result in the elevation of the intrinsic GTPase activity, thereby reducing signal transducing activity including oncogenicity of the Ras proteins. In order to clarify these points, we are currently examining the oncogenicity and anti-oncogenicity of the Glu64 mutants of the Ras and Rap1A proteins, as well as the effect of the Thr66 or Val66 mutations on the intrinsic GTPase activity of the Ras proteins.

ACKNOWLEDGEMENT

We are grateful to Dr. A. Hammacher and Mr. L. Fabri for their preprartion of GAP3 from bovine brain, to Dr. A. W. Burgess for his constant encouragement and invaluable advice throughout this work, to Dr. R. Simpson, Ms J. Discolo and Mr. Mathews for their constant supply of oligonucleotide primers for PCR.

REFERENCES

1. Barbacid, M. (1987) Ann. Rev. Biochem. 56, 779-827.
2. Pizon, V., Chardin, P., Lerosey, I., Olofsson,, B. and Tavitian, A. (1988) Oncogene 3, 201-204.

3. Kitayama, H., Sugimoto, Y., Matsuzaki, T., Ikawa, Y. and Noda, M. (1989) Cell 56, 77-84.
4. Pizon, V., Lerosey, I., Chardin, P. and Tavitian, A. (1988) Nucleic Acids Res. 16, 7719.
5. Ohmstede, C.A., Farrell, F.X., Reep, B.R., Clemetson, K.J. and Lapetina, E.G. (1990). Proc. Natl. Acad. Sci. USA 87, 6527-6531.
6. Kitayama, H., Matsuzaki, T., Ikawa, Y. and Noda, M. (1990) Proc. Natl. Acad. Sci. USA 87, 4284-4288.
7. Trahey, M. and McCormick, F. (1987) Science 238, 542-545.
8. Kikuchi, A., Sasaki, T., Araki, S., Hata, Y. and Takai, Y. (1989) J. Biol. Chem. 264, 9133-9136.
9. Adari, H., Lowy, D.R., Willumsen, B.M., Der, C.J., and McCormick, F. (1988) Science 240, 518-521.
10. Cales, C., Hancock, J.F., Marshall, C.J. and Hall, A. (1988) Nature 332, 548-551.
11. Quilliam, L.A., Der, C.J., Clark, R., O'Rourke, E.C., Zhang, K., McCormick, F. and Bokoch, G.M. (1990) Mol. Cell. Biol. 10, 2901-2908.
12. Maruta, H., Holden, J., Sizeland, A. and D'Abaco, G. (1991) J. Biol. Chem. in press.
13. Sigal, I.S.,, Gibbs, J.A., D'Alonzo, J.S., and Scolnick, E.M. (1986). Proc. Natl. Acad. Sci. USA 83, 4725-4729.
14. Willumsen, B.M., Papageorge, A.G., Kung, H.F., Bekesi, E., Robins, T., Johnsen, M., Vass, W.C. and Lowy, D.R. (1986). Mol. Cell Biol. 6, 2646-2654.
15. Gibbs, J.B., Schaber, M.D., Schofield, T.L., Scolnick, E.M. and Sigal, I.S. (1989) Proc. Natl. Acad. Sci. USA 86, 6630-6634.
16. Fasano, O., Aldrich, F., Tamanoi, E., Taparowsky, M., Furth, M. and Wigler, M. (1984) Proc. Natl. Acad. Sci. USA 81, 4008-4012.
17. Baldwin, G.S., Stanley, I.J. and Nice, E.C. (1983) FEBS Letter 153, 257-261.
18. Srivastava, S.K., Donato, D.A. and Lacal, J.C. (1990) Mol. Cell. Biol. 9, 1779-1783.
19. Lowe, D.G., Capon, D.J., Delwart, E., Sakaguchi, A. Y., Naylor, S.L. and Goeddel, D.V. (1987) Cell 48, 137-165.
20. Lowe, D.G., Ricketts, M., Levinson, A.D. and Goeddel, D.V. (1988) Proc. Natl. Acad. Sci. USA Proc. Natl. Acad. Sci. USA 85, 1015-1019.
21. Garret, M.D., Self, A.J., Van Oers, C. and Hall, A. (1989) J. Biol. Chem. 264, 10-13.
22. Smith, D.B. and Johnson, K.S. (1988) Gene 67, 31-40.
23. Maruta, H. (1989) In ras Oncogenes (Spandidos, D.A. ed.), Plenum Publishing Co., New York, pp255-260.
24. Brandt, D.R., Asano, T., Pedersen, S.T. and Ross, E.M. (1983) Biochemistry 22, 4357-4362.
25. Bradford, M.M. (1976) Anal. Biochem. 72, 248-254.
26. Beitel, G.J., Clark, S.G. and Horvitz, H.R. (1990) Nature 348, 503-509.
27. Holden, J., Nur-E-Kamal, M.S.A., Fabri, L., Nice, E., Hammacher, A. and Maruta, H. (1991) J. Biol. Chem. 266, In Press.

PROGRESSIVE FACTORS IN ONCOGENE TRANSFECTED RODENT EMBRYO

FIBROBLASTS

Heather Marshall[1], Przemyslaw Popowicz[1], Georg Engel[1],
Catharina Svensson[2], Göran Akusjärvi[2] ,and Stig Linder[1]

1. Division for Experimental Oncology, Radiumhemmet
Karolinska Hospital, S-10401 Stockholm, Sweden
2. Department of Microbial Genetics, Karolinska
Institute S-10401 Stockholm, Sweden

INTRODUCTION

Activated *ras* oncogenes are inefficient in the transformation of primary rodent embryo fibroblasts (REFs). However,
in unison with viral or cellular genes such as the adenovirus E1A, polyoma large-T, mutant p53 or *myc* genes, transformation will occur. No common biochemical activity has
been described for these cooperating oncogenes. We and
others have demonstrated that treatment of T24-*ras* oncogene
transfected REF cells with the glucocorticoid dexamethasone
(DEX) facilitates transformation (Martens *et al.*, 1988;
Yamashita *et al.*, 1988; Marshall *et al.*, Exp. Cell. Res, in
press;). During a critical period between 1-3 months after
transfection, cellular growth was found to be dependent on
glucocorticoid. Hormone independence invariably develops
during subsequent culture. These observations raise the
possibility of hormonal promotion and progression of rat
embryo fibroblasts expressing activated *ras* genes, and may
offer a convenient model for studies of promotion and
progression *in vitro*.

Transfection with the activated *ras* oncogene has been
reported to induce a metastatic phenotype in some cells
(Liotta *et al.*, 1991). The metastatic propensity of *ras*

The Superfamily of ras-Related Genes
Edited by D.A. Spandidos, Plenum Press, New York, 1991

transformed cells has been shown to be repress by adeno-
virus-2 E1a (Pozzatti *et al.*, 1986). The mechanisms whereby
these genes affect the metastatic phenotype have not been
fully elucidated.

RESULTS

Glucocorticoid facilitates the transformation of T24-*ras*
transfected REF cells

REF cells cotransfected with adenovirus-2 E1a + T24-*ras* are
efficiently established into cell lines. In contrast, the
growth of cells transfected with T24-*ras* slows down at 2-3
weeks after transfection. Continuous growth of T24-*ras*
transfected REF cells in glucocorticoid results in the
outgrowth of cells with a transformed morphology. However, a
period of slow growth is observed in such cultures at 3-4
weeks after transfection. Omission of hormone following
transfection results in a flat phenotype in the majority of
the transfected cell clones and arrested cellular growth.
Addition of glucocorticoid to such flat cells does not
affect cellular morphology. However, at later passages, the
cells become hormone independent and acquire an increasingly
higher plating efficiency in soft agar (Table 1).

Analysis of p21-*ras* expression in T24-*ras* transfected REF
cells revealed that glucocorticoid treated cells expressed
elevated levels of this protein. Similarly, *ras* mRNA levels
were elevated. Similar to the effects on cell morphology
described above, cellular *ras* expression decreased after
removal of glucocorticoid. Readdition of DEX did not result
in restoration of expression. However, we observed that
cells which developed hormone independent growth during
continuous passage also expressed high levels of p21-*ras*.
These observation suggest that high levels of p21-*ras*
expression are necessary for the induction of a transformed
phenotype in REF cells, as previously suggested (Spandidos &
Wilkie, 1984; Land *et al.*, 1986). Furthermore, it appears
that glucocorticoid treatment represents one route to
accomplish stable, high expression of p21-*ras* in REF cells.

How does facilitated transformation of *ras* transfected cells
by glucocorticoid relate to the mechanism of transformation
by adenovirus E1a?

The adenovirus E1a oncogene encodes multifunctional proteins
which are able to cooperate with *ras* in transformation

Table 1

Growth properties of REF cells transfected with T24-*ras*.

Treatment	Morphology (4 wks[1])	p21	Agar plating eff. (%)[2]			
			51	79	96	114
–	flat	+	<1			
0.5μM DEX	rounded	+++	4	7	22	30

[1] Morphology of representative colonies of G418 resistant cells transfected with a T24-*ras*-NEO[r] plasmid.
[1] Plating efficiency in soft agar of G418 resistant cells at different times (in days) after transfection.

Table 2

Plasmid	Treatment	Formation of foci
pmt	–	–
pmt	DEX	+
pmt + G5/3	–	–
pmt + G5/3	DEX	+
pmt + E1A		+++

The size and number of foci developing after transfection with *pmt* + DEX was smaller than that observed in *pmt* + E1A transfected cultures.

(Ruley, 1983). E1a proteins associate with cellular proteins such as the retinoblastoma susceptibility gene product (p105-RB) (Whyte *et al.*, 1988). Furthermore, they are positive and negative regulators of gene transcription. The ability to negatively regulate transcription depends on region in E1a also necessary for cell transformation (Schneider *et al.*, 1987). This has lead to speculations of whether repression of enhancer driven transcription is required for E1a to transform cells. The repression function partially overlaps that of the glucocorticoid receptor concerning the ability to repress the AP-1 transcription factor stimulated expression of the genes encoding the metalloproteases stromelysin and collagenase I (Offringa *et al.*, 1988). We have attempted to address the question of whether whether transcriptional repression contribute to the oncogenic effect of E1A by studying the transforming capacity of E1A mutants in the presence and absence of glucocorticoid.

We have attempted to correct the transformation defective E1a mutant which is deficient in the enhancer repression function (mutant G5/3 (Schneider *et al.*, 1987); deficient in conserved region 1 of E1A) by addition of glucocorticoid. We used both T24-*ras* and polyomavirus MT (*pmt*) in these experiments. In Table 2, results using *pmt* are presented. Our results show that the defect of the E1a mutant can not be complemented by hormone. The interpretation of this result is, however, complicated by the fact that the region deleted in the G5/3 mutant also is involved in binding the p105-RB protein and other cellular proteins (Whyte *et al.*, 1989). In order to address the question of the necessity of the AP-1 repression function for cell transformation further, we have attempted to cotransfect E1A genes lacking conserved region 2 (CR2) with polyomavirus. E1A proteins deficient in this region show decreased p105-RB binding but maintained enhancer repression.These experiments have shown that the E1A CR2 mutant GCX (Schneider *et al.*, 1987) does facilitate polyomavirus mediated REF cell transformation (Marshall, unpublished observations), suggesting that transcription repression of E1A may facilitate transforma-tion of REF cells.

The enhancer repression function of adenovirus-2 E1a is not required for cell transformation

Another approach to the problem of a common activity between

Ela and glucocorticoid in cell transformation is to determine whether the AP-1 repression activity of Ela in fact is
necessary for transformation by this gene. To this end,
adenovirus Ela genes carrying deletions in the carboxy-
terminal exon were cotransfected with T24-*ras* into primary
REF cells. These mutants found to be able to cooperate with
ras to induce foci of transformed cells. The expression of
stromelysin was examined in clonal lines derived from these
foci. Stromelysin is a secreted metalloprotease which is
expressed at high levels in transformed fibroblasts
(Matrisian *et al.*, 1986). Deletion of a region in the E1A C-
terminal exon resulted in high levels of stromelysin
expression in transformed cells. Transformed cell lines
expressing high levels of stromelysin showed invasive
properties in an *in vitro* invasion assay. During culture,
some clones became progressively more able to grow in soft
agar. These data suggest that the ability of E1A to repress
metalloprotease expression is dispensable for its transforming activity. The repression activity rather appears to
have a role in determining phenotypic properties of Ela
transformed cells

We have further characterized the effects of Ela and
glucocorticoid hormones on the expression of cellular
proteases. *Ras* transformed REF cells expressed high levels
of stromelysin, 72kDa and 92kDa type IV collagenases. Whereas
the expression of both stromelysin and 92kDa type IV
collagenase was repressed by Ela and glucocorticoid,
expression of 72kDa type IV collagenase was unaffected.
Since both the expression of E1A and treatment with
glucocorticoid resulted in repression of invasive
properties, these results show the significance of these
enzymes in determining cellular invasive properties.

We have observed progressive phenomena also during growth of
Ela + *ras* transfected cells. Of 5 cell lines studied in
detail, 2 showed an increased plating efficiency in soft
agar from passage 6 to passages > 40. Progression to
increased anchorage independence was not paralleled by an
increased invasive propensity in that the ability to invade
a basal membrane remained unaltered or even decreased during
passage. One of cell line which showed a decreased invasive
propensity during passage was studied in some detail. The
loss of invasive capacity correlated to a loss of stromelysin and 92kDa collagenase type IV expression in this line.
Stromelysin activity in late passage cells was not inducible
with TPA, suggesting that some component of the transcrip-

tion machinery was altered in these cells. Preliminary data suggest that invasive cells can be recovered by treatment with 5-azacytidine. Our data, then, suggest that the cellular invasive capacity does not necessarily increase during passage of cells in culture, and that changes in invasive potential are independent of and separable from transformation as measured by growth in semi-solid medium.

DISCUSSION

The term "immortalization" has been keyed to describe the ability of some viral and cellular oncogenes to facilitate the establishment of cell lines *in vitro*. The relevance of immortalization in relation to phenomena such as promotion and progression described in *in vivo* systems such as the mouse skin model is unclear. We have demonstrated that REF cells expressing a large-T antigen deficient mutant of polyomavirus can be efficiently established into cell lines in the presence of glucocorticoid (Martens *et al.*, 1988) and that T24-*ras* transfected REF cells are stimulated by this hormone (Marshall *et al.*, in press). Therefore, it is possible that immortalization by polyoma large-T may be related to hormonal or phorbol ester promotion in other models. Findings of progression to hormone independence and increased growth in soft agar of cells expressing *ras* oncogenes are consistent with such views.

We have not as yet been able to understand the mechanism leading to hormone promotion of *ras* transfected REF cells in culture. Our studies of the E1a oncogene in fact suggest that the ability of this oncogene to repress AP-1 stimulated transcription is not essential for cooperation with *ras*. Other groups have observed that E1A mutants deficient in the C-terminal region, which we have found to be required for efficient transcription repression, are deficient in immortalization of primary epithelial cells (Quinlan *et al.*, 1988). It is therefore quite possible that transcription repression may contribute to the transforming activity of E1a, but that this activity is not required for cooperation of E1a + *ras*. Our studies clearly show that E1a transcription repression diminishes the invasive properties of *ras* transformed cells. In the REF cell system, E1A appears to affect the expression of stromelysin and 92kDa type IV collagenase, whereas two other genes involved in the control of extracellular proteolysis, 72kDa type IV collagenase and TIMP-1 are not affected.

In some tumor types, the timing or progressive order of mutation has to some extent been identified. In human colon cancer, the DCC gene, the *ras*-gene, and the p53 gene appear to be affected during progression of adenomas to carcinomas (Fearon & Vogelstein, 1990). p53 mutations have been uncovered from a number of different tumor types. It is conceivable that changes in the expression of genes such as p53 might be involved in the progression during culture in our systems.

CONCLUSIONS

We have shown that similarly to adenovirus E1a, glucocorticoid hormones may facilitate transformation of REF cells transfected with activated *ras*-genes. In both cases, the resulting clones progress towards an increased growth in soft agar. T24-*ras* containing clones derived from DEX dependent cultures progress in time towards DEX independence. The molecular mechanisms underlying hormonal promotion and progression are currently under investigation.

ACKNOWLEDGEMENTS

This work has been supported by grants from the Swedish Cancer Society and from the King Gustav V:s Jubilee Foundation.

REFERENCES

Fearon, E. R. and Vogelstein, B. (1990). A genetic model for colorectal tumorigenesis. <u>Cell</u>. 61, 759-767.

Land, H., Chen, A. C., Morgenstern, J. P., Parada, L. F. and Weinberg, R. A. (1986). Behavior of myc and ras oncogenes in transformation of rat embryo fibroblasts. <u>Mol. Cell. Biol</u>. 6, 1917-1925.

Liotta, L. A., Steeg, P. S. and Stetler-Stevenson, W. G. (1991). Cancer metastasis and angiogenesis: an imbalance of positive and negative regulation. <u>Cell</u>. 64, 327-336.

Marshall, H., Martens, I., Svensson, C., Akusjärvi, G. and Linder, S. Glucocorticoid hormones may partially substitute for adenovirus E1A in cooperation with ras. <u>Exp. Cell Res</u>. in press.

Martens, I., Nilsson, M., Magnusson, G. and Linder, S. (1988). Glucocorticoids facilitate the stable transformation of embryonal rat fibroblasts by a polyomavirus large tumor antigen deficient mutant. Proc. Natl. Acad. Sci USA. 85, 5571-5575.

Matrisian, L. M., Leroy, P., Ruhlmann, C., Gesnel, M.-C. and Breathnach, R. (1986). Isolation of the oncogene and epidermal growth factor-induced transin gene: complex control in rat fibroblasts. Mol. Cell. Biol. 6, 1679-1686.

Offringa, R., Smits, A. M. M., Houweling, A., Bos, J. L. and van der Eb, A. J. (1988). Similar effects of adenovirus E1A and glucocorticoid hormones on the expression of the metalloprotease stromelysin. Nucl. Acids Res. 16, 10973-10984.

Pozzatti, R., Muschel, R., Williams, J., Padmanabhan, R., Howard, B., Liotta, L. and Khoury, G. (1986). Primary embryo cells transformed by one or two oncogenes show different metastatic potentials. Science. 232, 223-227.

Quinlan, M.P., Whyte, P., and Grodzicker, T. (1988). Growth factor induction by the adenovirus type 5 E1A 12S protein is required for immortaliztion of primary epithelial cells. Mol. Cell. Biol. 8, 3191-3203

Ruley, H. E. (1983). Adenovirus early region 1A enables viral and cellular transforming genes to transform primary cells in culture. Nature. 304, 602-606.

Schneider, J. F., Fisher, F., Goding, C. R. and Jones, N. C. (1987). Mutational analysis of the adenovirus E1A gene: the role of transcriptional regulation in transformation. EMBO J. 6, 2053-2060.

Spandidos, D. A. and Wilkie, N. (1984). Malignant transformation of early passage rodent cells by a single mutated human oncogene. Nature. 310, 469-475.

Whyte, P., Buchkovitch, K. J., Horowitz, J. M., Friend, S. H., Raybuck, M., Weinberg, R. A. and Harlow, E. (1988). Association between an oncogene and an antioncogene: the adenovirus E1A proteins bind to the retinoblastoma gene product. Nature. 334, 124-129.

Whyte, P., Williamson, N. M. and Harlow, E. (1989). Cellular targets for transformation by the adenovirus E1A proteins. Cell. 56, 67-75.

Yamashita, T., Kato, H. and Fujinaga, K. (1988). Conditional immortalization and/or transformation of rat cells carrying v-abl or EJras oncogene in the presence or absence of glucocorticoid hormone. Int. J. Cancer. 42, 930-938.

THE YPT-BRANCH OF THE RAS SUPERFAMILY OF GTP-BINDING PROTEINS
IN YEAST: FUNCTIONAL IMPORTANCE OF THE PUTATIVE EFFECTOR
REGION

D. Gallwitz, J. Becker, M. Benli, L. Hengst,
C. Mosrin-Huaman, M. Mundt, T.J. Tan, P. Vollmer
and H. Wichmann

Department of Molecular Genetics
Max-Planck-Institute for Biophysical Chemistry
D-3400 Göttingen, Germany

INTRODUCTION

Both in unicellular and multicellular organisms, a large
number of structurally related small GTP-binding proteins,
collectively known as ras superfamily of proteins, has been
identified within the past few years. These proteins are
similar in size and they share highly conserved sequence
motifs which, in the case of H-ras p21, have been demonstrated to be contact regions for the bound nucleotide (see
1,2 for review). It is thought, and in very few cases it is
known, that the diverse members of this superfamily of
proteins are regulators of basic cellular processes,
including the transmission of proliferation signals, the
vectorial transport and the fusion of vesicles, the proper
functioning of the cytoskeleton and cell differentiation (see
1-3 for review).

According to characteristic features of their primary
structure, four subfamilies of small GTP-binding proteins can
be distinguished: ras, rho, ypt/rab and arf proteins (1-4).
Evolutionary highly conserved members of all four sub-
families have been identified in eukaryotic species as
distant as yeast and man. One is led to conclude, therefore,
that many of these proteins might even perform an identical
or very similar function in uni- and multicellular organisms.
This is, however, not true for the Ras proteins in yeast and
mammals: although H-ras p21 can, at least in part, compensate
the loss of the essential RAS1 and RAS2 gene products in the
budding yeast Saccharomyces cerevisiae, yeast Ras proteins
regulate the activity of adenylyl cyclase, but mammalian ras
proteins seem to be integrated into other regulatory pathways
(5-7).

From studies with yeast mutants it has become clear that
members of the ypt and the arf family, including Ypt1p (8-
11), Sec4p (12,13), Sar1p (14) and Arf1p (15), participate in
guiding vesicular traffic in the secretory pathway. Notably,
several mammalian proteins belonging to the ypt/rab subfamily

The Superfamily of ras-Related Genes
Edited by D.A. Spandidos, Plenum Press, New York, 1991

Subfamily	Gene	Effect of Gene Disruption	Mammalian Counterpart
RAS	*RAS1* *RAS2*	Lethality (when both genes are disrupted)	H-, K-, N-ras
	RSR1	Randomized budding sites	rap1,2
YPT	*YPT1*	Lethality	rab1A,B
	SEC4	Lethality	?
	YPT3A *YPT3B*	?	rab11
	RYH1	t^s Phenotype	rab6
	YPT7	?	rab7
RHO	*RHO1* *RHO2*	Lethality none	rhoA,B,C
	CDC42	Lethality	cdc42 (Hs) rac1,2

Fig. 1.   Members of the ras superfamily in *S. cerevisiae* and their mammalian homologues. Arf proteins are not listed. Gene disruption data are from the following sources: *RAS1/2* (35), *RSR1* (36), *YPT1* (37), *SEC4* (12), *RYH1* (L.H. and D.G., submitted), *RHO1,2* (38), *CDC42* (39).

have been localized to different compartments of the exocytic and endocytic pathway (16,17), and the nonhydrolyzable GTPγS has been found to inhibit various *in vitro* protein transport reactions (18-20), suggesting that several of these small GTPases perform comparable functions in yeast and mammals. This assumption is indeed strengthened by the finding that several of the ypt/rab proteins are functionally interchangeable between evolutionary widely separated species (21-24).

Yeast as a genetically tractable organism offers many advantages for investigating the function of proteins. In particular, mutations can be inserted into defined gene segments to generate mutant proteins, and due to an efficient system of homologous recombination, the mutant genes can easily be used to replace the corresponding wild-type genes and search for phenotypic alterations. Conditional-lethal mutants allow to screen for suppressor genes whose protein products might act in the same pathway as the gene under study. The latter approach has already been successfully applied to the study of Ypt1p (25) and Sec4p (26) function, for instance.

In this contribution, we describe molecular genetic and biochemical experiments aimed to get inside into the functional importance of the putative effector region of yeast Ypt proteins, the corresponding protein segment in H-ras p21 taking part in the interaction with a GTPase-activating protein, rasGAP (27, for review).

RAS - FAMILY

		EFFECTOR REGION		PROTEIN	
30	D E	Y D P T I E D S Y	R K	Human	H-ras
37	D G	Y D P T I E D S Y	R K	S.cerev.	Ras1
37	D E	Y D P T I E D S Y	R K	S.cerev.	Ras2
35	D E	Y D P T I E D S Y	R K	S.pombe	ras1

YPT/RAB - FAMILY

38	E S	Y I S T I G V D F	K I	Human	rab1
38	E S	Y I S T I G V D F	K I	Mouse	ypt1/rab1
35	N D	Y I S T I G V D F	K I	S.cerev.	Ypt1
35	E S	Y I S T I G V D F	K I	S.pombe	ypt1
47	P S	F I T T I G I D F	K I	S.cerev.	Sec4
36	P S	F I T T I G I D F	K I	S.pombe	ypt2
42	P S	F I T T I G I D F	K I	Dictyost.	sas1
38	L E	S K S T I G V E F	A T	Canine	rab11
40	M D	S K S T I G V E F	A T	S.cerev.	Ypt3A
40	I E	S K S T I G V E F	A T	S.cerev.	Ypt3B
37	I E	S K S T I G V E F	A T	S.pombe	ypt3
40	N T	Y Q A T I G I D F	L S	Human	rab6
37	D H	Y Q A T I G I D F	L S	S.cerev.	Ryh1
38	N T	Y Q A T I G I D F	L S	S.pombe	ryh1
35	N Q	Y K A T I G A D F	L T	Canine	rab7
	Q Q	Y K A T I G A D F	L T	S.cerev.	Ypt7
35	S T	F I S T I G I D F	K I	Canine	rab8
36	T T	F I S T I G I D F	K I	Canine	rab10
33	P V H D L T I G V E F G A			Human	rab2
49	P A F V S T V G I D F K V			Human	rab3A
35	D D S N H T I G V E F G S			Human	rab4
47	E F Q E S T I G A A F L T			Human	rab5
	T Q L F H T I G V E F L N			Canine	rab9

Fig. 2.   Comparison of effector domain sequences of ras and ypt/rab proteins. Identical effector regions of proteins from different eukaryotes are boxed. Sequences are from the following sources: H-ras (40), *S.c.* Ras1,2 (35), *S.p.* ras1 (41), human rab (42), canine rab (16,43), *S.c.* Ypt1 (44), mouse ypt1/rab1 (45), *S.p.* ypt1,3 (22), *S.p.* ypt2 (23), *S.p.* ryh1 (24), *S.c.* Ypt3A,B (M.B., H.W. and D.G., unpubl.), *S.c.* Ryh1 (L.H. and D.G., subm.), *Dictyostelium* sas1 (46).

RESULTS

A compilation of the presently known members of the ras superfamily of proteins (excluding Arf proteins) in the yeast *S. cerevisiae* is given in Fig. 1. Their mammalian homologues, identified either by functional interchangeability of the proteins or by their extensive sequence identity (usually larger than 60 %) are also shown. As can be seen from gene disruption studies, most of these yeast proteins have an essential function.

From a comparison of the putative effector regions (Fig. 2) it is evident that ras and different groups of ypt/rab proteins can be clearly distinguished. In this context, it is of special interest that proteins with an identical effector region, where tested, can functionally replace each other. For instance, *ras1⁻ras2⁻* yeast mutants can be partially

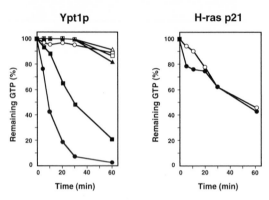

Fig. 3.   Intrinsic and yptGAP-accelerated GTPase activity of
yeast Ypt1p wild-type and mutant proteins and H-ras
p21. The proteins loaded with $5'-(\gamma-{}^{32}P)GTP$ were
incubated with or without partially purified yeast
yptGAP at 30°C, and labelled GTP remaining bound to
protein was assayed by a filter test at different
times of incubation (11). H-ras p21 and wild-type
Ypt1p (●,○), ypt1(D44N)p (■,□), ypt1(I41M)p (▲,Δ).
Open symbols: incubations without, closed symbols:
incubation with yptGAP.

rescued by H-ras p21 (28,29). We have found that in yeast,
the loss of Ypt1p and Ryh1p function can be complemented by
mammalian ypt1p/rab1p and rab6p, respectively (21,23) and
that a temperature-sensitive *S. cerevisiae* sec4 mutant can be
rescued with ypt2p from the evolutionary very distant fission
yeast *Schizosaccharomyces pombe* (23). In contrast, GTP-
binding proteins of the same family having only slightly
different effector regions are unable to replace each other
functionally: *S. pombe* ypt2p cannot compensate the loss of
function of either *S. cerevisiae* Ypt1p (21) or Ryh1p (L.H.
and D.G., unpublished results), and mammalian rab8p or rab10p
and *S. cerevisiae* Sec4p, whose effector regions differ by
only one amino acid residue (Fig. 2), are functionally not
interchangeable (M.M. and M. Zerial, unpublished results).
These results seem to indicate that the effector region is an
important functional determinant of GTP-binding proteins.

     Small GTP-binding proteins possess low intrinsic GTPase
activity. This activity is even lower in Ypt1p than in H-ras
p21 (30). GTPase-activating proteins specific for mammalian
(27) and yeast Ras proteins (31) have been characterized. We
have recently reported on the isolation of a GAP activity
from porcine liver that uses mammalian and yeast Ypt1p/rab1p
as the preferred substrates but is unable to accelerate
GTPase activity of H-ras p21 (11). A similar GAP activity was
also identified in and partially purified from yeast extracts
(Fig. 3).

     To examine the functional importance of the effector
region in *S. cerevisiae* Ypt1p, a large collection of mutant

37	38	39	40	41	42	43	44	45	Phenotypic Alterations
Y	I	S	T	I	G	V	D	F	
E	-	-	-	-	-	-	-	-	None
Q	-	-	-	-	-	-	-	-	None
F[1]	-	-	-	-	-	-	-	-	None
-	-	D	-	-	-	-	-	-	None
-	-	A[1]	-	-	-	-	-	-	None
-	-	-	E	-	-	-	-	-	cs
-	-	-	R	-	-	-	-	-	cs*
-	-	-	N	-	-	-	-	-	cs
-	-	-	V	-	-	-	-	-	cs
-	-	-	S[1]	-	-	-	-	-	None
-	-	-	-	M[1]	-	-	-	-	Lethal
-	-	-	-	-	E	-	-	-	cs
-	-	-	-	-	F	-	-	-	None
-	-	-	-	-	-	R	-	-	Lethal
-	-	-	-	-	-	N	-	-	Lethal
-	-	-	-	-	-	S	-	-	cs
-	-	-	-	-	-	E[1]	-	-	None
-	-	-	-	-	-	-	N[1]	-	ts
-	-	-	-	-	-	-	H	-	cs
-	-	-	-	-	-	-	-	Y	cs

Fig. 4. Amino acid substitutions of the Ypt1p effector region and the phenotypic alterations in haploids relying on the mutant proteins as the only source for Ypt1p. [1] (Data from ref.11).

proteins with single amino acid substitutions was generated and studied by the gene replacement method mentioned above (11 and C.M., unpublished results). As can be seen in Fig. 4, three substitutions (I41M, V43R and V43N) led to lethality, one resulted in a temperature-sensitive (D44N) and several others in a cold-sensitive phenotype. Surprisingly, several non-conservative amino acid exchanges did not interfere with the proper functioning of Ypt1p. As Ypt1p is known to act in ER to Golgi transport (8-11), it was not surprising to find that ypt1(D44N)p at the non-permissive temperature led to a defect in protein secretion and an accumulation of ER membranes. As expected from its presumed function in ER to Golgi vesicular transport, this mutant could also be shown to accumulate small vesicles (11).

Most importantly, ypt1(I41M)p causing a lethal phenotype did not serve as a substrate for yptGAP from either porcine liver (11) or yeast (Fig. 3; P.V., unpublished results). Although ypt1(I41M)p and ypt1(D44N)p had no significantly altered guanine nucleotide-binding capacity (11), the mutant protein with the D44N substitution resulting in a ts phenotype was less sensitive than wild-type Ypt1p to yptGAP from liver (11) and yeast (Fig. 3).

DISCUSSION

The ypt/rab subfamily of small GTP-binding proteins is the largest of the ras superfamily, and members of this group of proteins are evolutionary highly conserved. Of the six Ypt proteins presently known in the budding yeast S. cerevisiae, three have been shown to act as regulators in the secretory pathway and in protein sorting. Ypt1p seems to guide the vesicular traffic between the ER and the Golgi complex (8-11), Sec4p directs the transport of post-Golgi vesicles to the plasma membrane (12,13) and Ryh1p is likely to act in vacuolar protein sorting (L.H. and D.G., submitted).

Different members of the ypt/rab subfamily of proteins have different effector region sequences, structural features likely to signal functional specificity. It is clear, however, that other regions of the different GTP-binding proteins add to their functional specificity. Work with Ypt1p effector region mutants indicate, that the effector domain, similar to the situation in ras proteins, is important for the interaction of Ypt proteins with GTPase-activating protein(s). It seems perfectly possible that different GAPs with specificity for different Ypt proteins exist in eukaryotic cells. As from work with Ras proteins both in yeast (32,33) and mammals (34), the most likely role of GTPase-activating proteins is to negatively regulate the GTPases, a similar function might be envisaged for GTPase-accelerating proteins acting on Ypt1p or other members of this family. Future work in this direction will certainly be rewarding.

ACKNOWLEDGEMENTS

The work presented from this laboratory was generously supported by the Max-Planck-Society and by grants to D.G. from the Deutsche Forschungsgemeinschaft, the Bundesministerium für Forschung und Technologie and the Fonds der Chemischen Industrie. C. Mosrin-Huaman acknowledges the receipt of an EMBO fellowship. We thank colleagues from our Department for many fruitful discussions.

REFERENCES

1.   Bourne, H.R., Sanders, D.A., and McCormick, F. (1990) *Nature*, **348**, 125-132.
2.   Bourne, H.R., Sanders, D.A., and McCormick, F. (1991) *Nature*, **349**, 117-127.
3.   Hall, A. (1990) *Science*, **249**, 635-640.
4.   Gallwitz, D., Haubruck, H., Molenaar, C., Prange, R., Puzicha, M., Schmitt, H.D., Vorgias, C. and Wagner, P. (1989) In Bosch, L., Kraal, B., and Parmeggiani, A. (eds.) *The Guanine-nucleotide Binding Proteins*. Plenum Press, New York; NATO ASI Series A: Life Sciences, Vol 165, pp. 257-264.
5.   Birchmeier, C., Broek, D., and Wigler, M. (1985) *Cell*, **43**, 615-621.
6.   Beckner, S.K., Hattori, S., and Shih, T.Y. (1985) *Nature*, **317**, 71-72.
7.   Downward, J., Graves, J.D., Warne, P.H., Rayter, S., and Cantrell, D.A. (1990) *Nature*, **346**, 719-723.
8.   Segev, N., Mulholland, J., and Botstein, D. (1988) *Cell*, **52**, 915-924.
9.   Schmitt, H.D., Puzicha, M., and Gallwitz, D. (1988) *Cell*, **53**, 635-647.

10. Baker, D., Wuestehube, L., Scheckman, R., Botstein, D., and Segev, N. (1990) *Proc. Natl. Acad. Sci. USA*, **87**, 355-359.
11. Becker, J., Tan, T.J., Trepte, H.-H., and Gallwitz, D. (1991) *EMBO J.*, **10**, 785-792.
12. Salminen, A., and Novick, P.J. (1987) *Cell*, **49**, 527-538.
13. Walworth, N.C., Goud, B., Kabcenell, A.K., and Novick, P.J. (1989) *EMBO J.*, **8**, 1685-1693.
14. Nakano, A., and Muramatsu, M. (1989) *J.Cell Biol.*, **109**, 2677-2691.
15. Stearns, T., Willingham, M.C., Botstein, D., and Kahn, R.A. (1990) *Proc. Natl. Acad. Sci. USA*, **87**, 1238-1242.
16. Chavrier, P., Parton, R.G., Hauri, H.P., Simons, K., and Zerial, M. (1990) *Cell* **62**, 317-329.
17. Goud, B., Zahraoui, A., Tavitian, A., and Saraste, J. (1990) *Nature* **345**, 553-556.
18. Melancon, P., Glick, B.S., Malhotra, V., Weidman, P.J., Serafini, T., Gleason, M.L., Orci, L., and Rothmann, J.E. (1987) *Cell*, **51**, 1053-1062.
19. Goda, Y., and Pfeffer, S.R. (1988) *Cell*, **55**, 309-320.
20. Beckers, C.J.M., and Balch, W.E. (1989) *J. Cell Biol.* **108**, 1245-1256.
21. Haubruck, H., Prange, R., Vorgias, C., and Gallwitz D. (1989) *EMBO J.*, **8**, 1427-1432.
22. Miyake, S., and Yamamoto, M. (1990) *EMBO J.*, **9**, 1417-1422.
23. Haubruck, H., Engelke, U., Mertins, P., and Gallwitz, D. (1990) *EMBO J.*, **9**, 1957-1962.
24. Hengst, L., Lehmeier, T., and Gallwitz, D. (1990) *EMBO J.* **9**, 1949-1955.
25. Dascher, C., Ossig, R., Gallwitz, D., and Schmitt, H.D. (1991) *Mol. Cell. Biol.* **11**, 872-885.
26. Bowser, R., and Novick, P. (1991) *J. Cell. Biol.*, **112**, 1117-1131.
27. McCormick, F. (1989) *Cell*, **56**, 5-8.
28. DeFeo-Jones, D., Tatchell, K., Robinson, L.C., Sigal, I.S., Vass, W.C., Lowy, D.R., and Scolnick, E.M. (1985) *Science*, **228**, 179-184
29. Kataoka, T., Powers, S., Cameron, S., Fasano, O., Goldfarb, M., Broach, J., and Wigler, M. (1985) *Cell*, **40**, 19-26.
30. Wagner, P., Molenaar, C.M.T., Rauh, A.J.G., Brökel, R., Schmitt, H.D., and Gallwitz, D. (1987) *EMBO J.*, **6**, 2373-2379.
31. Tanaka, K., Lin, B.K., Wood, D.R., and Tamanoi, F. (1991) *Proc. Natl. Acad. Sci. USA*, **88**, 468-472.
32. Ballester, R., Michaeli, T., Ferguson, K., Xu, H.-P., McCormick, F., and Wigler, M. (1989) *Cell*, **59**, 681-686.
33. Tanaka, K., Matsumoto, K., and Toh, E.-A. (1989) *Mol. Cell. Biol.*, **9**, 757-768.
34. Zhang, K., DeClue, J.E., Vass, W.C., Papageorge, A.G., McCormick, F., and Lowy, D.R. (1990) *Nature* **346**, 754-756.
35. Powers, S., Kataoka, T., Fasano, O., Goldfarb, M., Strathern, J., Broach, J., and Wigler, M. (1984) *Cell*, **36**, 607-612.
36. Bender, A., and Pringle, J.R. (1989) *Proc. Natl. Acad. Sci. USA* **86**, 9976-9980.
37. Schmitt, H.D., Wagner, P., Pfaff, E., and Gallwitz, D. (1986) *Cell*, **47**, 401-412.

38. Madaule, P., Axel, R., and Myers, A.M. (1987) *Proc. Natl. Acad. Sci. USA*, **84**, 779-783.
39. Johnson, D.I., and Pringle, J.R. (1990) *J. Cell Biol.* **11**, 143-152.
40. Capon, D.J., Chen, E.Y., Levinson, A.D., Seeburg, P., and Goeddel, D.V. (1983) *Nature* **302**, 33-37.
41. Fukui, Y., and Kaziro, Y. (1985) *EMBO J.*, **4**, 687-691.
42. Zahraoui, A., Touchot, N., Chardin, P., and Tavitian, A. (1989) *J. Biol. Chem.*, **264**, 12394-12401.
43. Chavrier, P., Vingron, M., Sander, C., Simons, K., and Zerial, M., (1990) *Mol. Cell. Biol.* **10**, 6579-6585.
44. Gallwitz, D., Donath, C., and Sander, C. (1983) *Nature*, **306**, 704-707.
45. Haubruck, H., Disela, C., Wagner, P., and Gallwitz, D. (1987) *EMBO J.*, **6**, 4049-4053.
46. Saxe, S.A., and Kimmel, A.R. (1990) *Mol. Cell. Biol.* **10**, 2367-2378.

THE v-Ki-ras ONCOGENE UP-REGULATES MAJOR HISTOCOMPATIBILITY CLASS II ANTIGEN EXPRESSION IN EARLY-PASSAGE FIBROBLASTS

R.L. Darley and A.G. Morris

Cancer Research Campaign Interferon and Cellular Immunity
Research Group, Department of Biological Sciences
University of Warwick, Coventry CV4 7AL, U.K.

INTRODUCTION

A long-term goal of oncogene research is to understand how proto-oncogene products affect cell physiology and how their function is subverted in their mutationally activated form. Oncogene products are widely regarded as playing a role in cell growth and control, but they may also contribute to the malignant phenotype by influencing cell functions other than growth. The effect of oncogenes on the expression of antigens of the major histocompatibility complex (MHC) is one such example. These molecules are of particular interest because both experimental (1-4) and clinical evidence (5,6) suggests their expression on cancer cells can correlate with the ability of the host to reject the tumour. The function of these molecules is integral to the process of immune recognition, and it is thought that their expression decreases the tumorigenicity of cancer cells as a result of immunosurveillance by T cells. Modulation of MHC antigen expression by oncogenes, therefore, has a dual significance: not only may it provide information concerning the interaction of oncogenes with intracellular signalling pathways, but the outcome of this interaction may have indirect consequences for tumour growth.

Expression of class I and class II MHC antigens is controlled by intracellular as well as extracellular signalling agents. Interferons (IFN) are the principal up-regulators of MHC expression, but they are only effective if the relevant intracellular signals are also present. For MHC class I antigens, this is almost always the case and, consequently, interferons up-regulate the expression of MHC class I antigens on almost every cell type. MHC class II antigens are more selectively expressed and most cells are MHC class II antigen negative, even in the presence of interferon-γ (7). IFN-αβ does not induce MHC class II antigen expression.

The Superfamily of ras-Related Genes
Edited by D.A. Spandidos, Plenum Press, New York, 1991

The expression of both class I and class II MHC antigens is affected by aberrant signalling caused by oncogenes: c-myc amplification results in down-regulation of MHC class I antigen expression, whereas activation of ras has been shown to have varying consequences for the expression of MHC class II antigens. Observations of B cells and melanocytes (8,9) first indicated that the v-Ki-ras oncogene could increase constitutive MHC class II antigen expression. Work in this laboratory yielded quite different results for an immortalized mouse fibroblast line, C3H10T½ (10). This fibroblast line expresses high levels of MHC class II antigens upon induction with IFN-γ. In this case, transformation with v-Ki-ras largely abolished the expression of MHC class II antigens. Similar results were obtained with a MHC class II antigen inducible sub-line of Balb/c3T3 cells (unpublished data, AGM).

We report here our preliminary observations concerning the effects of v-Ki-ras on the MHC class II antigen expression of early-passage fibroblasts. Non-established cultures such as these probably represent better models for studying the effects of transformation by ras for two reasons. Firstly, they contain a heterogeneous population of cells whereas long-term culture selects for cells which most readily adapt to in vitro conditions and so may be less representative of the behaviour of normal cells. Secondly, evidence indicates that activation of ras may be an early event in carcinogenesis (11-13), thus the most appropriate model would be one where cooperative changes have not occurred. Early-passage lines approximate more closely to this condition than do established lines. As our results show, the effects of v-Ki-ras on MHC antigen expression on these early-passage fibroblasts is quite different to that observed for the established lines, C3H10T½ and Balb/c3T3.

MATERIALS AND METHODS

Cells

C3H kidney fibroblasts (CKF) were derived from adult kidney from C3H/He mice. Cultures were maintained in RPMI containing 20% foetal calf serum (FCS). Cells were transformed after 2 population doublings with a helper-free defective retrovirus, Ki MSV as previously described (10). The v-Ki-ras oncogene is the only gene expressed upon infection with this virus. Infection with virus itself (Ki MLV) had no effect on MHC class II antigen expression. The transformed line so derived, CKF RAS, was morphologically transformed and tumorigenic in nude mice. Experiments were carried out 3-6 cell doublings following transformation. Three further lines were similarly prepared: C3H embryo fibroblasts (CEF) derived from 16-day embryos and transformed after 24 cell doublings; Balb/c adult kidney fibroblasts (BKF) transformed after 4 cell doublings and Balb/c embryo fibroblasts (BEF) transformed after 15 cell doublings.

Interferon-γ and induction of MHC antigens

Recombinant murine interferon-γ (IFN-γ) was produced, purified and titrated as described elsewhere (14,15).

Fibroblasts were treated with 100 units/ml of IFN-γ and, 3 days later, harvested for staining for MHC antigens.

mAbs and antisera

mAbs used against murine MHC antigens were as follows; anti-H-2K^k: TIB 195/11.4.1 (ascites diluted to 1/100); anti-H-2A^k: TIB 92 (hybridoma supernatant); anti-H-2A^d (mismatched control): HB3 (hybridoma supernatant). For indirect immunofluorescence, the second-layer used was FITC-Fab₂ goat anti-mouse Ig (Cappel Lab., Malvern, PA) at a dilution of 1/60 in medium (HEPES-buffered RPMI containing 10% FCS, 0.2% sodium azide). Cells were stained and analysed by flow cytometry as previously described (10). mAbs and antisera used for analysis of p21 ras expression were as follows: anti-p21 ras; CRL1742 (hybridoma supernatant). Background staining was assessed using isotype-matched antibody, anti-IFN-γ:R46A2 (hybridoma supernatant). Second-layer reagents (from Amersham, UK) were biotinylated sheep anti-rat Ig (diluted to 1/400) followed by streptavidin-phycoerythrin complex (diluted to 1/50). Antisera were diluted in PBS containing 10mg/ml BSA and 0.05 % tween 20.

Analysis of p21 ras expression

Cells were fixed in suspension with phosphate-buffered 4% paraformaldehyde solution, permeabilized with 0.2% triton X100, then incubated with anti-p21 ras or control antibody for 60 minutes, with gentle agitation. Cells were washed 3 times with PBS containing 0.05% tween 20, then incubated with second antibody for a further 60 minutes. Cells were again washed and incubated with detection complex for 45 minutes. After final washing cells were analysed by flow cytometry as previously described (10). All steps were carried out at room temperature.

RESULTS AND DISCUSSION

p21 ras Expression

The data presented in figure 1 demonstrate that a five-fold increase in the level of p21 ras expression occurs following transformation with the v-Ki-ras oncogene. The single peak of expression demonstrates that all of the cells in the culture are transformed.

Expression of MHC class I antigens

None of the early-passage fibroblast lines or their ras-transformed counterparts expressed any detectable MHC class I antigen in the absence of IFN-γ (not shown). Treatment with IFN-γ induced high levels of MHC class I antigen expression. Following transformation by v-Ki-ras, the expression of MHC class I antigens remained unaffected (figure 2). This was also the case for CEF, BKF and BEF lines (not shown) and confirms our previous observations (10). The inability of v-Ki-ras to

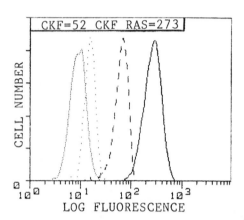

Figure 1. p21 ras expression . Solid line: p21 ras expression on CKF RAS cells; close dots: staining of control antibody on CKF RAS cells; dashed line: p21 ras expression on CKF cells; spaced dots: staining of control antibody on CKF cells. Inset: mean fluorescence above control levels in CKF and CKF RAS cells.

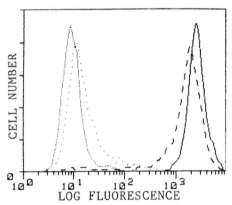

Figure 2. MHC class I antigen expression (H-2K^k). Solid line: (H-2K^k) expression on CKF RAS cells; close dots: staining of control antibody on CKF RAS cells; dashed line: (H-2K^k) expression on CKF cells; spaced dots: staining of control antibody on CKF cells.

affect either constitutive or inducible expression of MHC
class I antigens is of interest because loss of MHC class I
antigen expression is a feature of some carcinomas,
particularly colorectal tumours (16,17) where activated ras
genes are commonly found (18). This would be expected to
confer a selective advantage to the tumour through its ability
to escape immune destruction by cytotoxic T cells. Both our
experimental evidence and clinical observations (19) suggest
that there is no direct causative association between Ki-ras
mutations and the loss of MHC class I antigen expression.

Expression of MHC class II antigens

Expression of MHC class II antigens was undetectable on
CKF cells, either with or without induction by IFN-γ
(figure 3a). Transformation with v-Ki-ras did not affect
constitutive expression of MHC class II antigens; however,
these ras-transformed cells now strongly expressed MHC class II
antigen when induced with IFN-γ (figure 3b). We have also
found similar up-regulation following transformation of CEF and
BEF cells.

Previously, it has been reported that v-Ki-ras up-
regulates constitutive MHC class II antigen expression on B
cells (8) and melanocytes (9). This was not observed in this
case. However, it is well known that the requirements for MHC
class II antigen expression are cell-type dependent.
Therefore, it is not surprising that for fibroblasts, the co-
stimulatory activity of IFN-γ is required to manifest the MHC
class II antigen-expressing phenotype. All these data indicate
that signals induced by the v-Ki-ras oncogene contribute to the
up-regulation of expression of MHC class II antigens. These
data are, however, at odds with the previously observed effect

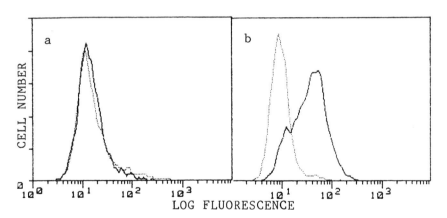

Figure 3. MHC class II antigen expression (H-2A^k). a) CKF
cells. Solid line: (H-2A^k) expression; close dots: staining of
control antibody. b) CKF RAS cells. Solid line: (H-2A^k)
expression; close dots: staining of control antibody.

for the established lines C3H10T½ (10) and Balb/c 3T3, where transformation with v-Ki-ras resulted in a loss of inducibility for MHC class II antigens. Together, these results indicate that, for fibroblasts, v-Ki-ras can exert both a positive and a negative regulatory function on MHC class II antigen expression depending on the nature of the cell line. As already mentioned, early-passage and established fibroblast lines behave differently upon transformation with v-Ki-ras, inducing, in established lines, full malignant transformation while ras-transformed early-passage cells display an intermediate morphology and transformed phenotype characteristic of premalignant cells. Ras induces transformation of fibroblasts, at least in part, by stimulating the expression of transforming growth factors (TGF) (20,21). Thus, the degree of transformation that these cells undergo due to the expression of a ras oncogene is determined by the production of, or responsiveness to these factors. It is of particular interest to this work that TGF-β is also an important immunomodulator and acts antagonistically to the action of IFN-γ in the induction of MHC class II antigens (22). Thus, ras-induced TGF-β , as well as mediating transformation, may also down-regulate MHC class II antigen expression and our work indicates this is indeed the case (manuscript in preparation). The loss of inducible MHC class II antigen expression for C3H10T½ and Balb/c3T3 cells is therefore explained by ras-induced production of TGF-β. This effect is not observed for early-passage fibroblasts (or other cell types of non-mesenchymal origin) probably because the level of production of, or sensitivity to TGF-β is more tightly regulated. Our work indicates that these cells are more responsive to ras-induced intracellular signals which oppose the effect of TGF-β by up-regulating MHC class II antigen expression (manuscript in preparation).

Previous work (1-3) has indicated that the expression of MHC class II antigens by tumour cells reduces their tumorigenicity. Thus, the expression of these molecules may in itself act indirectly as a mechanism for tumour suppression of ras-transformed cells.

ACKNOWLEDGEMENT

This work was supported by the Cancer Research Campaign.

REFERENCES

1. W. J. Bateman, R. Fiera, N. Matthews, A. G. Morris, Inducibility of class II major histocompatibility complex antigens by interferon-γ is associated with reduced tumorigenicity in C3H mouse fibroblasts transformed by v-Ki-ras, J. Exp. Med., 173: 193-196 (1991)
2. R. F. L. James, S. Edwards, K. M. Hui, P.D. Bassett, F. Grosveld, The effect of class II gene transfection on the tumorigenicity of the H-2K-negative mouse leukaemia cell line K36.16, Immunology, 72: 213-218 (1991)

3. S. O-Rosenberg, A. Thakur, V. Clements, Rejection of mouse sarcoma cells after transfection of MHC class II genes, J. Immunol., 144: 4068-4071 (1990)

4. G. J. Hammerling, D. Klar, W. Pulm, F. Momburg, G. Moldenhauer, The influence of major histocompatibility complex class I antigens on tumour growth and metastasis, Biochim. Biophys. Acta, 907:245-259 (1987)

5. T. P. Miller, S. M. Lippman, C. M. Spier, D. J. Slymen, T. M. Grogan, HLA-DR (Ia) antigen immune phenotype predicts outcome for patients with diffuse large cell lymphoma, J. Clin. Invest., 82: 370-372 (1988)

6. F. Esterban, F. Ruiz-Cabello, A. Concha, M. Perezayala, J. Sanchezrosas, F. Garrido, HLA-DR expression is associated with excellent prognosis in squamous cell carcinoma of the larynx, Clin. Exp. Metastasis, 8: 319-328 (1990)

7. P. Giacomini, P. B. Fisher, G. J. Duigou, R. Gambari, P. G. Natali, Regulation of class II MHC gene expression by interferons, Anticancer Research, 8: 1153-1162 (1988)

8. C. R. Hume, R. S. Accolla, J. S. Lee, Defective HLA class II expression in a regulatory mutant is partially complemented by activated ras oncogenes, Proc. Natl. Acad. Sci. USA, 84: 8603-8607 (1987)

9. A. P. Albino, A. N. Houghton, M. Eisinger, J. S. Lee, R. R. S. Kantor, A. I. Oliff, L. J. Old, Class II histo-compatibility antigen expression in human melanocytes transformed by Harvey murine sarcoma virus (Ha-MSV) and Kirsten MSV retroviruses, J. Exp. Med., 164: 1710-1722 (1986)

10. D. J. Maudsley, A. G. Morris, Kirsten murine sarcoma virus abolishes interferon-γ-induced class II but not class I major histocompatibility antigen expression in a murine fibroblast line, J. Exp. Med., 167: 706-711 (1988)

11. A. Balmain, M. Ramsden, G. T. Bowden, J. Smith, Activation of the mouse cellular Harvey-ras gene in chemically induced benign skin papillomas, Nature (Lond.), 307: 658-660 (1984)

12. K. Brown, M. Quintanilla, M. Ramsden, I. B. Kerr, S. Young, A. Balmain, v-ras genes from Harvey and Balb murine sarcoma viruses can act as initiators of two-stage mouse skin carcinogenesis, Cell, 46: 447-456 (1986)

13. E. Liu, B. Hjelle, R. Morgan, F. Hecht, J. M. Bishop, Mutations of the Kirsten-ras proto-oncogene in human preleukaemia, Nature (Lond.), 330: 186-188 (1987)

14. A. G. Morris, G. Ward, Production of recombinant interferons by expression in heterologous mammalian cells, in: Interferons and Lymphokines: a practical approach, M. Clemens, A. G. Morris, A. Gearing, editors. IRL Press: Oxford, 61-71 (1987)

15. G. J. Atkins, M. D. Johnston, L. M. Westmacott, D. C. Burke, Induction of interferon in chick cells by temperature-sensitive mutants of Sindbis virus, J. Gen. Virol., 25: 381 (1974)

16. M. E. F. Smith, J. G. Bodmer, A. P. Kelly, J. Trowsdale, S. C. Kirkland, W. F. Bodmer, Variation in HLA expression on tumours, in: Cold Spring Harbor Symposia on Quantitative Biology, vol. LIV Cold Spring Harbor Laboratory Press (1989)

17. F. Momburg, A. Ziegler, J. Harpprecht, P. Moller, G. Moldenhauer, G. J. Hammerling, Selective loss of HLA-A or HLA-B antigen expression in colon carcinoma, J. Immunol., 142: 352-358 (1989)

18. J. L. Bos, E. R. Fearon, S. R. Hamilton, M. Verlaan-de Vries, J. H. van Boom, A. J. van der Eb, B. Vogelstein, Prevalence of ras gene mutations in human colorectal cancers, Nature (Lond.), 327: 293-297 (1987)
19. M. R. Oliva, T. Cabrera, J. Esquivias, M. Perez-Ayala, M. Redondo, F. Ruiz-Cabello, F. Garrido, K-ras mutations (codon 12) are not involved in down-regulation of MHC class I genes in colon carcinomas, Int. J. Cancer, 46: 426-431 (1990)
20. P. L. Kaplan, M. Anderson, B. Ozanne, Transforming growth factor(s) production enables cells to grow in the absence of serum: an autocrine system, Proc. Natl. Acad. Sci. USA, 79: 485-489 (1982)
21. R. D. Owen, M. C. Ostrowski, Transcriptional activation of a conserved sequence element by ras requires a nuclear factor distinct from c-fos or c-jun, Proc. Natl. Acad. Sci. USA, 87: 3866-3870 (1990)
22. C. W. Czarniecki, H. H. Chiu, G. H. W. Wong, S. M. McCabe, M. A. Palladino, Transforming growth factor-β₁ modulates the expression of class II histocompatibility antigens on human cells, J. Immunol., 140: 4217-4223 (1988)

INCREASES IN PHOSPHOLIPASE A2 ACTIVITY AND IN THE ADP
RIBOSYLATION OF G PROTEINS IN K-RAS-TRANSFORMED THYROID CELLS

Daniela Corda, Salvatore Valitutti, Luisa Iacovelli
and Maria Di Girolamo

Laboratory of Cellular and Molecular
Endocrinology, Istituto di Ricerche Farmacologiche
"Mario Negri", Consorzio Mario Negri Sud, 66030 S.
Maria Imbaro, Italy

INTRODUCTION

Epithelial cell lines derived from rat thyroid have been employed to
analyze the effects of ras-induced transformation on the signal pathways
regulating growth and differentiation.

The FRTL5 cells (a normal thyroid cell line) maintain in culture many
features of the original tissue: they grow in a thyrotropin dependent
manner, synthesize thyroglobulin and are able to concentrate iodide[1-3].
When transformed by the kiMSV, k-ras oncogene, FRTL5 cells loose their
differentiated properties and acquire a transformed phenotype[4-6]. Three
cell lines were originated from the FRTL5 cells by infection with the
Kirsten Sarcoma Virus: KiKi and KiMol cells and a temperature sensitive
clone, Ts cells[4-6].

The expression of the ras p21 does not affect the basal activity of the
adenylyl cyclase in KiKi and KiMol cells[7]. We have recently analyzed the
lipid metabolism and, in particular, the activity of phospholipase A2 and
phospholipase C in KiKi, KiMol and Ts cells[8]. The expression of the p21 was
found to be associated in these systems with an augmented activity of a
phosphoinositide-specific phospholipase A2 (see below)[8].

In these ras-transformed cells hormones which regulate normal thyroid
cells such as thyrotropin and norepinephrine, are unable to stimulate the
activity of the transducing enzymes in the transformed cell line[7,8]. The
lack of hormonal stimulation might be related to a GTP binding (G) protein
defect in the ras-transformed cells. We have analyzed this hypothesis by
characterizing the G proteins present in normal and ras-transformed cells
and by evaluating their sensitivity to pertussis and cholera toxin[9,10]. We
find that the ability of toxins to ADP ribosylate the G proteins from
transformed cells is strikingly increased[9,10].

PHOSPHOLIPASE A2 ACTIVITY

The activities of phospholipase C and phospholipase A2 were evaluated
by measuring the inositol phosphate accumulation and the arachidonic acid

release in both normal and ras-transformed thyroid cells[8]. Whereas the basal levels of inositol bis and trisphosphate (InsP2 and InsP3) were similar in FRTL5, KiKi and KiMol cells, a significant increase in inositol monophosphate (InsP1) accumulation was observed in the two transformed cell lines (Figure 1). This was paralleled by a marked increase in glycerophosphoinositol (GlyPIn), more evident in the KiKi cells (Figure 1)[8].

The formation of GlyPIn requires the activation of phospholipase A1 and phospholipase A2-like activity. It has been previously proposed that ras p21 could act at the level of phospholipase A2, since higher amounts of lysophosphatidylcholine and lysophosphatidylethanolamine have been measured in ras transformed cells[11].

The phospholipase A2 activity was evaluated in FRTL5 and KiKi cells by measuring the total arachidonic acid release, as well as the release stimulated by the Ca^{++} ionophore A23187 (Table 1)[8]. In both cases there was a significantly higher arachidonic acid release in ras transformed cells than in the normal FRTL5 cells, suggesting that phospholipase A2 activity is more pronounced upon cell transformation (Table 1)[8]. The response to the ionophore A23187 as percent stimulation of the basal enzymatic activity was not different in the two cell lines (Table 1).

The Ca^{++} ionophore, which stimulates the cellular phospholipase A2 activity, induced an increase in GlyPIn and InsP1 in ras transformed cells, whereas there was no difference in InsP2 and InsP3 accumulation supporting the possible involvement of a phospholipase A2 acting on the phosphoinositide pool in the formation of GlyPIn[8].

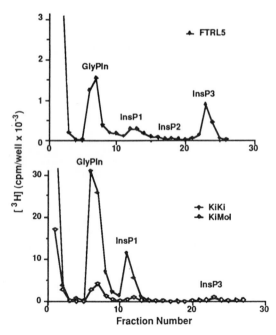

Figure 1. Elution pattern of inositol phosphates in FRTL5 (upper panel) and in KiKi and KiMol cells (lower panel). The cell extract and the elution were carried out as described in Ref.8. The peaks correspondent to GlyPIn, InsP1, InsP2 and InsP3 are identified in the Figure. Note the different scale between upper and lower panel. Taken from Ref.8.

Table 1. **Arachidonic acid release in normal and ras-transformed cells.**

	Control	A23187 (10μM)	
	% release[a]		
FRTL5 cells	1.4±0.2	3.6±0.8	[257±22][b]
KiKi cells	2.2±0.4	6.6±2.3	[300±35]
	(157±28)	(183±35)	

[a]Arachidonic acid release was measured as described in Ref.8 from cell monolayers incubated for 48 h in serum-free 199 medium containing 0.5 μCi/well [3H]-arachidonic acid. The release was evaluated over 30 min. Data are expressed as percent release (total [3H]-arachidonic acid released in 30 min/[3H]-arachidonic acid incorporated x100) and are mean±SD of 3 to 6 determinations. In parenthesis are the percent release of arachidonic acid in KiKi cells versus the normal FRTL5 cell line. [b]In brackets are indicated the percent increase induced by the Ca^{++} ionophore with respect to the basal release. The two values are not significantly different. Taken from Ref.8.

Support to the proposal that the phospholipase A2 activated in ras transformed cells is specific for the phosphoinositides, comes from the observation that the increase in GlyPIn is much larger (3-10 fold the normal cells value) than the increase in total arachidonic acid release (50-80%). Since the phosphoinositides represent a fraction (10%) of the total membrane lipids, the 50-80% increase in total arachidonic acid release might correlate well with the increase in phosphoinositide hydrolysis induced by the specific phospholipase A2 to form GlyPIn[8]. The formation of GlyPIn and InsP1 could be evaluated also in a temperature sensitive clone where the expression of the active p21 occurs at the permissive temperature of 33°C (Figure 2)[8]. Similarly to KiKi and KiMol cells, the Ts cell line was characterized by high levels of GlyPIn and InsP1 at the permissive temperature of 33°C (Figure 2). When these cells were kept for 24 h at 39°C, the GlyPIn and InsP1 decreased significantly (Figure 2). This phenomenon was time dependent, and required at least 12 h at 39°C. Ts cells kept for 24 h at 39°C (which showed low levels of GlyPIn and InsP1) were characterized by an increase in GlyPIn and InsP1 when switched back to 33°C for 48 h (Figure 2). This phenomenon required at least 8 h at 33°C. The GlyPIn levels of the 33°C Ts cells and of the cells which had the shift 33°C ->39°C ->33°C were virtually identical (Figure 3)[8]. Since GlyPIn and InsP1 were specifically increased upon cell transformation (i.e. at 33°C), whereas their levels were lower at the non permissive temperature (Figure 2), we hypothesize that the p21 could act as a regulatory component in their formation.

Other changes that occured upon ras-induced transformation included an enriched membrane pool of phosphoinositides (measured in the Ts cells maintained at the permissive temperature)[8]. All the changes reported occurred in parallel with morphological transformation[8].

On the basis of the above data, we propose that cell transformation by the k-ras oncogene might affect different steps of the membrane lipid metabolism, among which the most preeminent one is the activation of a phosphoinositide-specific phospholipase A2[8]. The activation of this enzyme

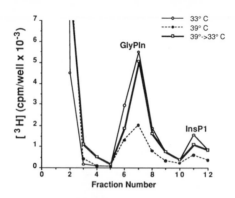

Figure 2. **GlyPIn and InsP1 formation in Ts cells at the permissive and non permissive temperatures.** The amounts of GlyPIn and InsP1 were evaluated under the following experimental conditions: 33°C (open dots), after 24 h at 39°C (full dots), and after 48 h at 33°, following the 24 h at 39°C (open squares).

leads to the formation of GlyPIn that in the epithelial cell systems analyzed in this study, can be considered a marker of malignancy[8].

ADP RIBOSYLATION OF G PROTEINS

FRTL5 cells are characterized by G proteins involved in the regulation of adenylyl cyclase and phospholipases A2 and C by thyrotropin, adrenergic, muscarinic and purinergic receptors[12-17]. In this cell line two G proteins with molecular masses of 40 and 41 kDa have been shown to be substrates for the ADP ribosylation induced by pertussis toxin (Figure 3)[10]. Using specific antisera raised against different αi subunits, the two proteins were identified as αi2 and αi3[9,10]. Cholera toxin-induced ADP ribosylation revealed two substrates with molecular masses of 42 and 45 kDa (Figure 3)[18].

In KiKi cells Gi and Gs proteins coomigrate on SDS-PAGE gels with the normal thyroid cell G proteins (Figure 3)[10]. A major difference between the G proteins of normal and transformed cells was that the latter appeared to be better substrates of the pertussis and cholera toxin-induced ADP ribosylation (the [32P]-ADP ribose labelling was at least ten fold that of normal cell membranes) (Figure 3). For example, the labelling of the αi subunit by pertussis toxin in 1 µg of KiKi cell membrane proteins was more pronounced than that of 50 µg of normal cell membrane proteins (50 fold increase) (Figure 3). This suggests that the ability of Gi and Gs to be ADP ribosylated was altered upon ras-induced cell transformation. The altered ADP ribosylation could be related either to an increase in the amount of G proteins present in ras-transformed cells or to a different modulation of the ADP ribosylation reaction in the two cell systems.

By western blot analysis employing the antibody AS58 which recognizes the αi2 subunit, it could be shown that this subunit was more aboundant in KiKi cells (about twice the amount of normal cells) (Figure 4)[10]. This difference, although statistically significant, does not explain the increased labelling by pertussis toxin of the G proteins in transformed cells (Figure 3)[10]. The phenomenon could therefore be due to some defect at the level of the ADP ribosylation reaction. Preliminary data tend to exclude that reduced hydrolysis of NAD by glycohydrolases in the transformed cells could play a role in the increase in G protein labelling, since the NAD available for the ADP ribosylation reaction was very similar in the two cell systems. A possible role of a poly-ADP-ribosylation reaction was excluded since inhibitors of this reaction did not affect the labelling of G proteins in both normal and transformed cell systems[10].

A difference in dissociated G proteins, which are not substrates of the pertussis toxin-induced ADP ribosylation in normal and transformed cells, could also be excluded[10].

Preliminary experiments indicate that endogenous modulators of the ADP ribosylation reaction are involved in the increased ADP ribosylation of G proteins in transformed cells[10].

In summary, we have shown (see above) that ras-transformed cells are characterized by an increased basal activity of a phosphoinositide-specific phospholipase A2[8]. The phospholipase A2 in thyroid cells is coupled to a pertussis toxin sensitive G protein[13,15-17]. It is conceivable that the increased amount and the modification of G proteins in ras transformed cells could play a role in altering the activity of transducing enzymes, such as phospholipase A2, which are known to be relevant in the control of growth and differentiation in the thyroid system.

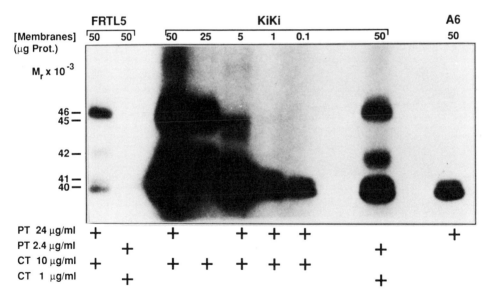

Figure 3. [32P]-ADP-ribosylation by pertussis (PT) and cholera (CT) toxin of membrane proteins from FRTL5, KiKi and Ts (A6) cells. The concentration of PT and CT are indicated. The assays were performed as previously described[10,18]. The additions to each sample are indicated, as well as the μg of membrane proteins. The molecular masses of the different bands are indicated.

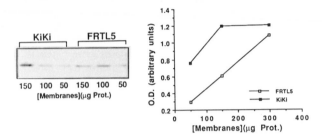

Figure 4. Densitometric analysis of the immunoblots of KiKi and FRTL5 cell membrane proteins (right panel). The immunoblots of a different experiment are shown in the left panel. The antibody AS58 which reacts with the αi2 subunit was employed. The experimental procedure have been previously described[10].

Acknowledgements: This study was supported by the Italian Association for Cancer Research (AIRC), and in part by Fidia S.p.A., by the Agenzia per la Promozione e lo Sviluppo del Mezzogiorno (PS.35.93/IND, DELIB.6168) and by the National Research Council (Convenzione CNR-Consorzio Mario Negri Sud).

REFERENCES

1. F.S. Ambesi-Impiombato, and H. Perrild eds FRTL5 today. Elsevier Science Publishers, Amsterdam, (1989).
2. F.S. Ambesi-Impiombato, R. Picone, and D. Tramontano, Influence of hormones and serum on growth and differentiation of the thyroid cell strain FRTL, Cold Spring Harbor Symp. Quant. Biol. 9:483-492 (1982).
3. W.A. Valente, P. Vitti, L.D. Kohn, M.L. Brandi, C.M. Rotella, R. Toccafondi, D. Tramontano, S.M. Aloj and F.S. Ambesi-Impiombato, The relationship of growth and adenylate cyclase activity in cultured thyroid cells: separate bioeffects of thyrotropin, Endocrinology 112:71-79 (1983).
4. A. Fusco, A. Pinto, F.S. Ambesi-Impiombato, G. Vecchio, and N. Tsuchida, Transformation of rat thyroid epithelial cells by kirsten murine sarcoma virus, Int. J. Cancer 28:655-662 (1981).
5. A. Fusco, A. Pinto, D. Tramontano, G. Tajana, G. Vecchio, and N. Tsuchida, Block in the expression of differentiation markers of rat thyroid epithelial cells by transformation with kirsten murine sarcoma virus, Cancer Res., 42:618-626 (1982).
6. G. Colletta, A. Pinto, P.P. Di Fiore, A. Fusco, M. Ferrentino, V.E. Avvedimento, N. Tsuchida, and G. Vecchio, Dissociation between transformed and differentiated phenotype in rat thyroid epithelial cells after transformation with a temperature-sensitive mutant of the kirsten murine sarcoma virus, Mol. Cell. Biol. 3:2099-2109 (1983).
7. G. Colletta, D. Corda, G. Schettini, A.M. Cirafici, L.D. Kohn, and E. Consiglio, Adenylate cyclase activity v-ras-k transformed rat epithelial thyroid cells FEBS Lett. 228:37-41 (1988).

8. S. Valitutti, P. Cucchi, G. Colletta, C. Di Filippo, and D. Corda, Transformation by the K-ras oncogene correlates with increases in phospholipase A2 activity, glycerophosphoinositol production and phosphoinositide synthesis in thyroid cells, _Cell. Signal._, in press (1991).

9. M. Di Girolamo, D. D'Arcangelo, P. Gierschik, and D. Corda, Characterization of trimeric G proteins in normal and k-ras transformed cells. _Cell Biol. Int. Rep._, 14S:45 (1990).

10. M. Di Girolamo, D. D'Arcangelo, P. Gierschik, and D. Corda, Increased ADP- ribosylation of trimeric G proteins in k-ras transformed cells, (submitted)

11. D. Bar-Sagi, and J.R. Feramisco, Induction of membrane ruffling and fluid-phase pinocytosis in quiescent fibroblasts by ras proteins _Science_ 233:1061-1068 (1986).

12. D. Corda, C. Marcocci, L.D. Kohn, J. Axelrod, and A. Luini, Association of the changes in cytosolic Ca^{2+} and iodide efflux induced by thyrotropin and by the stimulation of $\alpha 1$-adrenergic receptors in cultured rat thyroid cell, _J. Biol. Chem._ 260:9230-9236 (1985).

13. D. Corda, and L.D. Kohn, Role of pertussis toxin sensitive G proteins in the $alpha_1$ adrenergic receptor but not in the thyrotropin receptor mediated activation of membrane phospholipases and iodide fluxes in FRTL-5 thyroid cells, _Biochem. Biophys. Res. Commun._ 141:1000-1006 (1986).

14. C. Bizzarri, M. Di Girolamo, M.C. D'Orazio, D. Corda, Evidence that a guanine nucleotide-binding protein linked to a muscarinic receptor inhibits directly phospholipase C, _Proc. Natl. Acad. Sci. USA_ 87:4889-4893 (1990).

15. R.M. Burch, A. Luini, and J. Axelrod, Phospholipase A_2 and phospholipase C are activated by distinct GTP-binding proteins in response to α_1-adrenergic stimulation in FRTL5 thyroid cells, _Proc. Natl. Acad. Sci. USA_ 83:7201-7205 (1986).

16. D. Corda, L. Iacovelli, and M. Di Girolamo, Coupling of the α_1-adrenergic and thyrotropin receptors to second messenger systems in thyroid cells. Role of G-proteins. In: Maggi M, Johnston CA (eds) Horizons in Endocrinology. Raven Press, New York, 169-180 (1988).

17. D. Corda, C. Bizzarri, M. Di Girolamo, S. Valitutti, and A. Luini, G protein-linked receptors in the thyroid, _Adv. Exp. Med. Biol._ 261:245-269 (1989).

18. D. Corda, R.D. Sekura, and L.D. Kohn, Thyrotropin effect on the availability of Ni regulatory protein in FRTL-5 rat thyroid cells to ADP-ribosylation by pertussis toxin, _Eur. J. Biochem._ 166:475-481 (1987).

SUPPRESSION OF THE PHENOTYPE OF T24 H-RAS1 TRANSFORMED CELLS

Demetrios A. Spandidos[1,2]

[1]Institute of Biological Research and
Biotechnology, Hellenic Research Foundation
Athens, 11635, Greece
[2]Medical School, University of Crete, Heraklion
Greece

INTRODUCTION

Many tumors contain activated *ras* genes (Bos, 1989; Field and Spandidos, 1990). These genes are thought to be involved in the development of these cancers. It would be of great therapeutic value to find ways to prevent these cancer cells growing. Therefore it is of interest to explore methods by which the growth of *ras* transformed cells is controlled i.e. suppression of cancer phenotype. These studies are more easily undertaken *in vitro* i.e. in tissue culture systems than *in vivo* and in this paper we will concentrate on the *in vitro* studies. A recent review has classified the suppression of *ras* transformants into three groups (Kuzumaki, 1991). Viz. 1. Those that operate through genetic material (genetical). 2. Those that operate through biological response modifiers (biological) and 3. Drugs (pharmaceutical) (see Table 1).

Onco-suppressor genes are genes for which deletion or inactivation or loss-of-function mutations in both alleles are oncogenic (for review see Spandidos and Anderson, 1989). The presence of the wild type allele of such genes prevents or suppresses tumorigenesis. For example, introduction and expression of the normal retinoblastoma (RB) (Huang et al., 1988; Bookstein et al., 1990) or the p53 gene (Chen et al., 1990), suppressed the neoplastic properties of human tumor cells lacking endogenous wild type RB or p53 activities respectively. In a different set of studies a *ras*-related gene named K*rev*-1 with transformation suppressor activity has been described(Noda et al., 1989). Furthermore, expression of the normal H-*ras*1 gene can suppress the transformed and tumorigenic phenotypes induced by mutant *ras* genes (Spandidos and Wilkie,1988; Spandidos et al., 1990).

Although a consequence of oncogene activation may be the induction of cell proliferation (Mulcahy et al., 1985;), in some cases their activation may inhibit cell proliferation (Hirakawa and Ruley, 1988; Ridley et al, 1988). Oncogenes also

Table 1. Suppression of *ras* transformants

Type of suppression Reagents	Possible mechanisms
1. Genetical	
Normal cells	Onco-suppressor gene?
DNA from normal cells	An onco-suppressor gene
Krev1	*ras* p-21 antagonist
Normal *ras* genes	Mutated *ras* p21 antagonist?
Mutated *ras* genes	*ras* p21 antagonist
Poly (d G-m5dC)	Z-DNA formation
Antisense RNA	Hybridization arrest
Mutagens	Gelsolin? Tropomyosin?
2. Biological	
Interferon	DNA methylation
DIF	mRNA block, degradation
Antibodies	GTP binding inhibition
Arachidonic acid	Inhibition of palmitylation
cAMP	Increase of RII cAMP receptor
3. Pharmacological	
Methionine	Inhibition of glycolysis?
Swainsonine	Block of glycoprotein processing
Oxanonine	Inhibition of palmitylation
Compactin	Inhibition of isoprenylation
Azatyrosine	Unknown
Tamoxifen	Anti PKC activity
Genistein	Anti topoisomerase activity
Benzamide	Inhibition of poly (ADP-ribose) polymerase
Coumarin	Inhibition of poly (ADP-ribose) polymerase
Sodium butyrate	An onco-suppressor gene?

play a role in differentiation (Muller, 1986) in some cases blocking or in others inducing differentiation (Spandidos and Anderson, 1987).

It is now clear that members of the *ras* family can either block(Caffrey et al., 1987; Gambariand Spandidos, 1986; Olson et al., 1987) or induce(Bar-Sagi and Feramisco, 1985; Feramisco et al, 1987; Guerrero et al., 1988; Noda et al., 1985; Spandidos, 1989) differentiation depending on the cell system. We have previously noted these different effects of *ras* gene expression (Spandidos and Anderson, 1989) and discuss them further in this paper.

RESULTS AND DISCUSSION

The effect of normal cells on *ras*-transformants

We have investigated the effect of co-cultivated normal and *ras*-transformed mouse cells on the morphological phenotype of the malignant cells. Constant numbers of *ras* transformed cells were co-cultivated with increasing numbers of normal cells. It was found that the number of transformed foci was reduced but only in the presence of vast excess of normal cells. This effect is not attributed to overcrowding. Similar experiments were carried out to determine if injection of a mixture of normal and transformed cells would have an effect on tumor formation. When low numbers of transformed cells (10^3) were mixed with increased numbers of normal cells the time taken for the appearance of tumors was increased significantly. On the other hand when large numbers (1×10^6) of normal cells were used, the time for tumor formation was unaffected. Thus, here again the presence of excess normal cells ameliorated the effect of the tumorigenic cells (Kakkanas and Spandidos, 1989) Similar results have been previously obtained with Chinese hamster ovary lung cells transformed with the T24 H-*ras*1 oncogene which were suppressed *in vitro* by co-cultivation with primary hamster cells (Spandidos, 1986). The basis for this effect is not known but we are currently investigating three possibilities. The first is that through cell to cell contact growth inhibitory substances may pass through gap junctions from the normal to transformed cells. Secondly, normal cells may carry on their surface molecules which inhibit growth of the transformed cells. Both these possibilities require cell to cell contact. Thirdly, normal cells may release diffusible, growth inhibitory substances into the culture medium. This possibility would not require cell to cell contact, but might require excess of normal cells if the concentration of the inhibitor was low or the transformed cells were relatively insensitive to its effects.

The normal H-*ras*1 gene can act as an onco-suppressor gene

Normal and mutant T24 human H-*ras*1 genes on recombinant plasmids containing selectable markers were introduced into recipient rat 208F or human T24 cells. In some plasmids the normal H-*ras*1 gene was placed under the transcriptional control of the human metallothionin promoter which is inducible in the presence of ions. These recombinant· plasmids have been previously described(Spandidos and Wilkie, 1988; Spandidos et al, 1990). Transfection of 208F cells with a plasmid containing the T24 H-*ras*1 oncogene induced morphological alterations, anchorage independent growth and tumorigenicity in nude mice, while introduction of the normal H-*ras*1 gene showed none of these effects. This could not be attributed to absence of transfected plasmid in cells. However, transfection of 208F with a construct containing both the proto-oncogene and the T24 H-*ras*1 gene on the same plasmid gave rise to colonies the major proportion of which were morphologically normal, anchorage-dependent and non-tumorigenic.Thus, simultaneous transfer of the normal gene with the T24 H-*ras*1 gene apparently led to suppression of the cancer phenotype normally induced by the T24 gene. Similar results were obtained with the human T24 bladder carcinoma cells. The T24 cell line is tumorigenic in nude mice. It contains and expresses only the mutant oncogenic form of H-ras1 gene, since the normal allele has been deleted. Using an inducible metallothionin promoter in front of the exogenous normal H-*ras*1 gene a marked suppression of the tumorigenic phenotype was observed in the presence but not in the absence of Cd^{++}. The experiment suggested that normal H-*ras*1 gene behaved as an onco-suppressor gene in these human cancer cells. To determine if the level of proto-oncogene expression is important for the effect, we examined the relative levels of *ras* RNA and protein. Proteins in the transfected T24 cells were labelled with ^{35}S-methionine, extracted under non-denaturating conditions, immunoprecipitated using the anti-*ras* monoclonal antibody Y13 259 and subjected to electrophoretic separation as described previously (Spandidos and Wilkie, 1988; Spandidos et al, 1990).In each case it was found that the suppressed cells (flat transfectants) expressed both gene products and there was more normal p21 than mutant T24 p21. Expression was also measured by hybridizing labelled oligonucleotide specific for the normal and mutant genes to total cellular RNA. Furthermore, we observed that in a small proportion of the cell lines obtained by transfer of the normal and mutant H-*ras*1 genes on the same plasmid, tumors arose in nude mice after a very long lag period. Cell lines were derived from some of these tumors. It was found that while the parent line expresses more normal than mutant p21, these tumor derived cell lines express predominantly the mutant form. Similar results were obtained with differed parent and derived tumor cell lines and with tumorigenic cells obtained by growing suppressed cell lines in low serum conditions.

We have shown that the normal H-*ras*1 gene behaved as an onco-suppressor gene when transferred into recipient cells either simultaneously with or subsequent to transfection with mutant *ras* genes. Analysis of p21 expression showed that suppressed cells expressed more normal than mutant p21 while tumorigenic variants show reduced levels of the mutant p21. The results suggest that expression of normal H-*ras*1 can suppress the transformed and tumorigenic phenotypes induced by mutant *ras* genes and that this occurs when there is more normal than mutant protein. Results from the study of spontaneous cancer and established cell lines are consistent with these findings. Cancer cell lines often either contain and express only mutant alleles of *ras* genes, suppress expression of the normal allele or increase of the mutant allele (Feinberg et al.,1983; Sandos et al., 1984; Capon et al., 1983). The molecular mechanism behind the onco-suppression described in our study remain(s) to be determined. A possible explanation is competition by the normal H-*ras*1 gene product with the mutant gene product for cellular proteins or sites which interact with p21. Our results suggest that *ras* may be able to act as either an oncogene or an onco-suppressor gene depending on the cellular context. The mechanism by which *ras* can suppress the transformed and tumorigenic phenotypes of cells may be important in the analysis of the gene product which modulate onco-suppressor gene activity of cells *in vitro* and *in vivo*.

Expression of the T24 H-*ras* gene can induce differentiation

Using an electroporation technique we have introduced and expressed the recombinant plasmid pHO6T1 carrying the human T24 H-*ras* gene into the rat pheochromocytoma PC12 cells. We found that transfectants had undergone neuronal differentiation similar to that induced by the nerve growth factor(NGF). In control experiments the vector alone (Homer 6) or plasmid pHO6N1 carrying the normal H-*ras*1 gene had no effect (Spandidos, 1989). Our results are similar to those obtained transfecting the mutant HT1080 human N-*ras* (Guerrero et al, 1988) or retroviruses carrying the H-*ras* or K-*ras* genes (Noda et al, 1985). The same result could also be obtained by microinjecting the *ras* p21 oncoprotein but not the normal p21 (Bar-Sagi and Feramisco, 1985; Feramisco et al, 1985). On the other hand, we have previously found that the T24 H-*ras*1 gene inhibited adipocytic cell differentiation (Gambari and Spandidos, 1986). These results imply that *ras* oncogenes might play a physiological role in differentiation as well as in proliferation.

REFERENCES

Bar-Sagi, D. and Feramisco, J.R., 1985, Microinjection of the *ras* proteins into PC12 cells induces morphological differentiation. Cell 42:841.
Bookstein, R., Shew, J.-Y., Chen, P.-L., Scully, P. and Lee, W.-H., 1990, Suppression of tumorigenicity of human prostate carcinoma cells by replacing a mutated RB gene. Science 247:712.
Bos,J.L., 1989, *Ras* oncogenes in human cancer. Cancer Res., 49:4682
Caffrey, J.M., Brown A.M. and Schneider, M.D., 1987, Mitogens and oncogenes can block the induction of specific voltage-gated ion channels. Science 236, 570.
Capon, D.J., Chen E.Y., Levinson, A.D., Seeburg, P. and Goeddel, D.V.,1983, Complete nucleotide sequences of the T24 human bladder carcinoma oncogene and its normal homologue. Nature 309:33.
Chen, P.-L., Chen, Y., Bookstein, R. and Lee, W.-H., 1990, Genetic mechanisms of tumor suppression by the human p53 gene. Science 250:1576.
Feinberg, A.P., Vogelstein, B., Droller, M.J., Baylin, S.B. and Nelkin, B.D., 1983, Mutation affecting the 12th amino acid of the c-Ha-*ras* oncogene product occurs infrequently in human cancer. Science 220:1175.
Feramisco. J.R., Clark, R., Wong, G., Arheim, N., Milley, R. and McCormick, F. Oncogene-induced cell transformation by antibodies specific for amino acid 12 of *ras* protein. Nature 314:639.
Field, J.K. and Spandidos, D.A. 1990, The role of *ras* and *myc* oncogenes in human solid tumors and their relevance in diagnosis and prognosis. Anticancer Res., 10:1

Gambari, R. and Spandidos, D., 1986, Chinese hamster lung cells transformed with the human Ha-*ras*-1 oncogene: 5-azacytidine mediated induction of adipogenic conversion. Cell Biol. Int. Rep., 10: 173.

Guerrero, I., Pellicer, A. and Burstein, D.E., 1988, Dissociation of c-*fos* from ODC expression and neuronal differentiation in a PC12 subline stably transfected with an inducible N-*ras* oncogene. Biochem. Biophys. Res. Comm., 150:1185.

Hirakawa, T. and Ruley, H.E., 1988, Rescue of cells from *ras* oncogene-induced growth arrest by a second complementing oncogene. Proc. Natl. Acad. Sci. USA, 85:1519.

Huang, H.-J., S., Yee, J.-K., Shew, J.-Y., Chen. P.-L., Bookstein, R., Friedmann, T., Lee, E.Y.-H.P. and Lee, W.-H., 1988. Suppression of the neoplastic phenotype by replacement of the RB gene in human cancer cells. Science, 242:1563.

Kakkanas, A. and Spandidos, D.A., 1990. *In vitro* and *in vivo* onco-suppressor activity of normal cells on cells transfected with the H-*ras* oncogene. In Vivo, 4:109.

Kuzumaki, N. 1991, Suppression of *ras*-transformants, Anticancer Res. 11:313.

Mulcahy, L.S., Smith, M.R. and Stacy, D.W.,1985, Requirement for *ras* proto-oncogene function during serum-stimulated growth of NIH 3T3 cells. Nature, 318:73.

Muller, R., 1986, Proto-oncogenes and differentiation. TIBS, 11: 129.

Noda, M., Kitayama, H., Matsuzaki, T. Sugimoto, Y., Okayama, H., Bassin, R.H., and Ikawa, Y., 1989, Detection of genes with a potential for suppressing the transformed phenotype associated with activated *ras* genes. Proc. Natl. Acad. Sci. USA, 86:162.

Noda, M., Ko, M., Ogura, A., Liu D.-G., Amano, T. and Ikawa,Y., 1985, Sarcoma viruses carrying *ras* oncogenes induce differentiation-associated properties in a neuronal cell line. Nature, 318: 73.

Olson, E.N., Spizz, G. and Tainsky, M.A.,1987, The oncogenic forms of N-*ras* prevent skeletal myoblast differentiation. Mol. Cell Biol., 7:2104.

Ridley, A.J., Paterson, H.F., Noble, M. and Land, H.,,1988, *Ras*-mediated cell cycle arrest is altered by nuclear oncogenes to induce Schwan cell transformation. EMBO J., 7:1635.

Santos, E., Martin-Zanca, D., Reddy, E.P., Pierotti, M.A., Della Porta, G. and Barbacid, M., 1984, Malignant activation of a K-*ras* oncogene in lung carcinoma but not in normal tissues of the same patient. Science, 223:661.

Spandidos, D.A. 1986. The human T24 Ha-*ras*1 oncogene: A study of the effects of overexpression of the mutated *ras* gene product in rodent cells. Anticancer Res., 8:259.

Spandidos, D.A. 1989, The effect of exogenous human *ras* and *myc* oncogenes in morphological differentiation of the rat pheochromocytoma PC12 cells. Int. J. Devl. Neurosci., 7:1

Spandidos, D.A. and Anderson, M.L.M.,1987, A study of the mechanism of carcinogenesis by gene transfer of oncogenes into mammalian cells. Mutat. Res., 185:271.

Spandidos, D.A. and Anderson, M.L.M.,1989, Oncogenes and onco-suppressor genes: their involvement in cancer. J. Pathol., 157:1.

Spandidos, D.A., Frame M. and Wilkie, N.M.,1990. Expression of
    the normal H-*ras* gene can suppress the transformed and
    mutant *ras* genes. Anticancer Res. 10:1543.
Spandidos, D.A. and Wilkie, N.M., 1988, The normal human H-
    *ras*1 gene can act as an onco-suppressor. Br. J. Cancer
    58., Suppl IX: 67.

SEARCH FOR CORRELATIONS BETWEEN K-<u>RAS</u> ONCOGENE ACTIVATION AND

PATHOLOGY, IN SPORADIC AND FAMILIAL COLONIC ADENOMAS

J.C. Chomel[1], S. Grandjouan[2], D.A. Spandidos[3],
M. Tuilliez[4], C. Bognel[5], A. Kitzis[1],
J.C. Kaplan[1] and A. Haliassos[1,3]

1-Institut de Pathologie Moléculaire
2-Service de Gastroenterologie
4-Service de Pathologie, CHU COCHIN, 24 rue du
   du F[bg] St Jacques, 75014 Paris, France
3-National Hellenic Research Foundation
   Institute of Biological Research and
   Biotechnology, 48, Vas. Constantinou Ave.
   116 35 Athens, Greece
5-Institut Gustave Roussy 94805 Villejuif Cedex
   France

INTRODUCTION

The activation of K-<u>ras</u> oncogene by point mutation at its 12th codon has been reported as a molecular marker of tumor progression in colonic neoplasias (early event, correlated with the increasing size of adenomas, but not with dysplasia)[1]. The overall frequency of K-<u>ras</u> point mutations has probably been underestimated by the usual former techniques. Direct sequencing of PCR amplified K-<u>ras</u> oncogene demonstrated that 75% and 65% adenomas and adenocarcinomas, respectively, harbored a mutated allele of K-<u>ras</u> at codon 12[2]. We described a new sensitive method for an easier detection of this mutation, based upon the introduction of an artificial restriction site in a modified primer for selective *in vitro* amplification[3]. This sensitive method, connected with the feasibility of PCR amplification of DNA in formalin-fixed and paraffin-embedded tissues, prompted us to reevaluate the frequency of K-<u>ras</u> point mutations at codon 12 in both sporadic and familial colonic adenomas.

MATERIALS AND METHODS

DNA samples

DNA from adenomatous polyps and adjacent normal mucous was extracted according either to the classical method of phenol chloroform, or by the simplified guanidium method[4].

DNA from histological slides, fixed by formol and embeded in paraffin, was extracted by boiling in a lysis mixture after the dissolution of paraffin in chloroform.

Slices of paraffin with no embedded tissues, empty vials and normal lymphocytes were processed together with the colonic specimens and used as negative tests for DNA amplification and K-ras mutations.

As a positive control, we used DNA from the SW 80 cell line, which is known[5] to carry 2 mutated alleles at codon 12 of K-ras oncogene.

DNA from peripheral lymphocytes (50A), which are homozygous wild type at this same position as proved by allele specific oligodeoxynucleotide (ASO) probe hybridization, was used as a negative control.

Oligonucleotides primers and probes

The oligodeoxynucleotides were synthesised by the solid phase triester method. The primers were designed to introduce base substitution in the amplified fragments. The probe Kx-1, which hybridized with the 3' region of these fragments, was end labeled using $[\gamma^{32}P]ATP$ and T4-Polynucleotide kinase.

Polymerase chain reaction

In vitro enzymatic DNA amplification (PCR) was performed on an automated apparatus (DNA thermal Cycler from Perkin Elmer Cetus).

We performed 35 cycles of amplification. Each cycle includes 3 steps:
- denaturation of DNA at 94°C for 20 sec.
- annealing of the primers at 53°C for 30 sec.
- enzymatic extension at 72°C for 1 min.

Modified primers were designed to introduce a base substitution adjacent to the codon of interest in order to create an artificial restriction site with only one allelic form (wild type or mutated). We performed PCR with a modified

```
                   K-ras sequence with
                   wild codon 12 (GGT)
    5'..TAAACTTGTGGTAGTTGGAGCTGGTGGC......GACGAATATGATCCAACAATAGA..3'
      5'TAAACTTGTGGTAGTTGGAGCC 3' primers  3'CTTATACTAGGTTGTTATCT 5'
         K12Nm                                              KB12
                          PCR

    5'TAAACTTGTGGTAGTTGGAGCCGGTGGC.........GACGAATATGATCCAACAATAGA 3'
            Msp I site    (99 bp)

                       Msp I digestion
    TAAACTTGTGGTAGTTGGAGCCG and GTGGC.........GACGAATATGATCCAACAATAGA
        (21 pb)                                  (78 pb)
```

Figure 1. P.C.R. with modified primer creating artificial RFLP.

primer creating a Msp I recognition site only if codon 12 was of the wild type (Figure 1). This approach allowed us to screen for point mutations at codon 12 of K-_ras_ oncogene[6].

An aliquot of the PCR product was controlled in a 2-3% Nu Sieve gel for the presence of the amplified fragment (99 bp). The PCR product was digested by Msp I enzyme which gave two fragments of 21 and 78 bp in the case of a wild type sequence or left the product undigested in the case of a mutated sequence. The fragments were analyzed by electrophoresis on Nu Sieve agarose gels.

Detection of the minority fragments

After the electrophoresis, the fragments were transferred onto a nylon filter according to the standard method of Southern[7]. These filters were hybridized with the probe Kx-1 in a solution containing tetramethylammonium according to Verlaan de Vries _et al._[5] at 50°C for an hour. The filters were washed twice at 52°C in 0.1 SDS, 5X SSPE solution for 10 min. and autoradiographed for an hour at – 70°C using intensifying screens.

RESULTS AND DISCUSSION

The validity of our modified primer method is presented in Figure 2: 50A and SW 480 samples have been already described, SG3d1 DNA has a wild and a mutated allele as confirmed by allele specific radiolabeled probes. Lanes A and B show the PCR products of 50A DNA before and after Msp I digestion respectively; lanes C and D, those of SW80; lanes F and G, those of SG3d1.

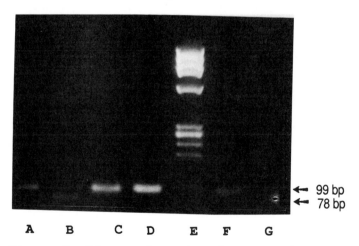

Figure 2. Electrophoresis of PCR products:
1 μg of DNA from 50A (A and B), SW80 (C and D) and SG3d1 (F and G) were amplified by PCR (35 cycles) using the primers described in Figure 1.
20 μl of the amplified samples were digested with Msp I (B, D and G) and electrophoresed on a 3% NuSieve agarose gel.
Lane E: 2μg of the ΦX 174/Hae III molecular weight marker.

The results of the electrophoresis correspond to the known genotype of these DNAs. The visualization of the results in the gel under U.V. light, without the use of radioactive probes, gives a sensitivity analogous to the previous method with the allele specific radiolabeled probes (i.e.≈ 5%). However, we can increase its sensitivity using an internal (i.e. non mutation specific) radiolabeled oligonucleotide probe (Kx-1), which reveals the presence of minority amplified and non digested fragments.

An example of this strategy is shown in Figure 3. SG3c DNA is negative for the mutation with the ASO probe method. In the stained gel after the PCR with the modified primer, we visualize only the 78bp fragment (corresponding to the wild type allele). But the hybridization with Kx-1 radiolabeled probe reveals, after autoradiography, another band (99 bp, corresponding to the presence of a minority mutation among the cells of the SG3c sample).

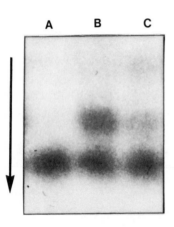

Figure 3. Hybridization of P.C.R. products with Kx-1 probe.
(5' - GCTAATTCAGAATCATTTTG - 3')
Lane A: 50A DNA (2 wild alleles)
Lane B: SG3d1 DNA (1 wild & 1 K-<u>ras</u> 12 mutated allele)
Lane C: SG3c DNA (1 wild & 1K-<u>ras</u> 12 mutated allele)

To establish the detection threshold of our method we performed a series of dilutions containing a decreased number of mutated alleles in a medium containing wild type DNA. Our dilution scale extended from 1/1 to 1/25000 of mutated alleles in normal ones. This allelic dilution corresponds to cell dilution, as mutated cells usually carry a normal and a mutated allele[8]. In Figure 4 the spot corresponding to the large (non digested) fragment (1) represents the mutated alleles of the heterozygous SG3d1 cells and the spot corresponding to the light (digested) fragment (2) represents the wild type alleles of the heterozygous SG3d1 and of the normal homozygous 50A cells. To interpret these results we must compare the intensities of spots (2) to those of spots (1) on the same lane and to those of spots (2) on the negative control lanes (B and G). This strategy leads to the detection of the mutated allele until the 1:25000 dilution.

We used the above technique to detect the mutations of codon 12 of the K-<u>ras</u> oncogene in cell populations from adenomas in 17 DNA samples from 14 patients with sporadic adenomas (3 with synchronous colon cancers), and in 11 DNA samples from 11 patients with Adenomatous Polyposis Coli (2 with synchronous colon cancers). Our results are represented in Table 1 (a & b) and depicted schematically in Figure 5.

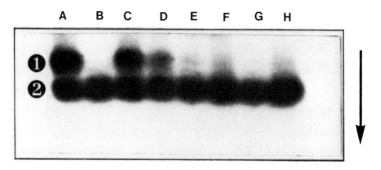

Figure 4. Dilutions of SG3d1 (heterozygous mutated-normal DNA) in 50A (homozygous wild type DNA)

LANE A:	1:1	50A/ SG3 d1
LANE B:	1:0	50A/ SG3 d1
LANE C:	10:1	50A/ SG3 d1
LANE D:	100:1	50A/ SG3 d1
LANE E:	1000:1	50A/ SG3 d1
LANE F:	5000:1	50A/ SG3 d1
LANE G:	1:0	50A/ SG3 d1
LANE H:	25000:1	50A/ SG3 d1

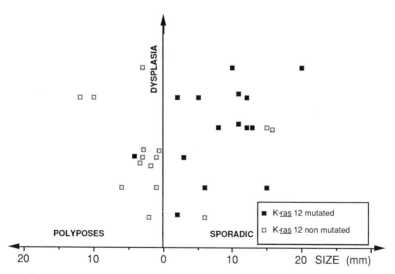

Figure 5. Graphical correlation between the presence of K-ras codon 12 point mutations the size of the corresponding polype and the grade of the dysplasia.

157

Table 1. Results of the detection of K-ras codon 12 point mutations in 17 DNA samples from 14 patients with sporadic adenomas and in 11 DNA samples from 11 patients with Adenomatous polyposis coli.

No pt	sexe	age	cancer + or -	diagnostic	number of polyps	size of polyp	grade of dysplasia	Tubulous or Villous	site	mutation + or -
POLYPOSES										
1	M	34	+	FAP	>1000	3 mm	2-3	TV	diffuse	-
2	M	25	-	FAP	> 100	3 mm	1	T	dffuse	-
3	M	23	-	FAP	> 100	3 mm	1	T	diffuse	-
4	F	27	+	FAP	> 100	12 mm	2	TV	mostly left	-
5	M		+	FAP		3 mm				-
6	F	26	-	FAP	> 100	2 mm	1	T	diffuse	-
7	M	26	-	FAP	> 100	2 mm	0	T		-
8	M	17	-	FAP	> 100	3 mm	1	T	diffuse	-
9	F	19	-	FAP	> 100	2 mm	1	T	diffuse	-
10	M	47	+	Lynch +polyposis	> 30	4 mm	1	T	diff (left)	+
11	F	68	+	Lynch +polyposis	> 50	10 mm	2	TV	diffuse	-
12	M	75	-	late polyposis	> 40	6 mm	2	TV	mostly right	-
13	M	61	ant	late polyposis	> 30	1 mm	1	TV	right	-
						1 mm	0-1	T	right	-
						1 mm	0-1	T	transv.	-
						1 mm	0-1	T	sigmoid	+
						1 mm	0-1	T	rectum	+

SPORADIC CASES

	Sex	Age		familial risk	Size			Location	
14	M	60	+	1	3 mm	1	T	left	+
15	M	67	+	4	20 mm	2-3	TV	left	+
16	F	58	+	4	2 mm	2	T		+
17	F	44	−	> 5	2 mm	0-1	TV	left angle	+
					15 mm	0-1	TV	left angle	+
					6 mm	0-1	TV	rectum	+
18	M	78	−	1	10 mm	2-3	TV	sigmoid	+
19	M	36	−	2	12 mm	2	T	rect.sigm.	+
					5 mm	2	T	left angle	+
20	F	88	−	1 or >	12 mm	1-2	TV	rect.sigm.	+
21	F	48	−	1	6 mm	0	T	legt angle	−
22	M	55	−	1	8 mm	1-2	T	rectum	+
23	M	76	−	2	12 mm	1-2	TV	transv.	+
24	M	64	−	1	15 mm	1-2	T	sigmoid	−
25	F	56	−	1	9mm	1-2	TV	left	+
26	F	66	−	1 or >	11mm	1-2	TV	transv.	+
27	M	62	−	1	16 mm	1-2	T	sigmoid	−

We found only one mutation in an adenoma from a patient with an atypical form of FAP (Lynch family). This patient had a colectomy some years ago after the diagnosis of a colon adenocarcinoma. In contrast, in another young patient with multiple villous and dysplastic polyps, adenocarcinomas and lymph node metastasis, neither samples from each lesions was positive.

Attempts were made to establish correlation between the presence of an activated K-_ras_ allele and Dukes stages in the above cases but our results do not permit this, because our sample was statistically insufficient.

Previous studies of 7[1] and 29[9] FAP patients show 5 mutations in 40 and 75 samples respectively. In those studies, the origin of DNA was either frozen tissues or cryostatically sliced paraffin embedded tissues.

CONCLUSIONS

We proved that our approach was sensitive and specific for the detection of point mutations despite the presence of many copies of wild type alleles. It is about 1000 times more sensitive than all the previous methods, and many times faster.

Our preliminary results indicate that the activation of K-_ras_ oncogene is a rare event among the FAP patients, even in the very aggressive forms of the disease, contrary to the sporadic adenomas. Final correlation between K-_ras_ activation and FAP pathology could be established only after a large study of patients using a sensitive method of detection in DNA samples extracted by a standard protocol.

Finally, this philosophy will permit to correlate clinical stages with data from histology and molecular biology.

REFERENCES

1. Vogelstein, B., Fearon, E.R., Hamilton, S.R., Kern, S.E., Preisinger, A.C., Leppert, M., Nakamura, Y., Whyte, R., Smits, A.M.M. and Bos, J.L.(1988) Genetic alterations during colorectal-tumor development. _N. Engl. J. Med.,_ **319**, 525-532.
2. Burmer, G.C. and Loeb, L.A. (1989) Mutations in the KRAS2 oncogene during progressive stages of human colon carcinoma. _Proc. Natl. Acad. Sci. USA,_ **86**, 2403-2407.
3. A. Haliassos, J.C. Chomel, S. Grandjouan, J.C. Kaplan & A. Kitzis.(1989) Detection of minority point mutations by modified PCR technique: a new approach for a sensitive diagnosis of tumor-progression markers. _Nucleic Acids Research._ **17**, 20, 8093-8100.
4. Jeanpierre, M. (1987) A rapid method for purification of DNA from blood. _Nucleic Acids Research,_ **15**, 9611.
5. Verlan-de Vries, M., Bogaard, M. E., Elst, H., Boom, J. H., Eb, A. J., & Bos, J. L. (1986) A dot-blot screening procedure for mutated _ras_ oncogenes using synthetic oligodeoxy-nucleotides. _Gene,_ **50**, 313-320.

6. Haliassos, A., Chomel, J.C., Tesson, L., Baudis, M., Kruh, J., Kaplan, J.C. and Kitzis, A. (1989) Artificial modification of enzymatically amplified DNA for the detection of point mutations. *Nucleic Acids Research,* **17**, 3606.

7. Southern, E. M. (1975) Detection of specific sequences among DNA fragments separated by gel electrophoresis. *J. Mol. Biol.* **98**, 503-517.

8. Cannon-Albright L. A., Skolnick M. H., Bishop D. T., Lee R. G., Burt R.W. (1988) Common inheritance of susceptibility to colonic adenomatous polyps and associated colorectal cancers. *New Engl. J. of Med.,* **319**, 533-537.

9. Farr, C.J., Marchall, C.J., Easty, D.J., Wright, N.A., Powell, S.C. and Paraskeva, C.(1988) A study of <u>ras</u> gene mutations in colonic adenomas from familial polyposis coli patients. *Oncogene,* **3**, 673-678.

APPENDIX

Cell lysis solution:

> 0,1M NaOH
> 2M NaCl
> 0,1% sodium dodecylsulfate

PCR mixture:

> 400ng of each primer
> 1mM of each dNTP
> 67mM Tris HCl pH 8,8
> 6,7μm EDTA
> 6,7mM $MgCl_2$
> 10mM β-mercaptoethanol
> 16,6mM ammonium sulfate
> 10% DMSO
> 1μg-100ng DNA to amplify
> 1 unit of Taq polymerase
> H_2O to 100 μl

Prehybridization solution:

> 3,0M tetramethylammonium Cl
> 50mM Tris HCl pH 8,0
> 2mM EDTA
> 100μg/ml sonicated, denatured herring sperm DNA
> 0,1% sodium dodecylsulfate
> 5x Denhart's solution

First and second wash solutions:

> 20mM sodium phosphate pH 7,0
> 0,36M NaCl
> 2mM EDTA
> 0,1% sodium dodecylsulfate

MOLECULAR EVOLUTION OF THE Ha-RAS-1 ONCOGENE: RELATIONSHIP BETWEEN DNA METHYLATION, FREQUENCY OF CpG DINUCLEOTIDES AND BINDING TO THE Sp1 TRANSACTING FACTOR

Roberto Gambari, Stefano Volinia, Giordana Feriotto,
Rafaella Barbieri, Roberta Piva and Claudio Nastruzzi

Istituto di Chimica Biologica, Università di Ferrara, Italy

One of the molecular events involved in the regulation of the expression of cellular oncogenes is the interaction between transcriptional factors and target DNA sequences present in the 5' genomic portion[1].
The recent cloning and sequencing of human oncogenes allows statistical analyses, with the aim to identify common putative regulatory boxes in their untrascribed regions and to study the molecular evolution of regulatory DNA sequences[1].
With respect to this issue, we review here some studies on the molecular organization of the human Ha-ras-1 oncogene.

DNA Methylation of the Ha-ras-1 oncogene

The Ha-ras-1 oncogene could be considered a typical example of eukaryotic gene displaying CpG rich portions in its 5' end.
Fig.1A shows that the distribution of CpG dinucleotides within the human Ha-ras-1 oncogene sequence is indeed not random, but exhibits a preferential accumulation of CpG in the 5' portion. Accordingly, the observed/expected (obs/exp) frequency ratio of CpG dinucleotides approaches the unity only in the 5' region (Fig.1B). On the contrary, obs/exp frequency ratios of TpG and CpA dinucleotides are higher than 1 throughout the Ha-ras-1 gene (Fig.1B).
Since methylated CpG dinucleotides tend to change to TpG or CpA following deamination of the methylcytosine[2], these data suggest that the molecular evolution of the Ha-ras-1 oncogene could be associated with methylation-dependent CG->TG and CG->CA mutations, with the exception of the 5' region. DNA methylation of the Ha-ras-1 proto-oncogene is, in this respect, an interesting topic. Accordingly, many studies on the pattern of DNA methylation of the Ha-ras-1 oncogene in human tumor cell lines have been

recently performed by different research groups[3-5] aiming to identify the role of methylation of this gene.

On one hand, it has been suggested that hypermethylated state of the Ha-ras-1 oncogene could inhibit its expression. "In vitro" methylation of the Ha-ras-1 oncogene indeed is associated with a sharp inhibition of transcription[4]. On the other hand, our group has demonstrated that the Ha-ras-1 oncogene exhibits a very peculiar pattern of DNA methylation in normal and tumor cell lines and tissues[6,7] (Fig.2), constantly displaying hypomethylated state of the 5' portion.

Taken together, these results suggest that, although the 5' unmethylated state of the Ha-ras-1 oncogene could be required for transcriptional activity, it is not sufficient[7].

Our hypothesis is that unmethylated state of the Ha-ras-1 gene is important for preserving CpG dinucleotides from CG -> TG or CA mutations.

Since the preservation of CpG dinucleotides could be related (a) to codon restrictions in coding DNA sequences[2] and (b) to an evolutionary pressure to maintain the capacity of non coding sequences to bind nuclear transcriptional factors, we analysed the sequences and the distribution of quintuplets containing CpG within different regions of the human Ha-ras-1 oncogene.

Distribution of the CpG containing 5-tuples in the human Ha-ras-1 oncogene

We have recently published computational methods to determine the frequency distribution of short sequences 3 to 10 nucleotides in length (k-tuples)[9]. When the number and the location of the 244 5-tuples containing CpG was considered in the human Ha-ras-1 oncogene (GenBank entry

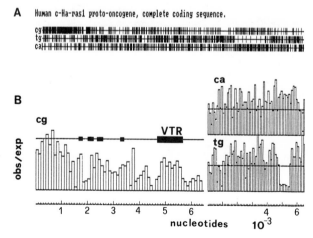

FIG.1. **A:** *Display of CpG, TpG and CpA dinucleotides throughout the human Ha-ras-1 oncogene (GenBank entry HUMRASH)[8].* **B:** *obs/exp frequency ratios of CpG (left), TpG and CpA (right) dinucleotides.*

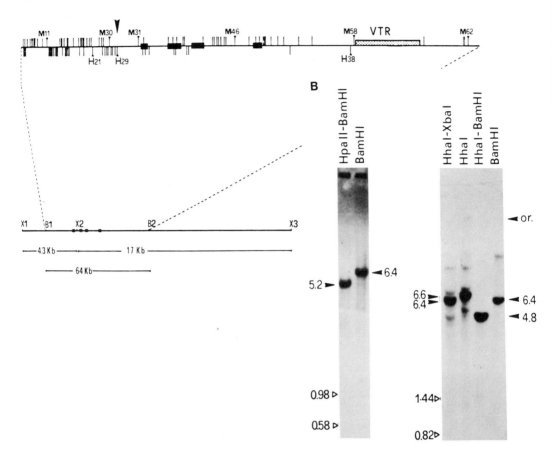

FIG.2. **A:** *Display of CCGG (M1-M62) and GCGC (H1-H38) sequences throughout the BamHI DNA fragment containing the human Ha-ras-1 oncogene.*

■ *= exons;* ▼ *= putative promoter region;* ☐ *= VTR segment; X1, X2, X3 = Xba I sites; B1, B2 = Bam HI sites.*

B: *Methylation pattern of CCGG and GCGC sites of the Ha-ras-1 oncogene in the human erythroleukemia K562 cell line[7]. DNA was digested with the indicated restriction enzymes, Southern blotted and hybridized with the T24-C3 Ha-ras-1 probe[7]. Black arrows indicate the sizes (Kb) of the Ha-ras-1 specific fragments. Open arrows (▷) indicate the position of DNA fragments which would be generated by the methyl-sensitive HpaII and HhaI restriction enzymes in the presence of unmethylated M57-M60 or H36-H38 sites.*

These results suggest that unmethylated state of the human Ha-ras-1 oncogene is restricted to its 5' portion. For experimental details, see ref.7.

HUMRASH) it was observed that these 5-tuples are present with sharply different frequencies (Fig.3). In Table I the frequencies of the 19 most represented CpG 5-tuples are listed, suggesting that the majority of them belong to the Sp1 consensus 5'-GGGGCGGGGC-3'. The data obtained by this computer-assisted analysis suggest that CpG quintuplets belonging to the Sp1 consensus sequence retain low probability to mutate.

In addition, when we compared the distribution of one of these CpG 5-tuples, GGCGG, with the distribution of the "mutated" GGTGG (CG->TG) and GGCAG (CG->CA) (Fig.4) a very striking behavior was observed. The distribution of these 5-tuples is sharply different, being the GGCGG sequence selectively present in the 5' portion of the gene. All the other CpG containing 5-tuples belonging to the Sp1 consensus exhibit the same behavior when their location is compared to their homologues containing TpG or CpA (data not shown).

Our hypothesis is that the methylation-related CG->TG and CG->CA mutations occurring during the molecular evolution of this gene have been generated mainly in the body of the gene and in its 3' portion, while unmethylated state of the 5' portion (Fig.2) prevents such mutational events.

Taken together, the data suggest that the molecular evolution of the human Ha-ras-1 oncogene leads to a selection of Sp1 binding motifs in the 5' portion. The frequently present ACGCC, GACGC and GGACG 5-tuples belong to the VTR sequence of the human Ha-ras-1 gene.

Relationship between CpG rich islands
and Sp1 binding sites: an hypothesis

In order to determine whether the methylation-dependent CG->TG or CA mutations lead to a selection in eukaryotic genes of other signals in addition to the Sp1 motif, we have analysed a set of 569 mammalian sequences (1.6 Mbases) from the BBN Laboratories GenBank, release 55.0.

In Fig.5 the frequency distribution of CpG containing 5-tuples is shown. We divided the 5-tuples in groups exhibiting frequency ratios between 0-0.1, 0.11-0.2, 0.21-0.3 ... 1.01-1.1. As it was expected, the majority of CpG containing 5-tuples in mammalian genes are rare. Among the 244 5-tuples, 181 display a obs/exp frequency ratio lower than 0.5. The distribution of the frequency ratios appears to be bimodal, and a peak of frequent CpG containing 5-tuples is evident.

Table II shows the sequences of these frequent 5-tuples. As it is evident, the sequences potentially belonging to Sp1 binding sites are found among the most frequent 5-tuples. All of the CpG 5-tuples belonging to the Sp1 consensus are present in Table II, therefore suggesting that the Sp1 binding activity played a role in the non random selection of specific CpG sequences in mammalian genes.

In addition, we like to point out that other 5-tuples present in Table I belong to degenerated Sp1 binding sites, such as the sequences CGTGG, which belongs to the Sp1 binding site of the HIV-1 LTR and the promoter of E2A of Adenovirus.

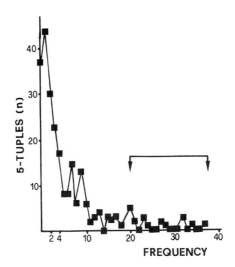

FIG. 3. *Frequency distribution of the 244 CpG containing 5-tuples within the human Ha-ras-1 oncogene. 37 5-tuples are not present in HUMRASH GenBank entry. 19 CpG 5-tuples are present more than 20 times (arrows). The sequences of these 5-tuples are listed in Table I.*

Table I. *Sequences of CpG 5-tuples present more than 20 times within the human Ha-ras-1 oncogene.*

5-tuple	frequency	5-tuple	frequency	5-tuple	frequency
ACGCC (+)	20	GCCGG	21	CGCCA (+)	27
GCGGG (*)	20	GACGC (+)	27	CGCCC (*)	37
GCCGC	20	GGACG (+)	34	CCGGG	32
GGCCG	20	CTCGC	20	CCGCC (*)	23
GGCGG (*)	20	CGGGC	23	CCCGG	28
GGGCG (*)	21	CGGGG (*)	24	CCCGC (*)	23
				CCCCG (*)	32

(+) *CpG 5-tuples belonging to the VTR;*
(*) *CpG 5-tuples belonging to the 5'-GGGGCGGGGC-3' Sp1 consensus motif.*

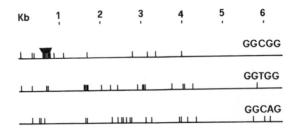

FIG.4. *Distribution of GGCGG, GGTGG and GGCAG 5-tuples within the human Ha-ras-1 oncogene.*

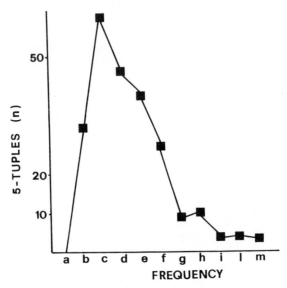

FIG.5. *Distribution of the obs/exp frequency ratios of the 244 CpG containing 5-tuples within 569 mammalian (1.6 Megabases) genomic sequences from GenBank. In this analysis we divided the CpG 5-tuples in groups exhibiting a obs/exp frequency ratio between 0-0.1(a), 0.11-0.2(b), 0.21-0.3(c), 0.31-0.4(d), 0.41-0.5(e), 0.51-0.6(f), 0.61-0.7(g), 0.71-0.8(h), 0.81-0.9(i), 0.91-1(l), 1.01-1.1(m).*
The sequences of the most frequent 5-tuples are listed in Table II.

Table II. *Sequences of the most frequent 5-tuples in 569 mammalian genomic sequences from GenBank.*

5-tuple	obs/exp	5-tuple	obs/exp	5-tuple	obs/exp
GGGCG (*)	0.888	GGCGG (*)	0.982	GGCCG	0.818
GCGGG (*)	0.948	GCGGC	0.804	GCCGC	0.826
GCCCG	0.849	CGTGG	0.814	CGGGG (*)	1.032
CGCTG	0.800	CGCCC (*)	0.914	CCGGG	0.997
CCGCC (*)	1.050	CCCGG	1.014	CCCGC (*)	0.957
CCCCG (*)	1.034				

(*) *CpG 5-tuples belonging to the 5'-GGGGCGGGGC-3' Sp1 consensus motif.*

DNA-binding compounds inhibiting the interaction of nuclear factor(s) with Sp1 motif: effects on Ha-ras-1 expression, cell growth and tumorigenicity of Ha-ras-1 transfected cell lines

The results shown in Table I and Table II suggest that the Sp1 binding site is required for biological functions of eukaryotic cells. Therefore, induced alterations of the interactions between nuclear factors and Sp1 motifs of eukaryotic genes could lead to inhibitory effects on cell growth.
We have recently reported the DNA-binding activity of aromatic polyamidines containing four benzamidine rings[10]. These drugs inhibit the interaction between a variety of nuclear factors and their target sequences. Despite these compounds are likely to retain a selectivity for AT/TA rich DNA segments, when they are administred at high concentrations inhibit also the binding of nuclear factors to double stranded oligonucleotides containing the Sp1 motif (Fig.6A). Interestingly, TAPP treatment of a Chinese Hamster cell line (FH06T1-1)[11] transfected with the activated human Ha-ras-1 oncogene leads to inhibition of the expression of the Ha-ras-1 oncogene (Fig.6B). This effect is obtained at concentrations that do not affect cell growth and cell cycle parameters[12].

Fig.7 shows that TAPP retains also inhibitory activity on the proliferation of FH06T1-1 cell xenografted into nude mice.
This inhibition of "in vivo" growth of FH06T1-1 cells was achieved either by intraperitoneal injection of TAPP or by local treatment with lecithin gels containing TAPP.
Taken together, these results suggest that DNA-binding drugs interfering with the binding of nuclear factor(s) to the Sp1 consensus motif could exhibit inhibitory effects on the expression of Ha-ras-1 oncogene, associated with both "in vitro" and "in vivo" antitumor activity.

Conclusions

The majority of CpG-containing 5-tuples (181 5-tuples) are rare (obs/exp frequency ratio <0.5) in eukaryotic genes. On the contrary, some of them

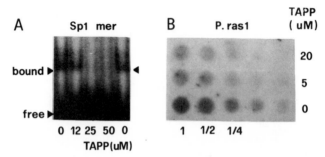

FIG.6. **A:** *Effects of the aromatic polyamidine TAPP on the binding of nuclear factors from a melanoma cell line to a synthetic 5' end labelled[13] Sp1 oligonucleotide. Gel retardation was performed as described elsewhere[13]. The intensity of the retarded band decreases when 12-50 uM TAPP is added to the reaction mixture, suggesting that this compound can inhibit interactions of nuclear proteins to Sp1 target DNA sequence.*
B: *Effects of TAPP on accumulation of Ha-ras-1 mRNA in FH06T1-1 cells transfected with the activated human Ha-ras-1 oncogene[11].Cytoplasmic dot-blot hybridization was performed as elsewhere described and suggests that TAPP inhibits accumulation of Ha-ras-1 mRNA[12].*

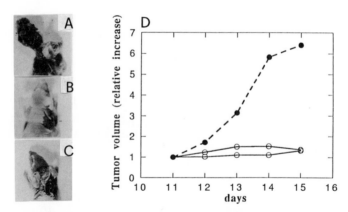

FIG.7. *"In vivo" antitumor activity of TAPP on FH06T1-1 cells xenografted into nude mice.* **A:** *untreated nude mice exhibiting growth of FH06T1-1 cells;* **B,C:** *nude mice xenografted with FH06T1-1 cells and treated with TAPP either by intraperitoneal injections (B) or by transdermal delivery using lecithin gels (C).*
D: *Kinetics of increase of the tumor mass generated by injections of FH06T1-1 cells in untreated (o) or transdermally treated with TAPP (o) nude mice. Details of this experiment are reported elsewhere[14]. Arrow indicates the onset of TAPP treatment.*

(14 5-tuples) are unexpectedly frequent (obs/exp frequency ratio = 0.81-1.1) and most of them belong to the 5'-GGGGCGGGGC-3' consensus Sp1 motif. Accordingly, all the CpG 5-tuples belonging to the Sp1 consensus motif are present among the most frequent CpG 5-tuples of the Ha-ras-1 oncogene. Chemical compounds exhibiting inhibitory effects on the interactions between nuclear factors and Sp1 motif were studied. It was found that they display also inhibitory activity on the expression of the Ha-ras-1 oncogene and on the "in vitro" and "in vivo" growth of Ha-ras-1 transfected tumorigenic cells.

We suggest that analysis of the effects of DNA-binding drugs on the interactions between nuclear factors and Sp1 target motifs could be a useful approach to identify new anti-tumor and anti-viral drugs exhibiting sequence selectivity.

Acknowledgements. *Work supported by AIRC and by Istituto Superiore di Sanità (AIDS-1990).*

References

1. B. Lewin, Oncogenic conversion by regulatory changes in transcription factors, Cell 64: 303 (1991).
2. L.P. Adams, DNA methylation, Biochem. J. 265: 309 (1990).
3. P. Feimberg and B. Vogelstein, Hypomethylation of ras oncogenes in primary human cancers, Biochem.Biophys.Res.Comm. 111: 47 (1983).
4. L.A. Chandler, Y.A. DeClerck, E. Bogenmann and P.A. Jones, Patterns of DNA methylation and gene expression in human tumor cell lines, Cancer Research 46: 2944 (1986).
5. S.E. Goelz, B. Vogelstein, S.R. Hamilton and A.P. Feimberg, Hypomethylation of DNA from benign and malignant human colon neoplasms, Science 228: 187 (1985).
6. R. Barbieri, R. Piva, D. Buzzoni, S. Volinia and R. Gambari, Clustering of unmethylated CCGG and GCGC sequences in the 5' region of the Ha-ras-1 oncogene of human leukemic K562 cells. Biochem.Biophys.Res.Comm. 145: 96 (1987).
7. R. Barbieri, C. Mischiati, R. Piva, C. Nastruzzi, P. Giacomini, P.G. Natali and R. Gambari, DNA methylation of the Ha-ras-1 oncogene in neoplastic cells, Anticancer Research 9: 1787 (1989).
8. H.S. Bilofsky, C. Burks, J.W. Fickett, W.B. Goad, F.I. Lewitter, W.P. Rindone, C.D. Swindell and C.S. Tung, The GenBank Gene Sequence Data Bank, Nucleic Acids Res. 14:1 (1986).
9. S. Volinia, F. Bernardi, R. Gambari and I. Barrai, Co-localization of rare oligonucleotides and regulatory elements in mammalian upstream gene regions, J.Mol.Biol. 203: 385 (1988).
10. R. Gambari, V. Chiorboli, G. Feriotto and C. Nastruzzi, TAPP (tetra-p-amidinophenoxyneopentane) inhibits the binding of nuclear factors to target DNA sequences, Int.J.Pharmaceutics, in press.
11. D.A. Spandidos and N.M. Wilkie, Malignant transformation of early passage rodent cells by a single mutated human oncogene, Nature 310: 469 (1984).

12. C.Nastruzzi, R. Barbieri, R. Ferroni, M. Guarneri and R. Gambari: Inhibition of "in vitro" tumor cell growth by aromatic poly-amidines exhibiting antiproteinase activity. Clinical and Experimental Metastasis, 7: 25 (1988).

13. R. Barbieri, P. Giacomini, S. Volinia, C. Nastruzzi, A.M. Mileo, U. Ferrini, M. Soria, I. Barrai, P.G. Natali and R. Gambari, Human HLA-DRα gene: a rare oligonucleotide (GTATA) identifies an upstream sequence required for nuclear protein binding, FEBS Letters 268: 51 (1990).

14. A. Bartolazzi, R. Barbieri, C. Nastruzzi, P.G. Natali and R. Gambari, Antitumor activity of tetra-p-amidinophenoxy neopentane in a mouse model of human melanoma, In vivo, 3: 383 (1989).

Ha-RAS-1 TRANSFECTED CELLS:
INDUCTION TO DIFFERENTIATION
BY NATURAL AND SYNTHETIC RETINOIDS

Claudio Nastruzzi[1], Demetrios Spandidos[2], Giordana Feriotto[3],
Stefano Manfredini[1], Daniele Simoni[1] and Roberto Gambari[3]

[1] Dipartimento di Scienze Farmaceutiche, Università di Ferrara, Italy
[2] National Hellenic Research Foundation, Athens, Greece
[3] Istituto di Chimica Biologica, Università di Ferrara, Italy

Summary: The in vitro activity of natural retinoids (retinoic acid and retinol) and new classes of synthetic analogues of retinoids on differentiation of chinese hamster fibroblasts transfected with either the normal (FH06N1-1) or the activated (FH06T1-1) Ha-ras-1 oncogene is reported. The most active synthetic retinoids were found to induce adipogenic conversion of both FH06N1-1 and FH06T1-1 cell lines in a range comprised between 10 to 50 µM, concentrations that are 2 to 10 times lower than those required in the case of natural retinoids.

1. Introduction

Malignant transformation is associated with high expression levels of cellular oncogenes, including those belonging to the ras oncogene superfamily[1,2]. Cell lines transformed with either the normal or the activated human Ha-ras-1 gene are therefore a useful experimental model to determine the ability of a variety of compounds, including natural and synthetic retinoids, to induce differentiated cellular functions. This information could be of great interest, since biological response modifiers able to induce at high efficiency terminal differentiation are currently proposed in the experimental therapy of cancer[3].

With respect to this issue, the biological activities of natural retinoids and of their synthetic analogues have stimulated a large number of chemical, biochemical and pharmacological investigations on the mechanisms of action and therapeutical possibilities of this class of compounds[4].

This interest is largely due to the ability of retinoids to induce differentiated functions in normal and neoplastic cells of different origin and histotype[5-7].

For instance, retinoids induce (a) the granulocyte-like and macrophage differentiation of the human promyelocytic leukemia cell line HL60[5], (b) the adipogenic conversion of fibroblasts[6] and (c) the differentiation of teratocarcinoma cells[7]. Because of their ability to control the epithelial cell differentiation, retinoids have been utilized for the treatment of different skin diseases[8], and the experimental treatment of cancer[9].

However, there are several limitations and disadvantages related to the clinical use of retinoids, such as the low stability shown by these compounds, their high toxicity and their poor solubility in aqueous solvents[8].

In order to avoid these disadvantages, a number of retinoid analogues have been synthetised with molecular structures mimicking vitamin A and natural related compounds[10]. In recent reports we have previously published about the synthesis and the antiproliferative activity of new retinoids incorporating an isoxazole or isoxazoline ring in the tetraene side chain or with an isoxazole ring substituting the cyclohexenyl ring of the natural retinoids[11, 12].

In this paper we report the differentiating activity of these new synthetic retinoids (their chemical structures are reported in Fig.1) on two cell lines obtained after transfection of chinese hamster lung fibroblasts with the Ha-ras-1 proto-oncogene or its activated counterpart.

2. Materials and Methods

2.1 Chemicals

The natural retinoids were purchased from Fluka (Switzerland).

Synthesis, purification procedures, yields, analytical and spectroscopical data of the synthetic retinoids have been reported elsewhere[11, 12].

2.2 Cell lines and culture conditions

Standard conditions for cell growth were α-medium (GIBCO, Grand Island, NY), 50 mg/l streptomycin, 300 mg/l penicillin, supplemented with 10-15% fetal calf serum (Flow Laboratories, Inc., McLean, VA) in 5% CO_2, 80% humidity.

FH06N1-1 and FH06T1-1 cells[13] were plated at initial density of 2-5 x 10^4 cells per 35-mm tissue culture dish. Before determinations of cell growth the unattached cells were usually rinsed from the dishes with phosphate-buffered saline.

The adherent cells were then removed by 0.05% trypsin/0.021% EDTA and counted electronically with a Model ZF Coulter Counter (Coulter Electronics, Inc., Hialeah, FL). Assays were carried out in triplicate and usually counts differed by <7%.

2.3. Adipogenic conversion

Fat droplets indicating adipocytic-like phenotype were detected by Oil red O staining as reported elsewhere[14].

3. Results

3.1. Ha-ras-1 transfected cells: differential ability to perform adipogenic conversion

In order to obtain informations on the influence of the Ha-ras-1 oncogene products on the capacity of cellular systems to undergo terminal

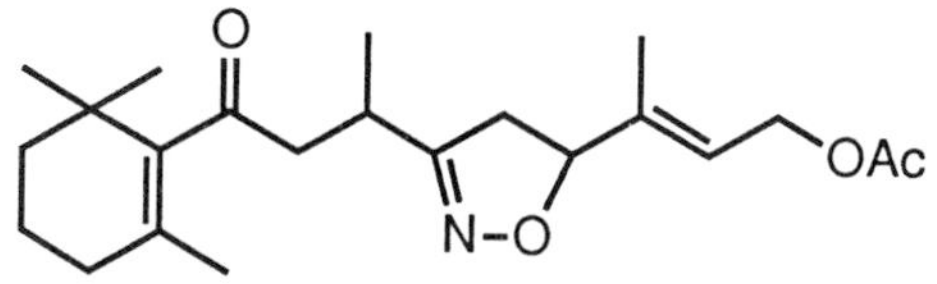

RET-VIa

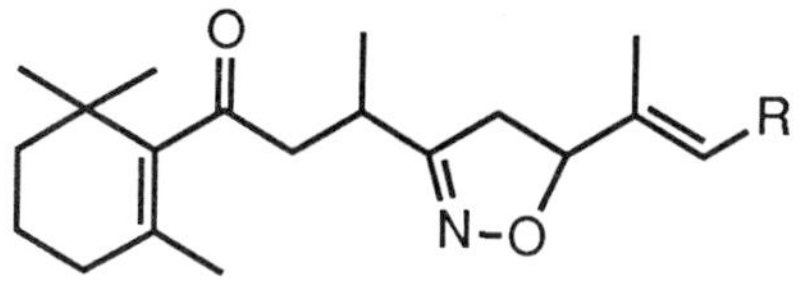

R = CH$_2$OH   RET-VIIa
R = COOH   RET-VIIc

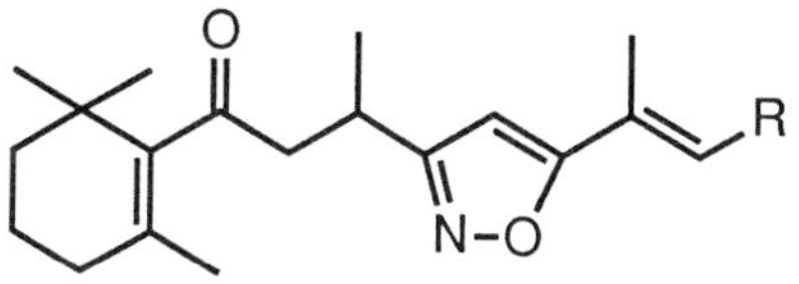

R = CH$_2$OH   RET-IXa
R = COOH   RET-IXc

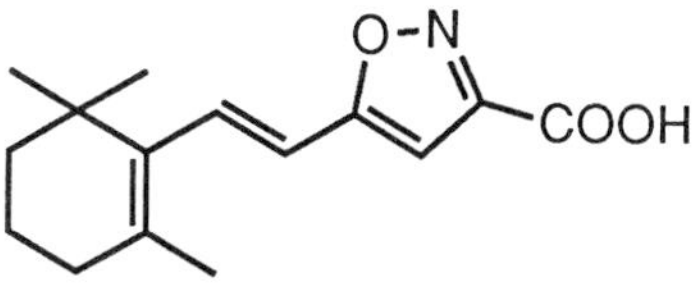

RET-XIIb

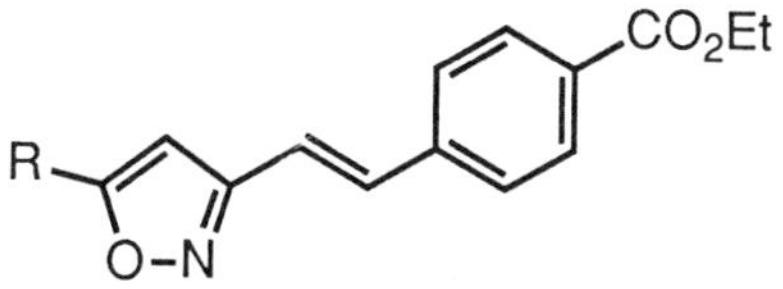

R = CH$_3$ RET-6a
R = nC$_6$H$_{13}$ RET-6b

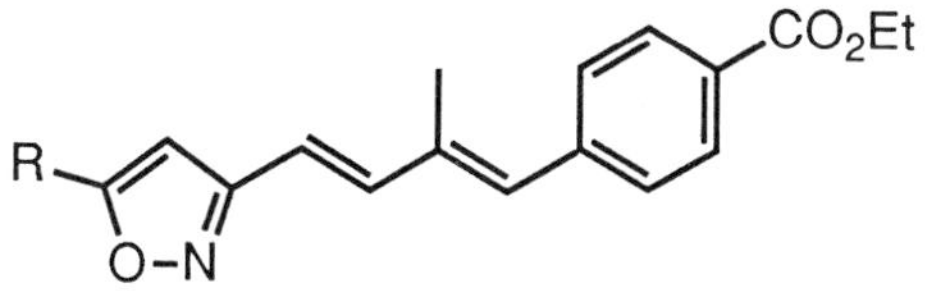

R = CH$_3$ RET-8a
R = nC$_6$H$_{13}$ RET-8b

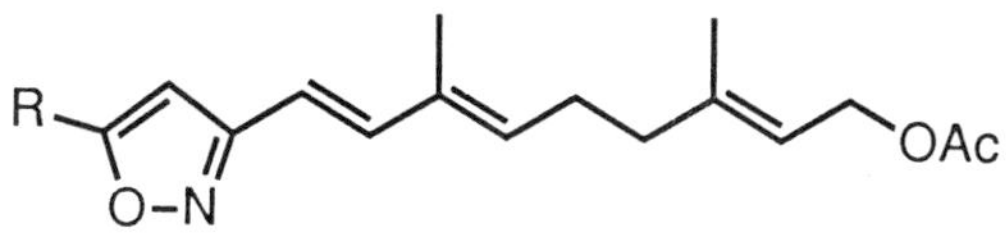

R = CH$_3$ RET-9a
R = nC$_6$H$_{13}$ RET-9b

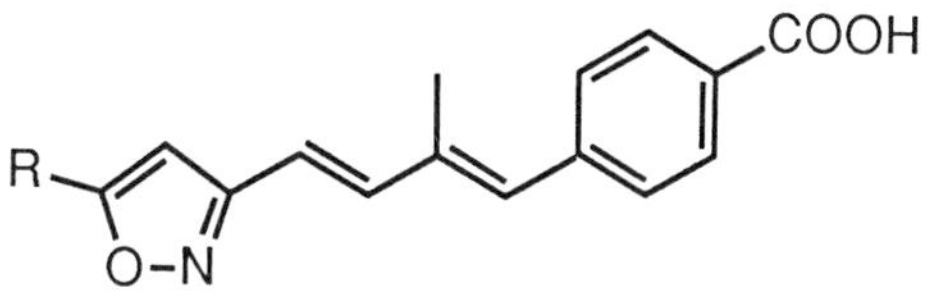

R = CH$_3$ RET-11a
R = nC$_6$H$_{13}$ RET-11b

Fig.1. *Chemical structures of the synthetic retinoids tested*

differentiation, we utilized two cell lines, FH06N1-1 and FH06T1-1, obtained after transfection of normal chinese hamster lung fibroblasts with plasmids carrying either the Ha-ras-1 proto-oncogene (pH06N1-1) or the activated T24 Ha-ras-1 oncogene (pH06T1-1) sequences, under the trascriptional control of the Moloney Sarcoma Virus and SV40 Virus enhancers. In Fig.2 and Table I are reported some of the morphological, cytological and biochemical differences between the two cell lines. With respect to their ability to be induced to differentiate to adipocyte-like cells it is to be underlined that after treatment with 5-azacytidine (a well known differentiating agent in other cellular systems), the FH06N1-1 cells but not the FH06T1-1 cells can be induced to adipogenic conversion (Fig.3A). In Fig.3B is reported a typical example of differentiated FH06N1-1 cells showing into the cytoplasms fat droplet inclusions, that can be specifically stained with Oil red O.

3.2. Retinoic acid induced differentiation of cells transfected with the activated Ha-ras-1 oncogene

In Fig.4 are reported experiments showing that when retinoic acid (RA) is added to FH06N1-1 and FH06T1-1 cell cultures in a concentration range between 12 and 100 μM also the cells transfected with the activated ras oncogene can be efficiently induced to differentiate to adipocyte-like cells. The microphotographs of FH06T1-1 cells after 7 days of RA treatment are shown in Fig.4A. The relationships between activity and concentrations are reported in Fig.4B.

Table I. *Biological features of FH06N1-1 and FH06T1-1 cell lines.*

	FH06N1-1	FH06T1-1	reference
Oncogenic sequence	Human Ha-ras-1 proto-oncogene	Human activated T24-Ha-ras-1	14
Growth in soft agar	NO	YES	14
Contact inhibition	YES	NO	14
Matrix Driven Translocation	NO	YES	16
Tumorigenicity	NO	YES	14
Adipogenic conversion induced by 5-azacytidine	YES	NO	15

3.3 Effects of synthetic retinoids on cell growth and differentiation of FH06T1-1 tumor cells

In order to avoid problems of stability and toxicity related to the use of natural retinoids, we have synthetized different series of retinoid analogues and tested their ability to inhibit tumor cell proliferation and to induce cellular differentiation of the FH06T1-1 cell line.

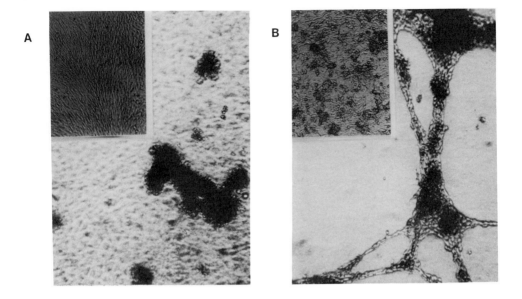

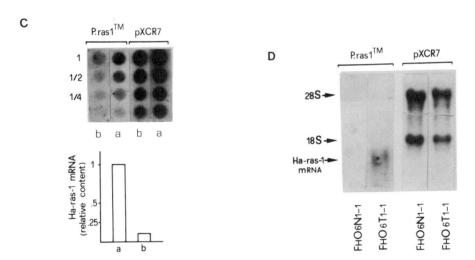

Fig.2. A, B: *Morphology of FH06N1-1 (A) and FH06T1-1 (B) cell lines cultured on plastic support (insert) or on the "in vitro" reconstituted basement membrane Matrigel* [15].

C, D: *Dot-blot (C) and northern (D) analyses of the Ha-ras-1 mRNA accumulation in FH06N1-1 (b) and FH06T1-1 (a) cells. Bars represent the relative Ha-ras-1/rRNA hybridization ratios. Hybridizations were carried out with P.ras-1*[TM] *and pXCR7 probes, specific for Ha-ras-1 and rRNA (28S and 18S) sequences, respectively* [16].

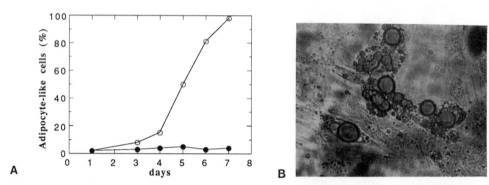

Fig.3 A: *Kinetics of the increase of the proportion of adipocyte-like FH06N1-1 (open circles) and FH06T1-1 (filled circles) cells after treatment with 25 μM 5-azacytidine. Results represent the proportion of Oil-red O positive cells[14].*
B: *Adipocyte-like FH06N1-1 cells stained with Oil red O. Cells were cultured for 6 days in the presence of 25 μM 5-azacytidine.*

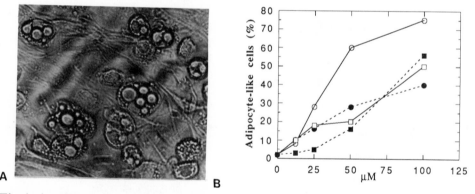

Fig.4 A: *Adipocyte-like FH06T1-1 cells stained with Oil red O. Cells were cultured for 6 days in the presence of 50 μM retinoic acid.*
B: *Effects of retinoic acid (solid lines) and retinol (dotted lines) on the adipogenic conversion of FH06N1-1 (squares) and FH06T1-1 (circles). Determinations (performed after 5 days of cell growth) are reported as percentage of Oil red O positive cells.*

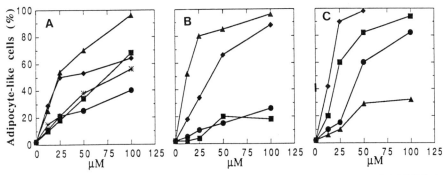

Fig.5. *Effects of synthetic retinoids on the adipogenic conversion of FH06T1-1 cells. Determinations were performed after 5 days of cell growth.*

A : ● = RET-VI;   ■ = RET-VIIa;   ♦ = RET-VIIc;   ▲ = RET-IXb;   * = RET-XIIb.

B : ● = RET-6a;   ■ = RET-8a;   ♦ = RET-9a;   ▲ = RET-11a.

C : ● = RET-6b;   ■ = RET-8b;   ♦ = RET-9b;   ▲ = RET-11b.

In Fig.5 are reported the data relative to induction to differentiation, suggesting that most of the synthetic compounds were able to induce adipogenic conversion of FH06T1-1 cells. Some of the most active compounds (for istance RET-11a, RET-9a and RET-IXb) can induce differentiation with low toxic effects, while other compounds (RET-6b and RET-9b) exhibit strong antiproliferative activity. This latter property of synthetic retinoids is summarized in Table II, where are reported the concentrations of retinoids able to cause 50% inhibition of cell growth.

Table II. *Effects of retinoids on the proliferation of FH06T1-1 cells.*

Compds.	ID_{50}	Compds.	ID_{50}	Compds.	ID_{50}	Compds.	ID_{50}
Retinoic Acid	85	RET-VI	130	RET-6a	80	RET-6b	50
Retinol	55	RET-VIIa	75	RET-8a	150	RET-8b	68
		RET-VIIc	70	RET-9a	75	RET-9b	15
		RET-IXb	80	RET-11a	200	RET-11B	200
		RET-XIIb	60				

4. Discussion

New therapeutical approaches based on the administration of differentiation inducers has been proposed for cancer treatment[3-5]. These approaches, in contrast with the widely used chemotherapy and immunotherapy, require the induction to differentiation of tumor cells minimalizing cytotoxic effects and leading to a reduction of the general toxicity.

With the aim to find new low molecular weight inducers exhibiting strong differentiation-inducing activity, we report in this paper data on the effects of natural and synthetic retinoids on the adipogenic conversion of Ha-ras-1 transfected chinese hamster lung fibroblasts.

The results obtained suggest the use of Ha-ras-1 transfected cells as "in vitro" cellular model to identify active compounds likely to exhibit efficient "in vivo" antitumor activity.

Unlike 5-azacytidine, retinoids were found to induce adipocytic-like differentiation of cells transformed with the activated human Ha-ras-1 oncogene.

The data shown in Fig.5 and Table II indicate that not all the synthetic retinois exhibit the same activity on FH06T1-1 cells. Compounds RET-9b and RET-11a are the most active in inducing differentiation at concentrations 2-10 fold lower that those of retinoic acid.

In conclusion, we suggest that both natural and synthetic retinoids could be used as differentiation inducers in anticancer experimental therapy.

5. Acknowledgements

This paper was supported by AIRC, Regione Emilia-Romagna and Regione Veneto.

6. References

1. D.A. Spandidos, Mechanisms of carcinogenesis: the role of oncogenes, transcriptional enhancers and growth factors, Anticancer Res. 5:485 (1985).
2. G.M. Cooper, Cellular transforming genes, Science 271:801 (1982).
3. M. Hozumi, Fundamentals of chemotherapy of myeloid leukemia by induction of leukemia cell differentiation, Adv.Cancer Res. 38:121 (1983).
4. M.B. Sporn, A.B. Roberts and D.S. Goodman, "The retinoids," Vol 2, Academic Press, New York (1980).
5. S.J. Collins, The HL-60 promyelocytic leukemia cell line: proliferation, differentiation, and cellular oncogene expression, Blood 70:1233 (1987).
6. M.B. Sporn, G.H. Clamon, M.N. Dunlop, D.L. Newton, J.M. Smith and U. Saffiotti, Activity of vitamin A analogues in cell cultures of mouse epidermis and organ cultures of hamster trachea, Nature 253:47 (1975).
7. S. Strickland and V. Mahdavi, The induction of differentiation in teratocarcinoma cells by retinoic acid, Cell 15:393 (1978).
8. W. Bollag, Vitamin A and retinoids: from nutrition to pharmacotherapy in dermatology and oncology, Lancet 1:860 (1983).
9. J.S. Bertran, Structure-activity relationships among various retinoids and their ability to inhibit neoplastic transformation and to increase

cell adhesion in the C3H/10T1/2CL8 cell line, <u>Cancer Res</u> 40:3141 (1980).

10.   M.I. Dawson, P.D.Hobbs, K.A. Derdzinski, W.R. Chao, G. Frenking, G.H. Loew, A.M. Jetten, J.L. Napoli, J.B. Williams, B.P. Sani, J.J. Wille and L.J. Shiff, <u>J.Med.Chem.</u> 32:1504 (1989).

11.   P.G. Baraldi, M. Guarneri, S. Manfredini, D. Simoni, M.Aghazade Tabrizi, R. Barbieri, R.Gambari and C.Nastruzzi, Synthesis of some isoxazole analogues of retinoids: biological effects toward tumor cell lines, <u>Eur.J.Med.Chem.Chim.Ther.</u> 27:279 (1990).

12.   D. Simoni, S. Manfredini, M.Aghazade Tabrizi, R. Bazzanini, M. Guarneri, R. Ferroni, F. Traniello, C.Nastruzzi, G. Feriotto and R.Gambari, New isoxazole derivatives of retinoids: synthesis and activity on growth and differentiation of tumor cells, <u>Drug Design and Delivery</u> in press.

13.   D.A. Spandidos and N.M. Wilkie, Malignant transformation of early passage rodent cells by a single mutated human oncogene, <u>Nature</u> 310:469 (1984).

14.   R. Gambari and D.A. Spandidos, Chinese hamster lung cells transformed with the human Ha-ras-1 oncogene: 5-azacytidine mediated induction to adipogenic conversion, <u>Cell Biol.Int.Rep.</u> 10:173 (1986).

15.   C. Nastruzzi, R. Barbieri, R. Ferroni, M. Guarneri, D. Spandidos, D. R. Gambari, Tetra-amidines exhibiting anti-proteinase activity: effects on oriented migration and in vitro invasiveness of a chinese hamster cell line transfected with tha activated human T24-Ha-ras-1 oncogene, <u>Cancer Letters</u> 50:93 (1990).

16.   C. Nastruzzi, G. Feriotto, D. Spandidos, D. Anzanel, R. Ferroni, M. Guarneri, R. Barbieri and R. Gambari, Effects of benzamidine derivatives on Ha-ras-1 mRNA accumulation in a chinese hamster cell line transformed with the activated human T24 Ha-ras-1 oncogene, <u>Anticancer Res</u> 8:269 (1988).

THE THREE-DIMENSIONAL STRUCTURE OF P21 IN THE CATALYTICALLY ACTIVE

CONFORMATION AND ANALYSIS OF ONCOGENIC MUTANTS

Ute Krengel, Ilme Schlichting, Axel Scheidig,

Matthias Frech, Jacob John, Alfred Lautwein,

Fred Wittinghofer, Wolfgang Kabsch, and Emil F. Pai

Max-Planck-Institut für medizinische Forschung

Abteilung Biophysik

Jahnstraße 29

6900 Heidelberg, FRG

SUMMARY

The three-dimensional crystal structure of the catalytically active, GTP-analogue containing complex of H-ras encoded p21 (aa 1–166) has been determined at 1.35 Å resolution. It has the same topology as the G-binding domain of elongation factor Tu. The structure analysis revealed the binding sites of the nucleotide and of the essential cofactor Mg^{2+} in great detail and made it possible to propose a mechanism for GTP hydrolysis. In addition to the wild-type protein, the structures of several p21 mutants have been solved. While the overall structures of these proteins are not perturbed, there are small, but significant differences at the positions of the mutated amino acids. In the oncogenic mutants (Gly-12→Arg, Gly-12→Val, Gln-61→Leu, Gln-61→His), these mutations interfere with the proposed mechanism of catalysis and thus lead to a reduced rate of GTP hydrolysis.

INTRODUCTION

The prevalence of mutated forms of the ras oncogene in human tumours makes it an attractive target to study the function of the ras gene product p21 (Barbacid, 1987; Bos, 1989). These mutations

The Superfamily of ras-Related Genes
Edited by D.A. Spandidos, Plenum Press, New York, 1991

are usually point mutations in two amino acids of the p21 sequence. It may turn out that the malfunctioning of mutated p21 is involved in, if not responsible for, uncontrolled growth of certain human tumours. For understanding the role of p21 in these processes, and for designing therapeutical agents directed towards mutated p21, it is important to understand the structure of this protein and its transforming mutants in addition to the biochemical processes in which it is involved. A major focus of our present research is thus to investigate the structures of the p21 wild-type and several oncogenic p21 mutants.

RESULTS AND DISCUSSION

3D-structure of cellular p21

We have crystallized the complexes between p21 and GDP or slowly hydrolyzing analogues of GTP: GppNp, GppCp and GTPγS (Scherer et al., 1989). While the crystals of the GppNp- and GppCp-complexes are isomorphous and all belong to the trigonal space group $P3_221$, the p21·GTPγS crystals show a different morphology. Crystals of the GppNp- and GppCp-complexes crystallize readily, are stable both mechanically and in the X-ray beam, and diffract to very high resolution (at least 1.35 Å and 1.54 Å, respectively). The three-dimensional structure of the p21·GppNp complex has been solved using heavy atom isomorphous replacement techniques (Pai et al., 1989). The structure is very similar to the 1.54 Å structure of the p21·GppCp complex, which has been solved recently (U. Krengel, 1991), and is also similar to the structure of p21·GppCp reported by Kim and coworkers (Milburn et al., 1990).

The secondary structural elements are shown in Fig.1, together with the position of some amino acids that are important for GTP binding, for interaction with the GTPase activating protein GAP, and for oncogenic activation. It has been shown by mutational analysis that residues 32-40 are involved in the interaction with GAP. They are located in loop L2 and in the second ß-sheet. These residues are highly exposed to the solvent and are well defined in the structure. Since loop L4 (residues 59-65) is also located on the outside of the molecule and near to L2, we have postulated that this loop is also involved in the interaction with GAP. In confirmation of this we have shown that the product of the Krev gene, which has the same sequence in loop L2, but a different one in L4, is not activated by GAP, but binds very tightly to it (Frech et al., 1990).

The details of the interaction of the guanine nucleotide with the protein have been described (Pai et al., 1990). The guanine base is

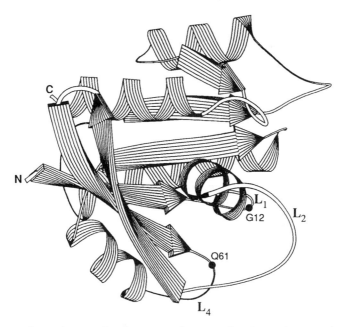

Fig.1. Schematic view of the topology of the three-dimensional structure of p21·GppNp. The two amino acids Gly-12 and Gln-61, which are often mutated in human tumours, are labelled. Loops L1, L2, and L4 are important for GTP and GAP binding and for the GTPase activity of p21.

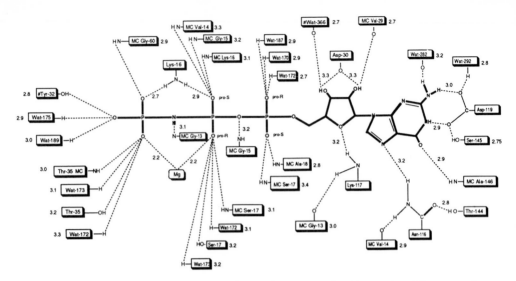

Fig.2. Scheme for the interactions between GppNp and p21 or water molecules. All dashed lines correspond to hydrogen bonding interactions (below 3.4 Å) between appropiate partners. The corresponding distances of the hydrogen bonds are given behind the residue, except for cases where one residue makes several bonds; then the distances are given adjacent to the dashed line. The values refer to the distances between the corresponding heteroatoms.

bound by interaction with the conserved elements NKXD (residues 116–119) and SAK (residues 145–147) (Fig.2 and Halliday, 1983; Dever et al., 1987; Gilman, 1987). The carboxylate group of Asp-119 interacts with the exocyclic amino group and the endocyclic nitrogen N1 of the base, and is therefore responsible for the discrimination between guanine- and adenine nucleotides, while the keto group at position 6 of the guanine base makes a hydrogen bond to the main chain nitrogen atom of Ala-146. The main function of Asn-116 seems to be to tie together the three elements that are involved in nucleotide binding: the phosphate binding loop [10]GXXXXGKS, the [116]NKXD- and the [145]SAK-motifs. Another element responsible for the tight binding of the guanine base is the hydrophobic interaction with the aromatic side chain of Phe-28 and the aliphatic side chain of Lys-117 on either side of the base. Phe-28 itself is held in place by another hydrophobic interaction between its aromatic ring and the aliphatic side chain of Lys-147, which is lying on top of the plane of the ring in a stretched-out conformation.

The ribose ring is in the 2'-endo conformation and the angle chi of
the N-glycosidic bond is -112^0, which is just at the border of an
anti conformation. As reported before (Pai et al., 1989 and 1990),
the 2'- and 3'-hydroxyl groups of the ribose are more or less
exposed to the solvent with only weak hydrogen bonds to the side
chain of Asp-30.

The phosphate binding site is characterized by a magnitude of
interactions. Each of the eight phosphate oxygens of GppNp has at
least two hydrogen bond donors or the Mg^{2+}-ion close enough for an
interaction. The hydrogen bond donors include the main chain NH
groups of residues 13-18, which create a strong electrostatic field
and act as a positive polarized pocket, in which the phosphates are
embedded. They also include Gly-60 and the hydroxyl groups of Ser-17
and Thr-35 as well as the phenolic hydroxyl group of Tyr-32 from a
neighboring p21-molecule, which contacts the γ-phosphate. Furthermore, the ε-amino group of Lys-16 is coordinated to the ß- and γ-

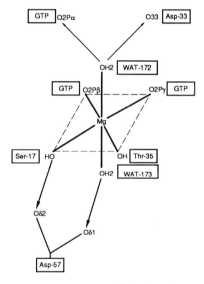

Fig.3. Schematic drawing of the Mg^{2+}-binding site showing the ligands
of the first coordination sphere of the metal ion and some of the
interactions of these ligands.

phosphate oxygens. There is also a hydrogen bond between the NH-
group of Gly-13 and the atom bridging the ß- and γ-phosphate, in
this case GppNp. Since this hydrogen bond should be stronger with a
bridging oxygen atom, it is likely that this interaction is
responsible for the lower affinity of GppNp as compared to GTP (see
also Schlichting et al., 1990).

As reported before (Pai et al., 1989 and 1990), Mg^{2+} in the three-dimensional structure of the p21·GppNp-complex is octahedrally coordinated to one oxygen of each the ß- and the γ-phosphate, to the side chain hydroxyl groups of Ser-17 and Thr-35, both of which are highly conserved in all nucleotide binding proteins, and to two water molecules. The high resolution structure proves that Asp-57, which is conserved as part of the DXXG motif in guanine-nucleotide binding proteins and serves as a direct ligand of Mg^{2+} in the p21·GDP-complex (Schlichting et al., 1990), is not in the first coordination sphere of Mg^{2+} in the triphosphate structure. Instead, it is hydrogen-bonded to one of the water molecules, Wat-173, which is directly liganded, as shown schematically in Figure 3. Asp-57 further binds to the side chain of Ser-17. The sixth ligand of the metal ion, Wat-172, is held in place by the interaction with the main chain oxygen of Asp-33 and the pro-R oxygen of the α-phosphate.

Mechanism of GTP hydrolysis

Guanine nucleotide binding proteins have an intrinsic GTPase activity, which is of central importance for their signal-switch function in the cell. In oncogenic p21 mutants GTP hydrolysis occurs at a reduced rate. This makes it very interesting to study the GTPase reaction in order to understand the mechanism of oncogenic activation.

It has been shown that the GTPase reaction, which (in chemical terms) is a phosphoryl transfer reaction of the γ-phosphate group from GTP to water, proceeds in a single step via a nucleophilic attack of a water molecule. Fortunately, as a result of the high resolution of our structure, we were able to see a water molecule in a proper position for nucleophilic attack. We have proposed a scheme for GTP hydrolysis, in which water is activated by the side chain of Gln-61, supported by Glu-63 and the side chain of Gln-63, through general base catalysis as shown schematically in Figure 4. GTP is probably activated by Mg^{2+} and Lys-16.

p21 mutants

How is the GTPase reaction disturbed in oncogenic p21? To answer this question, we crystallized four different mutants of p21 with the mutations Gly12→Arg, Gly12→Val, Gln-61→Leu and Gln61→His, all of which occur frequently in human tumours. The protein structures were solved by molecular replacement techniques.

The structural analysis turned out to be disappointing in some ways; no spectacular differences between wild-type and mutant proteins could be found that would allow "drug designers" to construct chemical inhibitors, which could potentially block oncogenic p21 specifically in human tumours. Instead, the differences were very subtle (see Fig.5A and Krengel et al., 1990). For the Gln-61 mutants they were very minor and led us to the conclusion that the mutants simply do not function properly because of their different side chains. The hydrophobic side chain of Leu-61 is obviously not able

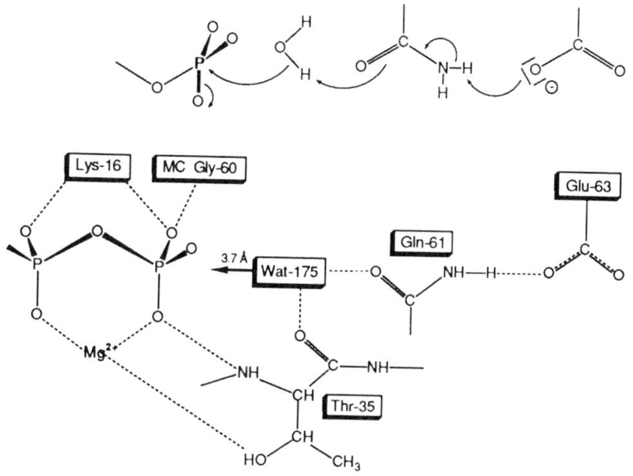

Fig.4. Schematic view of the GTPase reaction showing the nucleophilic water molecule attacking the γ-phosphate and the activation of this water molecule by the protein.

to perform the same function as Gln-61 of the cellular protein. For p21(Q61H), the case is not that simple. Usually, histidines are capable of functioning as reversible proton acceptors, and a histidine at the position of Gln-61 should therefore be a good activator of the nucleophilic water molecule Wat-175. However, for His-61 none of the three conformations identified correspond to the "catalytically active" conformation of Gln-61. Possibly, the "active" conformation is energetically unfavourable for a histidine residue at this position, or a rate limiting conformational change is less likely to occur in this mutant.

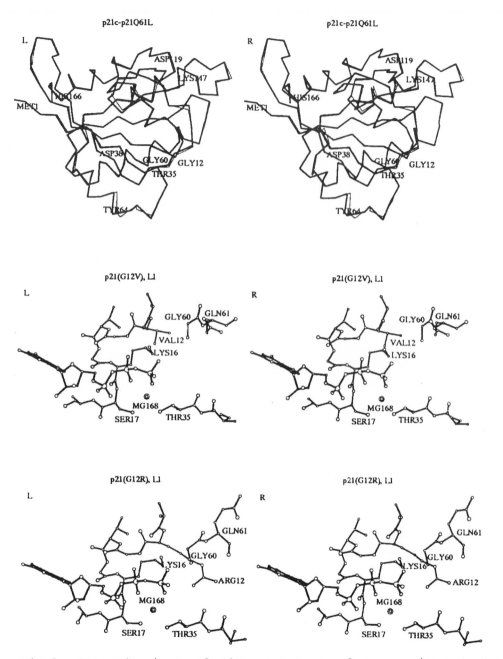

Fig.5. Stereodrawings of the structure of oncogenic mutants complexed to GppNp. A. Cα plot of p21(Q61L) (thick line) in comparison to the cellular protein (thin line). B. The phosphate binding region of p21(G12V). C. The phosphate binding region of p21(G12R) showing the Arg-12 side chain interacting with the γ-phosphate.

The most frequent ras mutations found in human tumours have a mutation of Gly-12, which is part of the phosphate binding loop that wraps around the ß- and γ-phosphates. It has also been shown by in vitro mutagenesis studies that the mutation of Gly-12 to any amino acid except proline renders the protein oncogenic (Seeburg et al., 1984). Before publication of the p21 structure, it was generally believed that Gly-12 had such unusual phi and psi angles that it cannot tolerate any side chain without destroying the geometry of the phosphate binding loop. However, as could be seen from the published structure, Gly-12 has normal dihedral angles and only Gly-10, -13 and -15 have unusual dihedral angles. In the position-12 mutants we observed that the phosphate binding loop is not (Arg-12, Pro-12), or only very slightly (Val-12) disturbed. The Val-12 mutant has a branched side chain in place of the "no side chain residue" glycine, which is bulky enough to push the catalytically important Gln-61 out of the way (see Fig.5B), whereas the large side chain of Arg-12 comes close to the γ-phosphate and makes a hydrogen bond to one of the phosphoryl oxygens (see Fig.5C). It thereby replaces the water molecule, which provides an obvious reason for the low intrinsic GTPase activity of this mutant ($1.1 \cdot 10^{-3}$ min^{-1}; John et al., 1988).

To summarize, all the oncogenic mutants analyzed somehow interfere with the catalytically active conformation of wild-type p21. This is not the case for the nontransforming p21 mutants D38E and G12P. While Glu-38 is located in the effector region and is therefore further away from the catalytic site, the Pro-12 side chain is so small that it does not interfere with loop L4, or the nucleophilic water molecule. In this mutant density is well defined for Wat-175 as well as for the "catalytically active" conformation of Gln-61 (U. Krengel, 1991).

GAP action

The GTPase activating protein GAP accelerates the GTPase rate of normal p21 approximately 10^5 fold (J. John, 1990). Since we have not crystallized the p21·GAP complex, we can only speculate on the way GAP accelerates the chemical cleavage of GTP. A model has been discussed, in which GAP supplies p21 with an element necessary for the GTPase reaction, possibly an arginine that is present in Gα proteins and in GAP in a similar sequence motif (McCormick, 1989). Another possibility would be a reaction scheme where GAP is responsible for bringing p21 into the proper conformation which is active in the enzymatic reaction. From our structural analysis, it appears likely that such a conformational change involves loop L4, which is the most mobile element in the structure and flips between different conformations even in the crystal (Pai et al., 1990).

REFERENCES

Barbacid, M. (1987). Ras genes. Ann. Rev. Biochem. 56, 779-827

Bos,J.L. (1989). Ras oncogenes in human cancer. Cancer Res. 49, 4682-4689

Dever, T. E., Glynias, M. J. and Merrick, W. C. (1987). GTP-binding domain: three consensus sequence elements with distinct spacing. Proc. Natl. Acad. Sci. U.S.A. 84, 1814-1818

Frech, M., John, J., Pizon, V., Chardin, P., Tavitian, A., Clark, R., McCormick, F. and Wittinghofer, A. (1990). Inhibition of GTPase activating protein stimulation of ras-p21 GTPase by the Krev-1 gene product. Science 249, 169-171

Gilman, A. G. (1987). G proteins: transducers of receptor-generated signals. Ann. Rev. Biochem. 56, 615-649

Halliday, K. R. (1983). Regional homology in GTP-binding proto-oncogene products and elongation factors. J. Cyclic Nucleotide and Protein Phosphorylation Res. 9, 435-44

John, J., Frech, M. and Wittinghofer, A. (1988). Biochemical properties of Ha-ras encoded p21 mutants and mechanism of the autophosphorylation reaction. J. Biol. Chem. 263, 11792-11799

John, J. (1990). phD thesis, Heidelberg

Krengel, U., Schlichting, I., Scherer, A., Schumann, R., Frech, M., John, J., Kabsch, W., Pai, E. F. and Wittinghofer, A., (1990)

Three-dimensional structures of H-ras p21 mutants: molecular basis for their inability to function as signal switch molecules. Cell 62, 539-548

Krengel, U. (1991). phD thesis, Heidelberg

McCormick, F. (1989). Gasp: not just another oncogene. Nature 340, 678-679

Milburn, M. V., Tong, L., DeVos, A. M., Brünger, A., Yamaizumi, Z., Nishimura, S., and Kim, S.-H. (1990). Molecular switch for signal transduction: structural differences between active and inactive forms of protooncogenic ras proteins. Science 247, 939-945

Pai, E. F., Kabsch, W., Krengel., U., Holmes, K. C., John, J. and Wittinghofer, A. (1989). Structure of the guanine-nucleotide-binding domain of the Ha-ras p21 oncogene product p21 in the triphosphate conformation. Nature 341, 209-214

Pai, E. F., Krengel, U., Petsko, G. A., Goody, R. S., Kabsch, W., and Wittinghofer, A. (1990). Refined crystal structure of the triphosphate conformation of H-ras p21 at 1.35 Å resolution: implications for the mechanism of GTP hydrolysis. EMBO J. 9, 2351-2359

Scherer, A., John, J., Linke, R., Goody, R. S., Wittinghofer, A., Pai, E. F. and Holmes, K. C. (1989). Crystallization and preliminary X-ray analysis of the human c-H-ras oncogene product p21 complexed with GTP analogues. J. Mol. Biol. 206, 257-259

Schlichting, I., Almo, S. C., Rapp, G., Wilson, K., Petratos, K., Lentfer, A., Wittinghofer, A., Kabsch, W., Pai, E. F., Petsko, G. A. and Goody, R. S. (1990). Time-resolved X-ray crystallographic study of the conformational change in Ha-ras p21 protein on GTP hydrolysis. Nature 345, 309-315

Seeburg, P. H., Colby, W. W., Capon, D. H., Goeddel, D. V. and Levinson, A. D. (1984). Biological properties of human c-Ha-ras 1 genes mutated at codon 12. Nature 312, 71-75

DETECTION OF N-RAS MUTATIONS IN ACUTE MYELOID LEUKEMIA

A.V. Todd, S. Yi, C.M. Ireland[1], and H.J. Iland

The Kanematsu Laboratories
Royal Prince Alfred Hospital
Camperdown, NSW, Australia, 2050

[1]The Children's Leukaemia and Cancer Research Unit
Prince of Wales Hospital
Randwick, NSW, Australia, 2031

INTRODUCTION

N-ras gene activation due to single base substitutions has been found in 20-40% of patients with acute myeloid leukemia (AML) depending on the technique used. Differential hybridization to allele specific oligonucleotide (ASO) probes detects mutations in about 22% of patients (1-6) whereas leukemic DNA assayed by NIH/3T3 transfection ± *in vivo* transformation in nude mice is associated with a mutation frequency of approximately 40% (7-11). We have developed a rapid screening method, termed allele specific restriction analysis (ASRA), for analysis of mutations at codons 12, 13 and 61 of the N-ras gene (12). A similar approach for the identification of Kirsten-ras gene mutations at codon 12 has also been reported (13,14). ASRA involves PCR amplification of DNA or RNA using a mismatched primer which introduces appropriately positioned base substitutions in N-ras and creates a restriction site provided the adjacent sequence is normal. Resistance of the amplified product to digestion indicates the presence of a mutation in the original template.

The major advantage of ASRA over other protocols is that it can be used as the first step in more sensitive methods for the analysis of ras mutations. Southern blotting of ASRA gels followed by hybridization with ASO probes increases the sensitivity and allows identification of specific mutations. Further increases in sensitivity can be achieved by allele specific enrichment (ASE).

MATERIALS AND METHODS

Patients, Cell Lines and Plasmids

Bone marrow or peripheral blood was obtained from 45 consenting patients with primary AML at the time of initial diagnosis. Additional samples were obtained from 5 of these patients at relapse following remissions achieved by induction chemotherapy. PCR was performed on purified DNA (15) or RNA (16) templates or on crude RNA extracts obtained by lysis of cell pellets (17). The K562 and HL-60 human leukemia cell lines were obtained from the American Type Culture Collection. Since DNA from K562 contains no N-ras mutations at codons 12, 13 or 61 it was included as a negative control in all of the following analyses. The limits of detection for ASRA were performed using dilutions of an N-ras cDNA plasmid with a codon 12 mutation (courtesy of E. Webb, Walter and Eliza Hall Institute, Melbourne) in pAT7.8 containing normal N-ras sequence (courtesy of A. Hall, Institute of Cancer Research, London).

Strategies for the detection of point mutations in N-ras

The general strategies used to detect single base substitutions in N-ras are illustrated for codon 12 in Figure 1.

ASRA

Amplification of DNA samples with mismatched oligoprimers introduced an Eco NI site at codon 12, a Pfl MI site at codon 13, and either an Msc I site or a Sca I site at codon 61 provided the gene was normal. The primer sequences and all PCR conditions have been published previously (12). When analyzed on 5% NuSieve GTG agarose gels, resistance to digestion with one of these 4 enzymes indicated the presence of a mutation in the corresponding codon. Furthermore, 13 of the 21 possible base substitutions at these 3 codons can be precisely identified by analysis of restriction enzyme sites which occur naturally or are induced by this primer system and which are specific for mutant N-ras alleles (18).

ASRA combined with ASO hybridization

DNA from 9 patients at initial presentation was further analyzed at codon 12 by a more sensitive protocol which combines ASRA with ASO hybridization (18). The samples were subjected to amplification which introduced an Eco NI site at codon 12. The PCR products were then digested with Eco NI and Southern blotted onto Hybond N+ nylon membrane. Filters were hybridized with the mutation specific oligonucleotides N12-a, N12-b, N12-c, N12-d, N12-e and N12-f under the conditions described by Verlaan-De Vries and co-workers (19).

Allele Specific Enrichment (ASE)

DNA from the patient no. 26 and from K562 cells was amplified with a mismatched primer which introduced an Eco NI site at codon 12. Digestion with Eco NI cleaved normal homoduplex PCR products rendering a digestion resistant, mutant allele enriched template for a second round of amplification. An aliquot of the digested PCR products was reamplified with primers which preferentially amplified these uncut mutant sequences and maintained the induced Eco NI site. Primer sequences used in these amplifications have been published previously (20). The secondary PCR product was digested with Eco NI, Southern blotted, and then analyzed by ASO hybridization.

For comparison with analysis by ASO hybridization alone a 125 base pair (bp) region of N-ras exon 1 was amplified from the DNA of patient no. 26 with fully matched primers prior to Southern blotting and ASO hybridization.

RESULTS

Analysis by ASRA

The limits of detection for ASRA were assessed by digestion of the PCR product of N-ras plasmid with a known codon 12 mutation diluted through pAT7.8, which contains the normal N-ras sequence. Visualization by ethidium bromide staining of the PCR products allowed detection of 5-10% mutant sequence, and therefore affords equivalent sensitivity to ASO hybridization (21) or direct sequencing (22). These results are in agreement with the limits of detection for ASRA of K-ras (13).

ASRA demonstrated 10 of the 45 patients were initially positive for N-ras mutations, including 1 patient with both codon 12 and 13 mutations (Table 1). Analysis of DNA from 19 of these patients has been published previously (18). A representative analysis of N-ras codon 12 by ASRA for 9 patients is shown in Figure 2A.

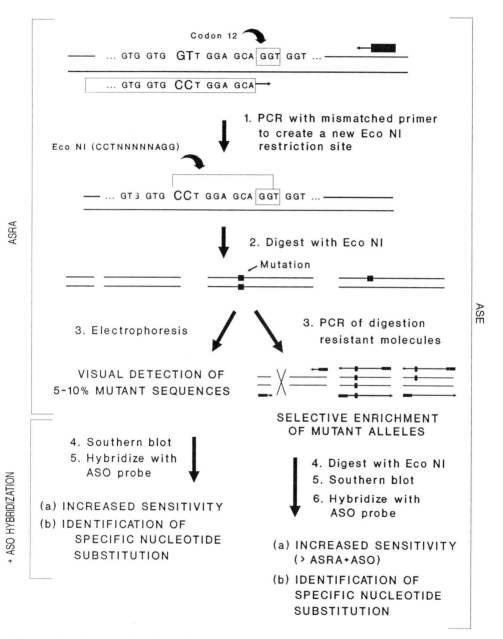

Figure 1. Strategies for the detection of N-ras mutations in codon 12.

Table 1. N-ras mutations in AML patients by ASRA. One patient (*) had mutations at both codons 12 and 13.

Codon	12	13	61	61
Position	1 & 2	1 & 2	1 & 2	3
Enzyme	Eco NI	Pfl MI	Msc I	Sca I
No. of patients with mutations (No. tested)	6* (45)	3* (45)	1 (45)	1 (19)

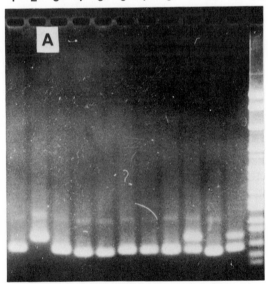

Figure 2A. Analysis by agarose gel electrophoresis of N-ras codon 12 DNA from the K562 cell line and from AML patients at initial presentation (P). DNA samples in lanes 1 and 3-11 were amplified around codon 12 and 20µl aliquots were digested with Eco NI. Amplified DNAs are as follows :- Lane 1: 26/P; lane 3: K562; lane 4: 1/P; lane 5: 2/P; lane 6: 8/P; lane 7: 15/P; lane 8: 30/P; lane 9: 38/P; lane 10: 41/P; lane 11: 42/P. The patient DNA in lane 2 was amplified but not digested. Lane 12: pBR322 digested with Msp I.

1 2 3 4 5 6 7 8 9 10 11 12

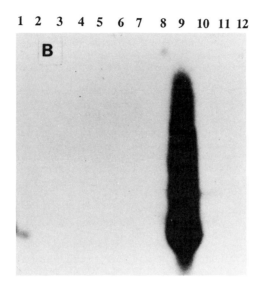

Figure 2B. Samples on the agarose gel in Figure 2A were Southern blotted, hybridized with N12-d, and autoradiographed.

1 2 3 4 5 6 7 8 9 10 11 12

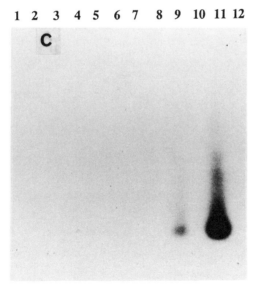

Figure 2C. A duplicate filter of samples on the agarose gel in Figure 2A was hybridized to N12-e, and autoradiographed.

K562 cell line DNA (lane 3) and DNA from patients negative by
ASRA for codon 12 mutations (lanes 1,4-8 and 10) digests completely
with Eco NI, whereas 2 patients (lanes 9 and 11 respectively) have
both wild type (83bp) and digestion resistant (99bp) mutant bands
indicating they were heterozygous at codon 12. The PCR product from 1
of these patients was sensitive to digestion with Pvu II indicating
this sample had a G-to-C transition at position 2 of codon 12 (12,18).
Analysis of the 5 patients at relapse showed that one patient had lost
a codon 13 mutation which had been detectable at presentation. The
other 4 patients were initially negative for N-ras mutations and did
not acquire mutations during the course of their disease.

Analysis of induced Msc I sites at codon 61 (positions 1 and 2)
indicated 1 of the patients in this study had a mutation at this
codon. HL-60 cells showed partial resistance to both Msc I and Mae I
confirming the well documented A-to-T substitution at codon 61
position 2. The pattern and relative intensities of bands after
analysis with Msc I was comparable for both DNA and RNA templates.
Digestion of the PCR product generated by the primer which induces a
Sca I site always resulted in a small amount of residual uncut
product. However the 3 possible base substitutions at codon 61
position 3 each result in sensitivity to another enzyme (Mae II,
Bst NI or Nla III) (18). The PCR product from 1 patient showed a
predominant Sca I resistant band and was sensitive to Mae II digestion
indicating an A-to-C substitution at this position.

Statistical analysis of the ASRA results for the first 19
patients studied indicated that N-ras mutations were more common in
elderly patients, particularly in those over 65 years of age (p<0.04,
Fisher's exact test), but showed no significant correlations with sex,
initial WCC, blast count or FAB subgroup. Patients with mutations were
less likely to achieve complete remission (2 of 4 compared with 12 of
15 mutation negative patients), but this difference was not
statistically significant. A similar trend against patients with N-ras
mutations was seen when overall survival was analyzed, but once again
statistical significance was not reached. In keeping with their older
age, patients with N-ras mutations generally received less intensive
chemotherapy, and these differences were sufficient to explain the
trend towards lower remission and shorter survival in the mutation
positive group.

Analysis by ASRA combined with ASO hybridization

In a preliminary study nine patients were further analyzed at
codon 12 by ASRA in combination with ASO hybridization (Figure 2A, 2B
& 2C). The mutation in DNA from patient no. 38 which was detected by
ASRA (2A:lane 9), and identified as a position 2 G-to-C substitution
by its sensitivity to Pvu II, was confirmed by ASO hybridization to
N12-d (2B:lane 9). Hybridization with N12-e identified the mutation
detected by ASRA in patient no. 42 (2A:lane 11) as a G-to-A mutation
at position 2 (2C:lane 11). The combination of techniques also allowed
identification of 2 additional mutations present at low levels in DNA
from patient no. 26 (2B:lane 1) and patient no. 38 (2C:lane 9). This
constitutes the third mutation found in patient no. 38 since a codon
13 mutation was also detected by ASRA.

Analysis by ASE and comparison with other methods

The results for analysis at codon 12 of DNA from patient no. 26
and K562 by ASRA, ASO, ASRA plus ASO, and ASE are shown in Figure 3a &
3b (20). Patient no. 26 DNA appeared negative by ASRA (3a:lane 4) and
by ASO hybridization with N12-d (3b:lane 2). However, when these two
protocols were combined, a low level mutation was detected (3b:lane
4). The intensity of the ASO signal was greatly enhanced following
analysis by ASE (3b:lane 3). K562 cell DNA was negative for codon 12
mutations by ASRA (3a:lane 6), by ASRA plus ASO hybridization
(3b:lane 6), and by ASE (3b:lane 1).

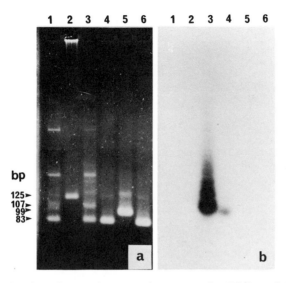

Figure 3. Analysis of DNA from patient sample 26/P and K562 cell line. Lane 5 is a 99bp marker. The PCR product of 26/P DNA using wild type primers for analysis by ASO hybridization alone is in lane 2. 26/P DNA amplified for ASRA (lane 4) or ASE (lane 3), and K562 DNA amplified for ASRA (lane 6) or ASE (lane 1), were digested with Eco NI prior to electrophoresis. Samples on the agarose gel (a) were Southern blotted, hybridized with N12-d, and autoradiographed (b).

DISCUSSION

ASRA is a rapid screening method suitable for detection of N-ras mutations in hematopoetic malignancies and other diseases. This method offers several advantages over other methods for the detection of point mutations. The method is non-radioactive and results for codons 12, 13 and 61 can be obtained in 24-48 hours. Double mutations on a single allele will usually not affect analysis, whereas double mutations can interfere with analysis by ASO hybridization (23). ASRA can provide insight into the relative expression of mutant and wild type alleles. Analysis of HL-60 templates showed that the relative intensities of the digestion sensitive and digestion resistant bands were approximately equivalent for DNA and RNA. This indicates that neither N-ras allele is differentially expressed in HL-60 cells.

ASRA was used to screen 45 samples from patients with AML for mutational activation of N-ras. Mutations were found in 10 of the 45 patients (22%) at the time of initial presentation. This result agrees with data pooled from a series of 6 studies using ASO hybridization which show N-ras mutations in 22% of 194 AML patients (1-6). Since the sensitivity of ASRA is comparable with that of ASO hybridization, with detection of 5-10% mutant sequences (12,13,21), it would be expected that both methods would detect mutations in a similar percentage of patients. Our finding that mutations detectable at presentation can be absent at relapse is also consistent with the findings of other authors (1,5).

The major advantage of ASRA over other detection methods is that it can be used as the first step in more sensitive protocols for the analysis of ras mutations. Haliassos et al. (24) combined ASRA with hybridization to an oligonucleotide probe complementary to both the normal and mutant PCR products. Their approach afforded a 1000-fold increase in sensitivity which allowed detection of 20% more K-ras mutations in colorectal cancers. Their protocol, however, has the potential for false positives due to hybridization of the probes to incompletely digested normal products. Our approach where ASRA has been combined with ASO hybridization not only eliminates this potential for false positives but also allows identification of specific mutations (18).

In a preliminary analysis of 9 patients at codon 12 of N-ras using ASRA in combination with ASO hybridization, the ASRA data were confirmed and the specific nucleotide substitutions were identified or confirmed. The increased sensitivity afforded by the combination of techniques also allowed identification of two more mutations present at low levels in patient samples which were not detectable by either ASRA or ASO hybridization alone. The presence of additional low level mutations demonstrated by this sensitive approach could explain the discrepancy found between data obtained by biochemical versus biological assays (1-11), and unequivocally establishes the existence of minor subclones in AML at the time of initial presentation. A further extension of the ASRA protocol termed allele specific enrichment (ASE) afforded even greater sensitivity than ASRA combined with ASO hybridization (20), and we have used ASE to confirm and facilitate recognition of the low level mutation in patient no. 26. The increased sensitivities of these methods may be useful in identifying minimal residual disease, or the emergence of new N-ras mutant clones in a subset of patients with AML. The ability to detect minimal residual disease by identification of cells containing N-ras mutations may have implications for autologous bone marrow transplantation, and could be used as a sensitive marker for comparing the efficiency of *in vitro* purging strategies.

ACKNOWLEDGEMENTS

Portions of this work were supported by the Australian National Health and Medical Research Council and the Kanematsu Research Fund. We thank Drs H. Kronenberg and Y. Kwan for providing patient samples. Portions of this work are currently in press (18).

REFERENCES

1. C.J. Farr, R.K. Saiki, H.A. Erlich, F. McCormick, C.J. Marshall, Analysis of *ras* gene mutations in acute myeloid leukemia by polymerase chain reaction and oligonucleotide probes. Proc. Natl. Acad. Sci. USA 85:1629-1633 (1988).
2. J.L. Bos, M. Verlaan-de Vries, A.J. van der Eb, J.W.G. Janssen, R. Delwel, B. Lowerberg, L.P. Colly, Mutations in N-*ras* predominate in acute myeloid leukemia. Blood 69:1237-1241 (1987).
3. H.P. Senn, Ch. Tran-Thang, A. Wodnar-Filipowicz, J. Jiricny, M. Fopp, A. Gratwohl, E. Signer, W. Weber, Ch. Moroni, Mutation analysis of the N-*ras* proto-oncogene in active and remission phase of human acute leukemias. Int. J. Cancer 41:59-64 (1988).
4. H. Mano, F. Ishikawa, H. Hirai, F. Takaku, Mutations of N-*ras* oncogene in myelodysplastic syndromes and leukemias detected by polymerase chain reaction. Jpn. J. Cancer Res. 80:102-106 (1989).
5. C.R. Bartram, W-D. Ludwig, W. Hiddemann, J. Lyons, M. Buschle, J. Ritter, J. Harbott, A. Frohlich, J.W.G. Janssen, Acute myeloid leukemia: Analysis of ras gene mutations and clonality defined by polymorphic X-linked loci. Leukemia 3:247-256 (1989).
6. J.J. Yunis, A.J.M. Boot, M.G. Mayer, J.L. Bos, Mechanisms of *ras* mutation in myelodysplastic syndrome. Oncogene 4:609-614 (1989).
7. D. Toksoz, C.J. Farr, C.J. Marshall. *ras* gene activation in a minor proportion of the blast population in acute myeloid leukemia. Oncogene 1:409-413 (1987).

8. J.W.G. Janssen, A.C.M. Steenvoorden, J. Lyons, B. Anger, Jv. Bohlke, J.L. Bos, H. Seliger, C.R. Bartram, *RAS* gene mutations in acute and chronic myelocytic leukemias, chronic myeloproliferative disorders, and myelodysplastic syndromes. Proc. Natl. Acad. Sci. USA 84:9228-9232 (1987).
9. S.W. Needleman, M.H. Kraus, S.K. Srivastava, P.H. Levine, S.A. Aaronson, High frequency of *N-ras* activation in acute myelogenous leukemia. Blood 67:753-757 (1986).
10. H. Hirai, J. Nishida, F. Takaka, Highly frequent detection of transforming genes in acute leukemias by transfection using *in vivo* selection assays. Biochem. Biophys. Res. Commun. 147:108-114 (1987).
11. J.L. Bos, D. Toksoz, C.J. Marshall, M. Verlaan-de Vries, G.H. Veeneman, A.J. van der Eb, J.H. van Boom, J.W.G. Janssen, A.C.M. Steenvoorden, Amino-acid substitutions at codon 13 of the N-*ras* oncogene in human acute myeloid leukemia. Nature 315:726-730 (1985).
12. A.V. Todd, H.J. Iland, Rapid screening of mutant N-ras alleles by analysis of PCR-induced restricition sites: Allele specific restriction analysis (ASRA). Leukemia and Lymphoma 3:293-300 (1991).
13. J. Wei, S.M. Kahn, J.G. Guillem, L. Shih-Hsin, I.B. Weinstein, Rapid detection of ras oncogenes in human tumors: application to colon, esophageal and gastric cancer. Oncogene 4:923-928 (1989).
14. A. Haliassos, J.C. Chomel, L. Tesson, M. Baudis, J. Kruh, J.C. Kaplan, A. Kitzis, Modification of enzymatically amplified DNA for the detection of point mutations. Nucleic Acids Res. 17:3606 (1989).
15. G.I. Bell, J.H. Karam, W.J. Rutter, Polymorphic DNA region adjacent to the 5' end of the human insulin gene. Proc. Natl. Acad. Sci. USA 78:5759-5763 (1981).
16. P. Chomczynski, N. Sacchi, Single-step method of RNA isolation by acid guanidinium thiocyanate-phenol-chloroform extraction. Analytical Biochemistry 162:156-159 (1987)·
17. F. Ferre, F. Garduno. Preparation of crude cell extract suitable for amplification of RNA by the polymerase chain reaction. Nucleic Acids Res. 17:2141 (1989)·
18. A.V. Todd, C.M. Ireland, T.J. Radloff, H. Kronenberg, H.J. Iland, Analysis of N-ras gene mutations in acute myeloid leukemia by allele specific restriction analysis. Am. J. Hematol. (in press) (1991).
19. M. Verlaan-de Vries, M.E. Bogaard, H. van den Elst, J.H. van Boom, A.J. van der Eb, J.L. Bos, A dot blot screening procedure for mutated *ras* oncogenes using synthetic oligodeoxynucleotides. Gene 50:313-320 (1986).
20. A.V. Todd, C.M. Ireland, H.J. Iland, Allele specific enrichment: A method for the detection of low level N-ras gene mutations in acute myeloid leukemia. Leukemia 5:160-161 (1991).
21. A. Neri, D.M. Knowles, A. Greco, F. McCormick, R. Dalla-Favera, Analysis of RAS oncogene mutations in human lymphoid malignancies. Proc. Natl. Acad. Sci. USA 85:9268-9272 (1988).
22. M. Bar-Eli, H. Ahuja, N. Gonzalez-Cadavid, A. Foti, M.J. Cline. Analysis of N-RAS Exon-1 mutations in myelodysplastic syndromes by polymerase chain reaction and direct sequencing. Blood 73:281-283 (1989).
23. J.W.G. Janssen and C.R. Bartram. Silent mutation at codon 15 interferes with the detection of a mutated N-ras codon 12 allele by oligonucleotide hybridization. Leukemia 3:235-235 (1989).
24. A. Haliassos, J.C. Chomel, S. Grandjouan, J. Kruh, J.C. Kaplan, A. Kitzis, Detection of minority point mutations by modified PCR technique: a new approach for a sensitive diagnosis of tumor-progression markers. Nucleic Acids Res. 17:8093-8099 (1989).

EVIDENCE FOR ras-DEPENDENT AND INDEPENDENT PATHWAYS IN NGF-

INDUCED NEURONAL DIFFERENTIATION OF PC12 CELLS

József Szeberényi[1], Hong Cai[2] and
Geoffrey M. Cooper[2]

[1]Department of Biology
University Medical School of Pécs
Hungary

[2]Dana-Farber Cancer Institute
Boston, MA 02115

PC12 cells expressing a dominant inhibitory mutant Ha-Ras protein [p21 (Asn-17) Ha-ras] were used to study signaling pathways involved in neuronal differentiation. Expression of the mutant protein completely blocks NGF-induced neurite outgrowth in these subclones. In contrast, NGF is able to potentiate the effect of dibutyryl cyclic AMP or ionomycin in inducing process formation in PC12 cells expressing the mutant Ras protein. These results suggest that (i) a Ras-independent signaling pathway exists in PC12 cells that is not sufficient for, but can contribute to NGF-induced neuritogenesis; (ii) the block of neuronal differentiation by the mutant Ras protein can be bypassed by either one of two second messengers, cyclic AMP and Ca^{++}.

INTRODUCTION

Mammalian ras genes (Ha-ras, Ki-ras, N-ras) encode small membrane bound proteins (p21s) that are able to bind guanine nucleotides, possess intrinsic GTPase activity and alternate between an active, GTP-bound and an inactive, GDP-bound form (Barbacid, 1987; Bar-Sagi, 1989; Hall, 1990). Although the function of Ras proteins is not understood, their strong resemblance to G proteins suggests that they might be involved in transmembrane signaling. There is increasing evidence for such a role of p21s in the process of neuronal differentiation of PC12 pheochromocytoma cells. PC12 cells respond to treatment with nerve growth factor (NGF) by shifting from a chromaffin cell-like phenotype to a neurite-bearing sympathetic neuronal phenotype (Greene, 1984). NGF-induced neuritogenesis can be mimicked by introducing oncogenic Ras proteins into PC12 cells (Bar-Sagi and Feramisco, 1985; Noda et al., 1985; Satoh et al., 1987; Der et al., 1988; Guerrero et al., 1988; Sassone-Corsi et al., 1989). The involvement of Ras proteins in PC12 cell differentiation is further supported by experiments, in which NGF-induced neurite outgrowth of PC12 cells was blocked by microinjection with anti-Ras monoclonal antibody (Hagag et al., 1986). Recently, similar results were obtained by expressing a dominant inhibitory Ha-Ras mutant protein in PC12 cells (Szeberényi et al., 1990). The protein encoded by the mutant gene (Ha-ras Asn 17; Feig and Cooper, 1988) has a preferential binding for GTP versus GDP and strongly inhibits serum or growth factor stimulated proliferation of HIH 3T3 cells (Cai et al., 1990). PC12 cells expressing the mutant p21 (M17-subclones) are resistant to NGF- or FGF-induced neuronal differentiation, while they retain responsiveness to direct stimulation by the second messenger analogs dibutyryl cyclic AMP (dbcAMP) and 12-0-tetradecanoylphorbol-13-acetate (TPA); Szeberenyi et al., 1990). Thus, M17-subclones provide a model system to study Ras-dependent and -independent signaling pathways in neuronal differentiation. In the present study we report the identification of a Ras-independent, growth factor-stimulated pathway that shows strong synergy with cyclic AMP- and Ca^{++}-mediated pathways in morphological differentiation of PC12 cells.

METHODS

Conditions for growing PC12 and M-M17-26 cells have been described earlier (Szeberényi et al., 1990). For details of inducing morphological differentiation, see the legends to the figures.

RESULTS

Wild type PC12 cells and an M17-subclone expressing high levels of the mutant p21 (M-M17-26; Szeberényi et al., 1990) were treated with NGF, dbcAMP, the Ca^{++} ionophore ionomycin (Balk et al., 1984) or the combination of these agents and scored for morphological differentiation (Figs. 1. and 2.). NGF induced neurite outgrowth (Fig. 1b.), while treatment with dbcAMP or ionomycin resulted in the appearance of shorter processes in PC12 cells (1c. an 1d.). dbcAMP strongly potentiated the effect of NGF (Fig. 1e. and 2.), in agreement with previous reports (Richter-Landsberg and Jastorff, 1986). A similar, but more delayed synergistic effect of ionomycin on NGF-induced neurite extension was also observed (Fig. 1f. and 2.). M-M17-26 cells were completely resistant to NGF-induced morphological differentiation (Fig. 1h.) and displayed a somewhat reduced response to dbcAMP or ionomycin (Fig. 1i. and 1j.) as compared to PC12 cells. However, the combination of NGF plus dbcAMP (Fig. 1k. and 2.) or NGF plus ionomycin

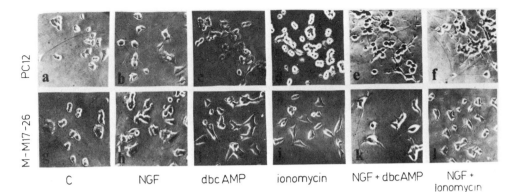

PC12

M-M17-26

| C | NGF | dbc AMP | ionomycin | NGF + dbcAMP | NGF + ionomycin |

Fig. 1. Morphological differentiation of PC12 and M-M17-26 cells. Cells were plated in 24-well dishes at a density of 10^4 cells per well in Temin's modified Eagle medium supplemented with 10% heat-inactivated fetal bovine serum and 5% horse serum. The next day the medium was replaced by a low-serum medium containing 0.5% horse serum and the cells were treated with NGF (10 ng/ml), dbcAMP (0.5 mM), and/or ionomycin (0.25 μM from a 0.1 mM stock solution in DMSO). PC12 (a to f) and M-M17-26 (g to l) were cultured without treatment (a and g) or in the presence of NGF (b and h), dbcAMP (c and i), ionomycin (d and j), NGF plus dbcAMP (e and k), or NGF plus ionomycin (f and l).Phase contrast micrographs were taken on day 2 (c, e, i, and k) or on day 4 (a, b, d, f, g, h, j, and l) of treatment.

(Fig. 1l. and 2.) evoked a strong differentiation response in these cells, with a maximal differentiation of about 50%. The potentiating effect of ionomycin was completely abolished by pretreating the cells with the Ca^{++}-chelator BAPTA/AM (not shown), indicating that the effect of ionomycin depends on the presence of free Ca^{++} in the medium. Note that the kinetics of the effect of dbcAMP and ionomycin are markedly different: the synergistic action of dbcAMP and ionomycin peaks at around day 2 and day 6, respectively, in both PC12 and M-M17-26 cells (Fig. 2.). Two additional NGF-resistant M17-subclones, Z-M17-5 and M-M17-2, that express lower levels of Ha-Ras Asn-17 protein (Szeberényi et al., 1990), displayed qualitatively similar synergistic responses to NGF plus dbcAMP and NGF plus ionomycin (data not shown). These experiments clearly show the existence of a Ras-independent, NGF-stimulated pathway that synergizes with both cAMP- and Ca^{++}-dependent mechanisms.

In an attempt to further characterize the NGF-stimulated, Ras-independent pathway, we studied the possible involvement of protein kinase C, an enzyme stimulated by NGF in PC12 cells (Hama et al., 1986; Haesley and Johnson, 1989). M-M17-26 cells, in which protein kinase C had been down-regulated by

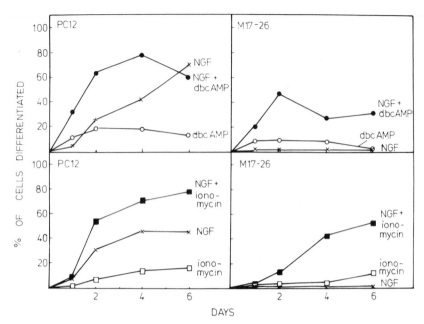

Fig.2. Quantitation of neuronal differentiation of PC12 and
M–M17-26 cells. Cells were treated as described in
the legend to Fig. 1. Approximately 200 cells were
counted in each well on day 1, 2, 4, and 6 of
treatment with NGF (x), dbcAMP (o), NGF plus dbcAMP
(●), ionomycin (□), or NGF plus ionomycin (■). Panels
A–C and B–D derive from separate experiments. Cells
carrying processes longer than their diameter were
scored differentiated. Untreated PC12 and M–M17-26
cells showed <1% differentiation at any time point.

prolonged treatment with TPA (Ballester and Rosen, 1985) were
treated with NGF plus dbcAMP: no inhibition of neurite
outgrowth by depletion of protein kinase C was observed (Fig.
3A). In control experiments process formation induced by
dbcAMP alone was not affected by protein kinase C down-
regulation (Fig. 3B); the potentiating effect of TPA on dbcAMP
treatment (Sugimoto et al., 1988) was, however, completely
abolished by chronic phorbol ester treatment of the cells
(Fig. 3C). This observation indicates that M–M17-26 cells were
in fact depleted of functional protein kinase C under our
experimental conditions. We can thus conclude that the NGF
inducible, Ras-independent pathway does not involve protein
kinase C.

DISCUSSION

In the studies presented here, we identified a Ras-
independent signal transduction pathway in PC12 cells that is

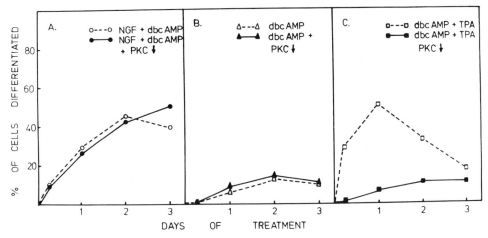

Fig. 3. The effect of protein kinase C down-regulation on the differentiation of M-M17-26 cells. Cells were treated and scored for differentiation, as described in the legend to Fig. 1., except that, when indicated (PK C↓), they were preincubated with 200 nM TPA for 48 h. After pretreatment with TPA, the medium was removed and treatment with NGF plus dbcAMP (o and ●), dbcAMP alone (△ and ▲), or dbcAMP plus TPA (20 nM, □ and ▣) was performed in fresh low-serum medium.

stimulated by NGF. This pathway (i) is not sufficient to induce neuronal differentiation by itself: PC12 subclones expressing detectable levels of Ha-Ras Asn-17 p21, encoded by a dominant inhibitory ras gene, are resistant to NGF-induced neuritogenesis (Figs. 1. and 2.; Szeberényi et al., 1990); (ii) is not sufficient to induce early response genes (fos, jun, zif268): in M17-subclones expressing high levels of the mutant p21 (such as M-M17-26) induction of these genes by NGF is strongly inhibited (Szeberényi et al., 1990); (iii) synergizes with Ca^{++} and cAMP agonists (Figs. 1. and 2.) in the stimulation of process formation, but not with TPA (data not shown); (iv) is not mediated by protein kinase C: down-regulation of protein kinase C does not affect the response of M-M17-26 cells to treatment with NGF plus dbcAMP (Fig. 3.); (v) is probably not mediated by protein kinase A or Ca^{++}/calmodulin dependent protein kinases: saturating concentrations of NGF plus dbcAMP or NGF plus ionomycin produce strong synergistic effects both in control and M17-subclones (Figs. 1 and 2) suggesting that these agents use separate signaling pathways.

Our present and previous observations (Szeberényi et al., 1990) indicate that the intracellular effect of NGF in PC12 cells is mediated by at least three signaling pathways: (i) a pathway required for NGF-induced neurite outgrowth and inhibited by low levels of Ha-Ras Asn-17 p21 ; (ii) a pathway that is indispensable for the full induction of early response genes and is inhibited by high levels of the mutant Ras protein only (Szeberényi et al., 1990); and (iii) the Ras-independent pathway described here.

The second messengers involved in the signal transduction from the NGF-receptor have not yet been identified. It has been suggested that the early phase of signaling is probably carried out by a single transducer and that the pathways diverge thereafter and involve multiple second messengers (Nairn et al., 1987). The signal transducing mechanisms may include protein tyrosine phosphorylation (Maher, 1988), protein serine/threonine phosphorylation (Rowland et al., 1987; Mutoh et al., 1988; Volonte et al., 1989; Damon et al., 1990; Tsao et al., 1990), activation of phospholipase C (Contreras and Guroff, 1987; Sugimoto et al., 1988), leading to increases in the intracellular concentration of diacylglycerol, inositolphosphates, Ca^{++} and other second messengers. An early step in the signaling process might be the divergence of Ras-dependent and -independent pathways. The signal transducer involved in the Ras-independent pathway has not yet been identified. Tyrosine or serine/threonine specific protein kinases (other than protein kinase A, C or Ca^{++}/calmodulin-dependent protein kinases) are reasonable candidates.

Our results also indicate that the block by Ha-Ras Asn-17 can be, at least partially, overcome by direct stimulation of cAMP- and Ca^{++}-dependent mechanisms. However, these experiments do not establish a direct relationship between NGF and cAMP- or Ca^{++}-mediated pathways. Earlier observations support the notion that cAMP is not a major factor in NGF-induced neuronal differentiation (Richter-Landsberg and Jastorff, 1986). The possible role of Ca^{++} in NGF-stimulated neurite outgrowth has been suggested: the concentration of inositol phosphates, second messengers that liberate Ca^{++} sequestered in intracellular pools, is increased in PC12 cells after NGF treatment (Contreras and Guroff, 1987). Similarly, NGF treatment is followed by a rapid and transient increase in Ca^{++} levels (Pandiella-Alonso et al., 1986). These results, together with our ionomycin-data are consistent with, but do not prove the notion that a Ca^{++}-dependent pathway may be involved in NGF-induced neuronal differentiation.

REFERENCES

Balk, S.D.,A. Morisi and H.S. Gunther, 1984, Proc.Natl. Acad.Sci. USA 81:6418
Ballester, R. and O.M. Rosen, 1985, J. Biol. Chem. 260: 15194
Barbacid, M., 1987, Annu.Rev. Biochem. 56:779
Bar-Sagi, D.,1989, Anticancer Research 9: 1427
Bar-Sagi, D. and J.R. Feramisco, 1985, Cell 42:841
Cai, H., Szeberényi and G.M. Cooper, 1990, Mol. Cell. Biol. 10:1314

Contreras, M.L. and G. Guroff, 1987, J. Neurochem. 48:1466

Damon, D.H., P.A. D'Amore and J.A. Wagner, 1990, J. Cell. Biol. 110:1333

Der, C.J., B. Weissman and M.J. MacDonald, 1988, Oncogene 3:105

Feig, L.A. and G.M. Cooper, 1988, Mol Cell. Biol. 8:3235

Greene, L.A., 1984, Trends Neurosci 7:91

Guerrero, I., A. Pellicer and D.E. Burstein, 1988, Biochem. Biophys. Res. Commun. 150:1185

Hagag, N., S. Halegoua and M. Viola, 1986, Nature 319:680

Hall, A., 1990, Science 249:635

Hama, T., K.-P. Huang and G. Guroff, 1986, Proc. Natl .Acad. Sci. USA 83:2353

Heasley, L.E. and G.L. Johnson, 1989, J. Biol. Chem. 264:8646

Hunter, T., 1987, Cell 50:823

Maher, P.A., 1988, Proc. Natl. Acad. Sci. USA 85:6788

Mutoh, T., B.B. Rudkin, S. Koizumi and G. Guroff, 1988, J. Biol. Chem. 263:15853

Nairn, A.C., R.A. Nichols, M.J., Brady and H.C. Palfrey 1987, J. Biol. Chem. 262:14265

Noda, M. et al., 1985, Nature 318:73

Pandiella-Alonso, A., A. Malgaroli, L.M. Vicentini and J. Meldolesi, 1986, FEBS Lett. 208:48

Richter-Landsberg C. and B. Jastorff, 1986, J. Cell. Biol. 102:821

Rowland, E.A., T.H. Muller, M. Goldstein and L.A. Greene, 1987, J. Biol. Chem. 261:7504

Satoh, T., S. Nakamura and Y. Kasito, 1987, Mol. Cell. Biol. 7:4553

Sassone-Corsi, P., C.J. Der and I.M. Verma, 1989, Mol. Cell. Biol. 9:3174

Sugimoto, Y., M. Noda, H. Kitayama and Y. Ikawa, 1988, J. Biol. Chem. 263:12102

Szeberényi, J., H. Cai and G.M. Cooper, 1990, Mol. Cell. Biol. 10:5324

Tsao, H., J.M. Aletta and L.A. Greene, 1990, J. Biol. Chem. 265:15471

Volonte, C., A. Rukenstein, D.M. Loeb and L.A. Greene, 1989, J. Cell. Biol. 109:2395

PROGNOSTIC IMPLICATIONS OF RAS ONCOGENE EXPRESSION

IN HEAD AND NECK SQUAMOUS CELL CARCINOMA

John Field

Department of Clinical Dental Sciences
School of Dentistry
University of Liverpool
P.O. Box 147
Liverpool L69 3BX

The ras gene family have been implicated in the development of a range of human cancers (Bos, 1989; Barbacid, 1990; Field and Spandidos, 1990). The ras gene family consists of three functional genes: H-ras, K-ras and N-ras, which encode for very similar proteins with a similar molecular weight of 21,000. Ras p21 proteins are homologous to G proteins, possess GTPase activity and are located on the internal part of the cytoplasmic membrane. A GTPase activating protein has been shown to take part in the hydrolyses of GTP by binding to the effector domain of the ras protein, and thereby takes part in signal transduction (Barbacid, 1987; McCormick, 1989; Schlichting et al., 1990).

It has been proposed that the normal p21 ras proteins when activated transduce a signal to an effector molecule and subsequently the ras protein is inactivated, whereas mutated ras proteins have lost their ability to be inactivated and may therefore stimulate growth and differentiation autonomously (Bos, 1989).

There are three recognised mechanisms of oncogene activation in human neoplasm, which include gene amplification, translocations and mutations. However, other mechanisms must exist as overexpression of oncogenes has been reported by many authors with no evidence of the aforementioned mechanisms. Activation of the ras genes has been demonstrated in a range of human neoplasm to be caused by specific base changes (Bos, 1989), whereas amplification or rearrangement of these genes is a rare finding in human tumours (Barbacid, 1990). It is also of note that c-Ha-ras restriction fragment length polymorphisms (RFLP) (Capon et al., 1983) have been found to be linked to susceptibility of individuals to cancer (Krontiris et al., 1985) and that deletions of the c-Ha-ras locus have been reported in a number of human neoplasia (Nordenskjold and Cavenee, 1988; Riou et al., 1988; Sheng et al., 1990). Elevated levels of ras expressions have been reported in a variety of human tumours and have been associated with tumour progression and prognosis in certain cases (Field and Spandidos, 1990).

The Superfamily of ras-Related Genes
Edited by D.A. Spandidos, Plenum Press, New York, 1991

Incidence and Risk Factors of Head and Neck Cancer

The incidence of head and neck cancer varies considerably worldwide. The highest reported rates are in India where oral cancers are the commonest sites of malignancy and account for over 40% of all cancer compared to 2% of all malignancies in the western world. When the worldwide figures are computed for mouth and pharynx cancers, this group accounts for six percent of all solid tumours (Parkin et al., 1988). In Great Britain it is disturbing to realise that the incidence and mortality rates of oral cancer are increasing, especially in younger men (Cancer Research Campaign Factsheet, 1990). In fact, increases in mortality from oral and pharyngeal cancer has been found in most E.C. countries over the last 30 years (Johnson, 1990).

There is overwhelming epidemiological evidence for a correlation between heavy smoking and lung cancer (Doll and Peto, 1976). Indeed, the incidence of head and neck cancer, and oral cancer in particular, appear to be directly related to tobacco use. Moreover the relative risk increases with duration and the quality of tobacco smoked (Silverman and Griffith, 1972; Stell, 1972; Wynder et al., 1977; Myers and Suen, 1989). There has been a relatively recent introduction by the tobacco industry of smokeless tobacco in the form of Skod Bandits, which are small paper pouches containing oral snuff. In fact, the use of these smokeless tobacco products was estimated at 12 million in the USA in 1985 and has led to an increase in oral lesions in young people (Cancer Research Campain Factsheet, 1990).

Also the relationship of alcohol consumption to squamous cell carcinoma has been well documented (Wynder et al., 1956; Krajina et al., 1975; Kurozumi et al., 1977; Schmidt and Popham, 1981). The mortality studies of alcoholics have consistently found that more than average numbers of deaths were caused by cancer of the head and neck and from cancer of the lung. There is also evidence that heavy drinking and smoking act synergistically to increase the risk of head and neck cancer and oral cancer in particular (Schmidt and de Lint, 1970; Nicholls et al., 1979; McCoy and Wynder, 1979).

A number of other risk factors have been proposed which include: _Candida albicans_ in candida leukoplakia, Syphilis in syphilitic leukoplakia, as well as HIV infection, and also individuals with low levels of Vitamin A, and C and iron deficiency are at a higher risk of developing oral cancer (Oral Cancer, Editorial, 1989).

Clinical Prognostic Variables in Head and Neck Cancer

The survival of any patients with cancer is determined by three factors: (i) Patient factors include age, sex, race, nutritional state and immunological status; (ii) Tumour factors include site of primary cancer, diameter of tumour, presence or absence of regional metastasis and of distant metastasis; (iii) Treatment modalities.

The most important prognostic indicator in head and neck cancer is the presence of regional lymph metastasis (Davis, 1985). TNM staging (UICC, 1987) of the disease; T (tumour diameter); N (presence or absence of palpable nodes); M (metastasis) is used by many head and neck oncologists and provides some prognostic information at stages I and IV; however, the ability to prognosticate stages II and III remain problematic (Davis, 1985). Other clinical indicators include histological differentiation and immunological staging which have been found to have limited prognostic value (Davis, 1985; Stell, 1988).

Genetic Risk Factors

We have previously demonstrated that there is elevated expression of several oncogenes, Ha-ras, Ki-ras, c-erbB-2 and c-myc in the development of squamous cell carcinoma of the head and neck (Field et al., 1986, Field and Spandidos, 1987, 1990; Field et al. 1989, 1990, 1991a,b; Field, 1991). We have previously shown that high levels of expression of the c-myc protein in tumour cells from patients with squamous cell carcinoma of the head and neck correlates with a poor prognosis (Field et al., 1989).

Among the cellular oncogenes, the ras gene family are considered to be involved in the development of malignant tumours and a number of studies have been undertaken to establish whether genetic alteration or aberrant expression of these genes are important in head and neck cancer. In the carcinogen induced mouse skin model system, A-T nucleotide transversions in codon 61 of the ras gene showed a high percentage of carcinomas and premalignant papilloma (Balmain et al., 1984; Quintanilla et al., 1986). The frequency of detecting ras mutations and their position maybe altered by the use of different chemical carcinogens (Zarb et al., 1985 Quintanilla et al., 1986, Brown et al., 1990) thereby indicating that specific mutagens create specific mutations in the ras gene. Balmain and co-workers concluded that the ras genes were directly responsible for the malignant activation of these tumours and that the Ha-ras gene was activated in the premalignant papillomas, the mutations must have occurred at an early stage of carcinogenesis. They also proposed that additional secondary changes are necessary to achieve the full malignant phenotype of skin carcinomas. Recently Bremmer and Balmain (1990) have investigated skin carcinomas in F1 hybrid mice with allele-specific restriction fragment length polymorphisms (RFLP) markers. On analysing these skin carcinomas for loss of heterozygosity and H-ras mutations, they showed an allelic imbalance at one or more loci on chromosome 7. The chromosomal alterations were only seen in tumours that had activated ras genes. In addition recent investigations have elucidated the timing of ras oncogene activation in the rat mammary carcinoma model system (Kumar et al., 1990). Chemically activated ras oncogenes have been shown to precede the onset of neoplasia and remain latent within the mammary gland until exposed to estrogen.

As previously mentioned there is epidemiological evidence to support the hypothesis that certain chemical carcinogens are important in the development of squamous cell carcinoma of the head and neck. For this reason there has been an effort by a number of groups to demonstrate a link between genetic alteration of ras genes and overexpression with the development and progression of this disease.

In oral squamous cell carcinoma patients from the Indian sub-continent, Saranath et al. (1989) reported amplification in 7/23 N-ras, 4/23 Ki-ras and 0/23 Ha-ras cases. However, no rearrangement or amplification of the ras gene family have been found in head and neck squamous cell carcinoma cases in western Europe (Field et al., 1986; Sheng et al., 1990; Merrit et al., 1990).

High levels of ras mutations have been reported in
adenocarcinomas of the pancreas (90%), colon (50%), thyroid
(50%), lung (30%) and in myeloid leukaemia (Bos, 1989). In
colon tissues the Ki-ras mutations were found in pre-cancer-
ous lesions which were assumed to progress into carcinomas
(Bos et al., 1987; Forrester et al., 1987). In addition,
mutations at codon 12 of the Ha-ras gene were shown to be
associated with cervical cancers of poor prognosis (Riou et
al., 1988; Sheng et al., 1990).

Ras mutations were analysed Sheng et al. (1990) in 54
head and neck tumours and they found only two tumours
contained a mutation (G-T transversion) in codon 12 in the
Ha-ras gene. The presence of mutations at position 12, 13
and 61 in Ki-ras, 12 and 61 in N-ras were also investigated
in 28 cases; however, none were found. Rumsby et al. (1990)
also investigated ras mutations in 37 head and neck cancers
and found 2 patients with (G-A transversions) at codon 12 in
Ha-ras gene. No mutations were found at positions 12, or 61
in Ki-ras or N-ras genes in these tumours. Similar low
levels (<5%) of ras mutations have been found by other
researchers (Chang, personal communication; Field et al.,
unpublished).

Again a dissimilar finding has been reported by the
Indian group working on oral carcinomas from those in western
Europe (Saranath et al., 1991). Thirty-five percent of the
oral carcinomas were found to have a Ha-ras mutation, thereby
suggesting that different environmental initiation agents are
involved in the Indian population. When considering the
prevalent use of betal quid chewing and its correlation with
oral cancer in India, it would appear to support the hypothesis of a "betal quid" carcinogen causing the ras mutations.

The loss of one Ha-ras allele was observed in 10 of 46
head and neck tumours from heterozygous patients for this
locus (Sheng et al., 1990). It is of particular interest
that all of the deletions were found in lymph node metastasis
with extracapsular rupture, and none were detected in any of
the primary tumours investigated. From this result it may be
argued that the Ha-ras allelic deletions are important in the
progression of head and neck cancer; however, as the patients
numbers are small, no statistical correlations were made.
This group have also examined deletions in the HBBC (gamma A
and gamma G of the ß globin cluster) PTH (parathyroid
hormone) CALC (calcitonin) and CAT (catalase) gene loci on
the short arm of chromosome 11p. They demonstrated deletions
of HBBC genes in 4 of 11 carcinomas, and these deletions were
accompanied with the loss of 1 Ha-ras allele in 3 tumours,
and with the loss one 1 calcitonin allele in 1 tumour (Sheng
et al., 1990). Ali et al. (1987) also found that the most
frequent loss of sequences on 11p in breast tumours was
between ß globin and the parathyroid hormone loci. In view
of these results it would suggest that genes other than
Ha-ras on the short arm of 11p are deleted in a number of
head and neck tumours and may play a role in the progression
of the disease.

However, there is also evidence from one source to
indicate that one tumour specimen derived from a gingival
squamous cell carcinoma had an activated K-ras oncogene using
the NIH 3T3 transfection assay (Howell et al., 1990). Also an
earlier investigation reported a high percentage of transforming genes from squamous cell carcinomas of the head and
neck; however, no specific oncogene was identified by these
investigators (Friedman et al., 1985).

These observations therefore leave open the hypothesis that genetic alterations in the _ras_ gene may have a role in the development of these head and neck tumours but are rarely found in head and neck patients in the western world.

Ras Overexpression in Head and Neck Cancer

Elevated expression of the _ras_ gene family has been demonstrated in head and neck cancers, using RNA hybridization analysis, in situ hybridization and immunocytochemical techniques (Field et al., 1986; Hollering and Shuler, 1989; Sheng et al.,.1990; Azuma et al., 1987; Tsuji et al., 1989; Field et al., 1991a). We initially studied the level of Ha-_ras_ and Ki-_ras_ RNA transcripts in 14 squamous cell carcinoma of the head and neck and found that a significant number of patients had elevated Ha-_ras_ and Ki-_ras_ levels of expression (Spandidos et al., 1985). Ha-_ras_ mRNA was detected in 5 squamous cell carcinomas of the oral cavity by in situ hybridization analysis (Hoellering and Shuler, 1989). These authors reported a distinctive pattern of Ha-_ras_ expression in the tumour mass and noted that the most intense hybridization with the Ha-_ras_ probe was at the invading edge or adjoining regions of the tumour into the underlying connective tissue. However, no correlation was found between p21 _ras_ expression and depth of invasion.

Sheng et al. (1990) found no significant difference in Ha-_ras_ RNA levels of expression in primary and metastatic head and neck squamous cell carcinoma. Also, it is of particular note that these authors found that the level of Ha-_ras_ expression was not influenced by genetic alterations of the Ha-_ras_ genes (i.e. loss of an allele, presence of a rare allele or a mutation) thereby giving weight to the hypothesis that a mechanism external to the _ras_ genes is responsible for elevated ras expression in these tumours.

The expression of p21 _ras_ oncoprotein has been studied with the use of the Y13-259 monoclonal antibody (Furth et al. 1982) in a range of human cancers (Field and Spandidos, 1990). We have studied the expression of p21 _ras_ in 69 squamous cell carcinomas of the head and neck (Field et al., 1991a) and have found a correlation between low levels of _ras_ expression and the disease-free survival period in patients with previously untreated tumours. The specimens were taken from oral, pharyngeal and laryngeal squamous cell carcinomas. Also a number of specimens were from lymph node metastasis in patients with primary squamous cell carcinoma of the head and neck. The level of p21 _ras_ expression was not found to correlate with whether the patient had been previously treated or not, site, TNM status, presence of pathological lymph node metastasis or extracapsular rupture. A correlation was found between p21 _ras_ staining and histopathology. Fifty-four percent (30/50) of the well or moderately well differentiated tumours had positive p21 _ras_ staining, compared to 93% (13/14) poorly differentiated tumours. The clinicopathological characterisation of the previously untreated squamous cell carcinomas were also investigated and no correlations were found.

Previous studies from two Japanese groups (Azuma et al., 1987; Tsuji et al., 1989) have also investigated _ras_ p21 expression using the Y13-259 monoclonal antibody. Azuma et al. (1987) reported that 14/23 _ras_ p21 positive squamous cell carcinomas had died, compared with 20/20 _ras_ p21 negative patients; however, these calculations were based on Fisher's exact probability test, which is of limited value for analysing survival data. Tsuji et al., (1990) found no correlation between ras expression and survival in 56 patients with oral cancer using the Log rank test. These two sets of data from Japan, even though disagreeing with each other, were also at variance with our findings. However, as we have previously argued, there may be a different set of environmental factors in western Europe compared to Japan (Field et al., 1991a).

We first of all calculated the survival period for all of the 73 patients, which was made up of 44 patients who have not been previously treated and 25 patients who have been previously treated. The survival period was calculated from the time of diagnosis for both the previously treated and untreated patients. The survival of this group of patients with elevated levels of p21 ras positive staining was greater than that of those with negative p21 ras staining, but this difference was not quite significant ($X^2 = 3.3$ $P = 0.07$). However, we also analysed the disease-free period for 44 patients whose specimens were taken from previously untreated tumours. A correlation was found between low levels of ras expression and the disease-free survival period in this group of patients ($X^2 = 4.2$ $P< 0.05$),(Figure 1) where three percent of the patients with ras negative staining were alive 60 months after diagnosis, compared with the 54 percent of patients with positive p21 ras expression who were still alive after the same time interval.

These results indicate that overexpression of the ras gene is important in the initiation stages of head and neck

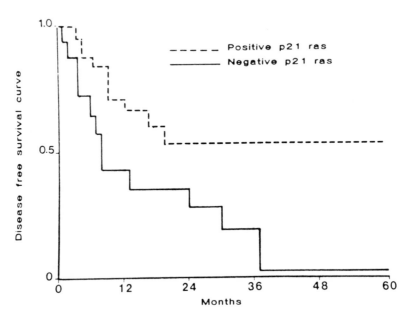

Figure 1.   Disease free survival curve of patients with squamous cell carcinoma of the head and neck, drawn up using the Kaplan and Meier (1958) product limit estimate with ras p21 positive staining and those with ras p21 negative staining.

carcinogenesis. However, in view of the studies on <u>ras</u> gene
amplification, rearrangements and mutations in head and neck
cancer, it would appear that other genetic mechanisms, as yet
unidentified, must exist to explain <u>ras</u> overexpression in
these tumours.

A similar observation has been made in human neuroblastomas using the Y13-259 monoclonal antibody to p21<u>ras</u> (Tanaka
et al., 1988). They reported that high levels of Ha-<u>ras</u> gene
product correlated with a favourable prognosis and early
stages of the disease.

We also considered investigating a possible link between
<u>ras</u> and p53 expression in head and neck cancer, as a previous
report indicated co-operation between <u>ras</u> and the p53 tumour
antigen in cellular transformation (Parada et al., 1984).
The p53 gene mutations have been demonstrated in a range of
human tumours (Crawford et al., 1981; Cattoretti et al.,
1988; Baker et al., 1989; Nigro et al., 1989; Takahashi et
al., 1989; Iggo et al., 1990; Chiba et al., 1990). In
particular, two studies demonstrated mutations in the p53
gene and abnormal expression of p53 in lung tumours which are
associated with smoking. Iggo et al. (1990) found abnormal
staining patterns in three cases of lung cancer which had
undergone p53 mutations. Normal p53 protein is known to have
a very short half life (6-20 minutes) and therefore it may be
assumed that detection of the p53 protein in tumour tissue is
synonymous with mutation, as the mutant form of p53 has a
half life of up to 6 hours, most likely due to stabilization
of the protein (Lane and Benchimol, 1990). Elevated p53
expression was found in 14/17 (82%) of squamous cell carcinomas of the lung compared to 8/21 (38%) new squamous cell
carcinomas (Iggo et al., 1990), while p53 mutations were

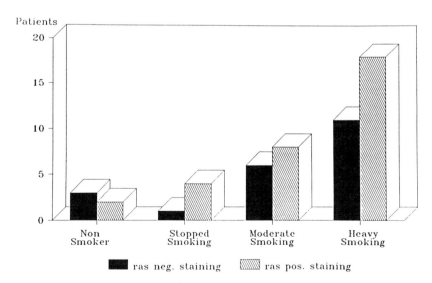

Figure 2. p21 ras expression and patients smoking history in
          head and neck cancer.
          Moderate smoking < 20 cigarettes per day,
          Heavy smoking > 20 cigarettes per day.

found in 6/11 (65%) squamous cell carcinomas and 12/33 (36%) new squamous cell carcinomas (Chiba et al., 1990). The squamous cell pathology of the lung is associated with a patient's history of smoking, and therefore provides evidence for an association between p53 mutations and smoking.

We have used the PAb 420 and PAb 1801 monoclonals to assess p53 expression in 73 SCC of the head and neck (Field et al., 1991b). A correlation was found between the patients smoking history and positive p53 staining. Six of seven non-smokers did not express p53, whereas 20 of 37 heavy smokers were found to have elevated p53 expression (P<0.005). It is also of particular interest that out of a group of ten patients who had given up smoking for more than 5 years (range 5-18 years), 9 had elevated levels of p53 expression. These results strongly suggest that mutagen(s) in tobacco cause a p53 mutation relatively early in the smoking history of these patients and that further event(s) are required for the development of the neoplasia.

These results also encouraged us to investigate whether the patients smoking history was related to p21 ras expression (Fig. 2); however, no correlations were found. When the p53 and p21 ras levels of expression were correlated with patients smoking history, it was found that 13/20 p21 ras positive, p53 positive staining squamous cell carcinomas were from patients that had a history of heavy smoking. Also, when the group of patients who had stopped smoking were included, 16/20 patients had positive staining for both of these gene products. When the data was further analysed by weighted logistic regression analysis, we found that p53 positive staining correlated with patients who had a history of heavy smoking or had stopped smoking. No correlation was found for p21-ras expression and smoking; however, the weighted logistic regression analysis of positive p21 ras and p53 staining correlated with the patient's history of heavy smoking (P <0.05) (Figure 3).

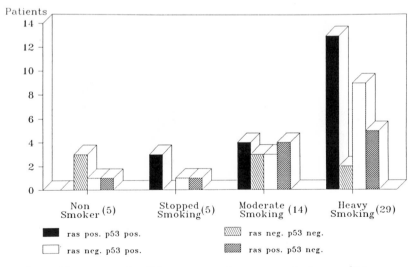

Figure 3. p21 ras and p53 expression correlates with a history of heavy smoking in squamous cell carcinoma of the head and neck.

These results, when considered in the light of the presently accepted model of p53 mode of action, may suggest that the carcinogens in tobacco cause mutations in the p53 gene and thereby increase its expression, and that there has been both a loss of suppressor gene function and a gain in the dominant transforming activity of the p53 gene. In addition, it may be argued that the head and neck tumour with no p53 staining may have either a normal p53 gene, or have lost p53 expression of both alleles, or may just contain very low levels of p53 mutant which have not been detected by either of the monoclonal antibodies we have used.

The presently accepted model for the action of ras oncogenes neoplasms, concerning the hypothesis that mutated ras genes have lost their ability to be inactivated and thereby stimulate growth in an uncoordinated manner, would seem to be inappropriate for head and neck cancer in the western world. As there is little evidence to indicate that ras mutations, rearrangements, amplification or loss of Ha-ras alleles are common in squamous cell carcinomas of the head and neck, therefore other, as yet, unidentified mechanisms must exist to explain ras overexpression in these tumours. However, as we have found a statistical correlation linking both overexpression of ras and p53 to the patient's smoking history, it may be suggested that co-operation between the abnormal expression of these genes is important in the initiation phases of head and neck cancer. Moreover, as our results demonstrate a correlation between elevated p21 expression and a favourable prognosis, it may be argued that ras is involved in the early stages of head and neck carcinogenesis.

Further studies are required to establish the incidence of p53 mutations in head and neck cancer, and if these results can be correlated with ras expression and smoking history in patients with premalignant lesions, then it maybe possible to ascertain whether aberrant expression of these two genes are the initiation events in squamous cell carcinoma of the head and neck.

Acknowledgements

Figure 1 was reproduced with kind permission from the European Journal of Surgical Oncology (Field et al., 1991a).

References

Ali, I. U., Lidereau, R., Theillet, C., and Callahan, R. Reduction to homozygosity of genes on chromosome 11 in human breast neoplasia. Science 238: 185-188 (1987).

Azuma, M., Furumoto, N., Kawamata, H., Yoshide, H., Yanagawa, T., Yura, Y., Hayashi, Y., Takegawa, Y. and Sato, M. The relation of ras oncogene product p21 expression to clinico-pathological status and clinical outcome in squamous cell head and neck cancer. The Cancer Journal 1: 375-380 (1987).

Baker, S. J., Fearon, E. R., Nigro, J. M., Hamilton, S. R., Preistinger, A. C., Jessup, J. M., vanTuinen, P., Ledbetter, D. H., Barker, D. F., Nakamura, Y., White, R. and Vogelstein, B. Chromosome 17 deletions and p53 gene mutations in colorectal carcinomas. Science 244: 217-221 (1989).

Balmain, A., Ramsden, M., Bowden, G. T. and Smith, J. Activation of the mouse cellular Harvey-ras gene in chemically induced benign skin papillomas. Nature 307: 658-660 (1984).

Barbacid, M. ras genes. An. Rev. Biochem. 56: 779-827 (1987).

Barbacid, M. ras oncogenes: their role in neoplasia. W. Eur. J. Clin. Inv. 20: 225-235 (1990).

Bos, J. L. ras oncogenes in human cancer: A review. Cancer Res. 49: 4682-4689 (1989).

Bos, J. L., Fearon, E. R., Hamilton, S. R., Verlaan-de Vries, M., Van Boom, J. H., Van der Eb, A. J., Vogelstein, B. Prevalence of ras gene mutations in human colorectal cancers. Nature 327: 293-297 (1987).

Bremmer, R. and Balmain, A. Genetic changes in skin tumour progression: correlation between presence of a mutant ras gene and loss of heterozygosity on mouse chromosome 7. Cell 61: 407-417 (1990).

Brown, K., Buchmann, A. and Balmain, A. Carcinogen-induced mutations in the mouse c-Ha-ras gene provide evidence of multiple pathways for tumor progression. Proc. Natl. Acad. Sci. USA 87: 538-542 (1990).

Cancer Research Campaign Factsheet. Oral Cancer, The Lancet 14: 1-5 (1990).

Capon, D. J., Chen, E. Y., Levison, A. D., Seeburg, P. H. and Goeddel, D. V. Complete nucleotide sequences of the T24 human bladder carcinoma oncogene and its normal homologue. Nature 302: 33-37 (1983).

Cattoretti, G., Rilke, F., Andreola, S., D'Amato, L. and Delia, D. p53 expression in breast cancer. Int. J. Cancer 41: 178-183 (1988).

Chiba, I., Takahashi, T., Nau, M., D'Amico, D., Curiel, D., Mitsudomi, T., Buchhagen, D., Carbone, D., Piantadosi, S., Koga, H., Reissman, P., Slamon, D., Holmes. E. C. and Minna, J. Mutations in the p53 gene are frequent in primary, resected non-small cell lung cancer. Oncogene 5: 1603-1610 (1990).

Crawford, L. V., Pim, D. C., Gurney, E. G., Goodfellow, P. and Taylor-Papadimitriou, J. Detection of a common feature in several human tumor cell lines - a 53,000 Dalton protein. Proc. Natl. Acad. Sci. USA 78: 41-45 (1981).

Davis, R. K. Prognostic variables in head and neck cancer. Otolaryng. Clin. N. Amer. 18: 411-419 (1985).

Doll, R. and Peto, R. Mortality in relation to smoking: 20 years observations on male doctors. Br. Med. J. 2: 1525-1536 (1976).

Field, J. K. The biology of oncogenes and their role in malignant transformation. In: Risk Markers of Oral Disease 2. Oral Cancer. N. W. Johnson (Ed.), Cambridge University Press, pp. 257-293, 1991.

Field, J. K. and Spandidos, D. Expression of oncogenes in
human tumours with special reference to the head and
neck region. J. Oral Pathol. 16: 97-107 (1987).

Field, J. K. and Spandidos, D. A. Expression of ras and
myc oncogenes in human solid tumours and their relevance
in diagnosis and prognosis. A review. Anticancer Res.
10: 1-22 (1990).

Field, J. K., Lamothe, A. and Spandidos, D. A. Clinical
relevance of oncogene expression in head and neck
tumours. Anticancer Res. 6: 595-600 (1986).

Field, J. K., Spandidos, D. A., Stell, P. M., Vaughan, E. D.,
Evan, G. I. and Moore, J. P. Elevated expression of the
c-myc oncoprotein correlates with poor prognosis in head
and neck squamous cell carcinoma. Oncogene 4: 1463-
1468 (1989).

Field, J. K., Spandidos, D. A., Yiagnisis, M., Reed, T.,
Papadimitriou, K. and Stell, P. M. Immunocytochemistry
of the C-erb-2 product in head and neck squamous cell
carcinoma. In Vivo 4: 88 (1990).

Field, J. K., Yiagnisis, M., Spandidos, D.A., Gosney, J.R.,
Papadimitriou, K., Vaughan E.D. and Stell, P. M. Low
levels of ras p21 oncogene expression correlates with
outcome in head and neck squamous cell carcinoma. Eur.
J. Surg. Oncol. (in press) (1991a).

Field, J. K., Spandidos, D. A., Malliri, A., Yiagnisis, M.,
Gosney, J. R. and Stell, P. M. Elevated p53 expression
correlates with a history of heavy smoking in squamous
cell carcinoma of the head and neck. Br. J. Cancer
(in press) (1991b).

Forrester, K., Almogvera, C., Han, K., Grizzle, W. E.,
Perucho, M. Detection of high incidence of Ki-ras
oncogenes during human colon tumorigenesis. Nature 327:
298-303 (1987).

Friedman, W. H., Rosenblum, B. N., Loewenstein, P., Thornton,
A., Katsantonis, G. and Green, M. Oncogenes: their
presence and significance in squamous cell cancer of the
head and neck. Laryngoscope 95: 313-316 (1985).

Furth, M. E., Davis, I. J., Fleurdelys, B. and Scolnick, E.
M. Monoclonal antibodies to the p21 products of the
transforming gene of Harvey murine sarcoma virus and of
the cellular ras gene family. J. of Virol. 43: 294-304
(1982).

Hoellering, J. and Shuler, C. F. Localization of H-ras mRNA
in oral squamous cell carcinomas. J. Oral Pathol. Med.
18: 74-78 (1989).

Howell, R. E., Wong, F. S. H., Fenwick, R. C. Activated
Kirsten ras oncogene in an oral squamous carcinoma. J.
Oral Path. Med. 19: 301-305 (1990).

Iggo, R., Gatter, K., Bartek, J., Lane, D. and Harris, A. L.
Increased expression of mutant forms of p53 oncogene
in primary lung cancer. Lancet 335: 675-679 (1990).

Johnson, N. W. Oral-facial Neoplasms: Global Epidemiology,
Risk Factors and Recommendations for Research. Final

Report of Working Group 2 of the Commission on Oral Health, Research and Epidemiology, Federation Dentaire Internationale. London, (1990).

Kaplan, E. L. and Meier, P. Nonparametric estimation from complete observation. J. Amer. Statis. Assoc. 53: 457-481 (1958).

Krajina, Z., Kucar, Z. and Konic-Carnelutti, V. Epidemiology of laryngeal cancer. Laryngoscope 85: 1155-1161 (1975).

Krontiris, T. G., Dimartino, N. A., Colb, M. and Parkinson, D. R. Unique allelic restriction fragments of the human Ha-ras locus in leukocyte and tumour DNAs of cancer patients. Nature 313: 369-374 (1985).

Kumar, R., Saraswati, S. and Barbacid, M. Activation of ras oncogenes preceding the onset of neoplasia. Science 248: 1101-1104 (1990).

Kurozumi, S., Harada, Y., Sugimoto, Y. and Sasaki, H. Airway malignancy in poisonous gas workers. Laryngol Otol. 91: 217-223 (1977).

Lane, D. P. and Benchimol, S. p53: oncogene or anti-oncogene? Genes and Develop 4: 1-8 (1990).

McCormick, F. ras GTPase activating protein: signal transmitter and signal terminator. Cell 56: 5-8 (1989).

McCoy, D. G. and Wynder, E. L. Etiological and preventive implications in alcohol carcinogenesis. Cancer Res. 39: 2844-2850, 1979.

Merrit, W. D., Weissler, M. C., Turk, B. F. and Gilmer, T. M. Oncogene Amplification in Squamous Cell Carcinoma of the Head and Neck. Arch. Otolaryngol. Head Neck Surg. 116: 1394-1398 (1990).

Myers, E. N. and Suen, J. Y. Cancer of the Head and Neck. 2nd ed., Churchill Livingstone Inc., NY (1989).

Nicholls, P., Edwards, G., Kyle, E. Alcoholics admitted to four hospitals in England. II. General and cause-specific mortaility. Quart. J. Stud. Alc. 35: 841-855, (1974).

Nigro, J. M., Baker, S. J., Preisinger, A. C., Jessup, J. M., Hostetter, R., Cleary, K., Bigner, S. H., Davidson, N., Baylin, S., Devilee, P., Glover, T., Collins, F. S., Weston, A., Modali, R., Harris, C. C. and Vogelstein, B. Mutations in the p53 gene occur in diverse human tumour types. Nature 342: 705-708 (1989).

Nordenskjold, M. and Cavenee, W. K. Genetics and the etiology of solid tumors. In: Important Advances in Oncology. DeVita, V. T. Jr., Hellman, S. & Rosenberg, S. A. (eds), p. 83. J. B. Lippincott: Philadelphia (1988).

Oral Cancer, Editorial Lancet 2: 311-312 (1989).

Parada, L. F., Land, H., Weinburgh, R. A., Wolf, D. and Rotter, V. Cooperation between gene encoding p53 tumour antigen and ras in cellular transformation. Nature 312: 649-654 (1984).

Parkin, D. M., Laara, E. and Muir, C. S. Estimates of the worldwide frequency of sixteen major cancers in 1980. Int. J. Cancer 41: 184-187 (1988).

Quintanilla, M., Brown, K., Ramsden, M. and Balmain, A. Carcinogen-specific mutation and amplification of Ha-ras during mouse skin carcinogenesis. Nature 322: 78-80 (1986).

Riou, G., Barrois, M., Sheng, Z. M., Duvillard, P. and Lhomme, C. Somatic deletions and mutations of c-Ha-ras gene in human cervical cancers. Oncogene 3: 329-333 (1988).

Rumsby, G., Carter, R. L. and Gusterson, B. A. Low incidence of ras oncogene activation in human squamous cell carcinomas. Br. J. Cancer 61: 365-368 (1990).

Saranath, D., Panchal, R. G., Nair, R., Mehta, A. R. and Deo, M. G. Oncogene amplification in squamous cell carcinoma of the oral cavity. Jpn. J. Cancer Res. 80, 430. (1989).

Saranath, D., Chang, S. E., Bhoite, L. T., Panchal, R. G., Kerr, I. B., Mehta, A. R., Johnson, N. W. and Deo, M. G. High frequency mutation in codons 12 and 61 of H-ras oncogene in chewing tobacco-related human oral carcinoma in India. Br. J. Cancer 63, 573-578 (1991).

Schlichting, I., Almo, S. C., Rapp, G., Wilson, K., Petratos, K., Lentfer, Ar., Wittinghofer, A., Kabsch, W., Pai, E., Petsko, G. A. and Goody, R. S. Time-resolved X-ray crystallographic study of the conformational change in Ha-Ras p21 protein on GTP hydrolysis. Nature 345: 309-315 (1990).

Schmidt, W. and de Lint, J. Estimating the prevalence of alcoholism from alochol consumption and mortality data. Quart. J. Stud. Alc. 31: 957-964 (1970).

Schmidt, W. and Popham, R. E. The role of drinking and smoking in mortality from cancer and other causes in male alcoholics. Cancer 47: 1030-1041 (1981).

Sheng, Z. M., Barrois, M., Klijanienko, J., Micheau, C., Richard, J. M. and Riou, G. Analysis of the c-Ha-ras-1 gene for deletion, mutation, amplification and expression in lymph node metastases of human head and neck carcinomas. Br. J. Cancer 62: 398-404 (1990).

Silverman, S. and Griffith, M. Smoking characteristics of patients with oral carcinoma and risk for second oral primary carcinoma. J Am Dent Assoc. 85: 637-642 (1972).

Spandidos, D. A., Lamothe, A. and Field, J. K. Multiple transcriptional activation of cellular oncogenes in human head and neck solid tumours. Anticancer Res. 5: 221-224 (1985).

Stell, P. M. Smoking and laryngeal cancer. Lancet 1: 617-618 (1972).

Stell, P. M. Prognostic factors in laryngeal carcinoma. Clin. Otol. 13: 399-409 (1988).

Suarez, H. G., Du Villard, J. A., Caillou, B., Schlumberger, M., Tubiana, C., Parmentier, C., Monier, R. Detection of activated ras oncogenes in human thyroid carcinomas. Oncogene 2: 403-406 (1988).

Takahashi, T., D'Amico, D., Chiba, I., Buchhaben, D. and Minna, J. Identification of intronic mutations as an alternative mechanism for p53 inactivation in lung cancer. J. Clinical Invest. 86: 363-369 (1989).

Tanaka, T., Slamon, D. J., Shimoda, C. W., Kawaguchi, Y., Tanaka, Y. and Ida, N. Expression of Ha-ras oncogene in human neuroblastomas and the significant correlation with patient prognosis. Cancer Res. 48: 1030-1034 (1988).

Tsuji, T., Sasaki, K., Hiraoka, F. and Shinozaki, F. The immunohistochemical detection of ras p21 and its correlation with differentiation in oral cancers. Journal of Tumour Marker Oncology 4: 415-419 (1989).

Union Internationale Contre le Cancer. TNM Classification of malignant tumours. Eds. P. Hermanek and L. Fobin. Springer Verlag, Heidelberg (1987).

Wynder, E.L. and Stellman, S.D. Comparative epidemiology of tobacco-related cancers. Cancer Res. 37: 4608-4622 (1977).

Wynder, E. L., Bross, I. J., Day, E. Epidemiological approach to the etiology of cancer of the larynx. JAMA 160: 1384-1389 (1956).

Zarbl, H., Sukumar, S., Arthur, A. V., Martin-Zanca, D and Barbacid, M. Direct mutagenesis of Ha-ras-1 oncogenes by N-nitroso-N-methylurea during initiation of mammary carcinogenesis in rats. Nature 315: 382-385 (1985).

POSSIBLE ROLE FOR HA-*ras* EXPRESSION IN INDUCIBLE

STEROIDOGENESIS IN IMMORTALIZED GRANULOSA CELL LINES

Abraham Amsterdam, Lea Eisenbach*, Byung Sun Suh,
Debora Plehn-Dujowich, Iris Keren Tal and Ada Dantes

Departments of Hormone Research and *Cell Biology
The Weizmann Institute of Science
Rehovot, 76100, Israel

ABSTRACT

Primary granulosa cells cotransfected with SV40 DNA and the Ha-*ras* oncogene can be induced to produce progestins (progesterone and 20α–dihydroprogesterone) when incubated with 8-Br-cyclic AMP and substances elevating intracellular cyclic AMP (cAMP) such as forskolin, choleratoxin and the *Bordetella pertussis* invasive adenylate cyclase (BPAC). In contrast, cells transfected with SV40 DNA alone show only traces of steroidogenic activity under similar stimulation. The steroidogenic capacity of the cotransfected lines was correlated with the epithelioid appearance of the cells and low expression of actin and actin binding proteins in these cells. Expression of isoforms 2 and 3 of tropomyosins which possess high affinity for actin filaments was extremely low in these cells compared to cells transfected with SV40 DNA alone. Expression of p21 in cotransfected individual lines was correlated to the steroidogenic capacity. Primary granulosa cells and luteinized cells also express modestly but significantly p21 precipitable by monoclonal antibodies against the proto/mutated oncogene product. The cotransfected cells were highly tumorigenic when injected to nude mice but pretreatment of the cells with BPAC, which resulted in prolonged intracellular accumulation of cAMP, prevented metastatic spread of the tumor cells. Therefore, high levels of intracellular cAMP may arrest proliferation of the transformed cells both *in vivo* and *in vitro*. It is suggested that the expression of the Ha-*ras* oncogene may be involved in inducible steroidogenesis in immortalized granulosa cell lines, while the product of the protooncogene may be implicated in this process in normal cells.

INTRODUCTION

Cellular and viral oncogenes are defined by their ability to elicit neoplastic transformation (for review see [1-4]). In a variety of systems, oncogenes have also been implicated in controlling differentiation[5-7]. *Ras* oncogenes are one of the most prevalent oncogenes in human[8-10] and carcinogen-induced animal tumors[11,12]. The identified *ras* genes in the mammalian genome[13-15] Ha-*ras*, K-*ras* and N-*ras*, encode proteins that bind guanine nucleotides[16], have GTPase activity[17,18] associate with the plasma membrane[19,20], and are homologous to G-proteins[21,22]. These properties suggest that *ras* proteins participate in signal transduction across the cell membrane[23,24]. In mammals, the *ras* proteins have been implicated in cellular proliferation and terminal differentiation[25,26]. However, the role of this protein in these processes remained enigmatic.

To investigate further the effects of *ras*-oncogene expression on the differentiated phenotype, the well characterized rat granulosa cell system was utilized[27-32]. These cells nurse the mammalian egg during each reproductive cycle, and differentiate to become the main site of progesterone production in the body[33], which is essential for the regulation of the menstrual cycle and the maintenance of pregnancy. Progesterone synthesis in granulosa cells is induced by gonadotropic hormones via a cAMP dependent pathway[30,31]. Cultured primary granulosa cells produce progesterone in response to elevated cAMP levels[34]. However, recent studies suggest that the gonadotropic hormones may also exert their effect in a cAMP independent manner by modulation of inositol triphosphate and cytosolic free Ca^{2+} [35].

In this review we demonstrate that when highly steroidogenic granulosa cells from preovulatory follicles are co-transfected with SV40 DNA and the Ha-*ras* oncogene, immortalized cell lines can be obtained which preserve cAMP stimulated progesterone production at levels similar to those of the highly steroidogenic primary cells. The lack of steroidogenic response in immortalized granulosa cells transfected with SV40 alone suggests that the cAMP stimulation of steroidogenesis may be modulated by the *ras* oncogene expression product p21.

RESULTS AND DISCUSSION

Establishment of immortalized granulosa cell lines

Granulosa cells cease to divide in primary cultures. Therefore, several attempts were made by different laboratories to immortalize these cells. Unfortunately, in preliminary attempts, although transfection of the cells with SV40 DNA produced immortalized cell lines, the cells lost almost completely their steroidogenic activity[36].

We have succeeded in immortalizing rat granulosa cells by cotransfection with SV40 and Ha-*ras* oncogene[37-41]. This combination produced immortalized lines which acquire a pronounced steroidogenic activity subsequent to stimulation with substances elevating cAMP within the cells. Interestingly, the transfected cells from preovulatory follicles produce a much higher amount of progesterone than those obtained from preantral follicles, similar to the steroidogenic activity of the normal cells from which they were derived[37,41].

The transformed cells however differ from primary cells, since induction of differentiation and suppression of cell growth in these cells is achieved only after prolonged incubation with cAMP[37,41] and the effect of cAMP seems to be reversible, unlike that in primary cells.

Role of the ras Protein Expression

Only cell lines transfected with the Ha-*ras* oncogene as well as with SV40 DNA retain their steroidogenic capacity while cells transfected with SV40 alone lose it. This is true for cells derived from both preantral and preovulatory follicles. In general, lines that show more characteristic epithelioid morphology express higher amounts of the oncogene product p21 and also show higher steroidogenic potential (Fig. 1; Suh and Amsterdam, in preparation). This would suggest that expression of the *ras* oncogene may be important for the expression of the steroidogenic enzymes under cAMP stimulation. Indeed, we also find that primary granulosa cells and luteinized cells produce significant amounts of p21 probably as a product of the proto-oncogene[37,41]. Expression of the *ras* oncogene in the transformed granulosa cells is probably a prerequisite for their differentiation since we did not find any modulation of p21 expression upon cAMP stimulation[38] (Fig.2) . In contrast, expression of p21 coded by the proto-oncogene could be modulated since it was modestly enhanced in luteinized cells compared to immature granulosa cells. However, since the antibodies used for the detection of the p21 oncogene can also recognize the products of the K-*ras* and N-*ras* proto-oncogenes[37], modulation of the p21 expression of any of the three members of this super family could occur. The mechanism by which expression of the *ras* oncogene affects differentiation is not yet clear and there are several possibilities:
(i)   The *ras* p21 participates in a signal transduction pathway leading to differentiation;
(ii) It antagonizes the anti-differentiating effect of SV40. Indeed, we found that in cotransfected granulosa cells a high molecular weight isoform of the T antigen is not expressed compared to cells transfected with SV40 alone[37,41] (Fig. 2);
(iii) *ras* expression affects the organization and expression of cytoskeletal proteins in reaching the right configuration of the cytoskeleton which is essential for the development of steroidogenesis in the cells[38]
Investigating the role of the non-mutated p21 in normal cells may illuminate the role of proto-oncogenes and suppressor genes in differentiation and luteinization of normal granulosa cells[6,30,31].

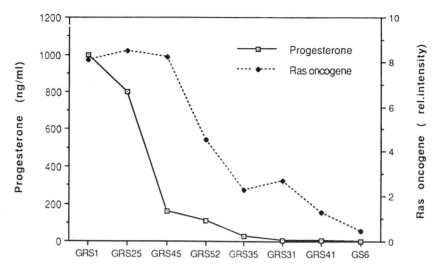

Fig. 1. **Comparison of steroidogenesis and Ha-*ras* oncogene expression (p21) in different rat granulosa cell lines obtained after cotransfection with SV40 and the Ha-*ras* oncogene (GRS lines).** Expression of the oncogene product p21 was measured after immunoprecipitation using monoclonal antibodies[41] and densitometer scanning of the electrophorized p21. Progesterone was measured subsequent to 48 h incubation at 37°C with 1 mM 8-Br cAMP. Only traces of progesterone (< 0.5 ng/ml/2x10^6 cells/48 h) were detected in non-stimulated cells[41] GS6; line established subsequent to transfection of primary rat granulosa cells with SV40 DNA alone.

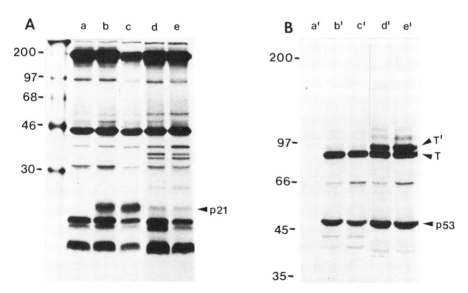

Fig. 2. **Expression of T-antigen and *ras* protein in established granulosa cell lines.** Granulosa cell lines from preovulatory follicles established by transfection with SV40 (d,e,d',e') and SV40 plus Ha-*ras* DNA (a-c, a'-c') were cultured for 48 hr in the absence (a,b,d,a',b',d') and presence (c,e'c',e') of 1 mM 8-Br-cAMP. During the last 12 hr in culture the cells were labeled with [35S]methionine. Equal cpm of radioactive proteins from each culture were immunoprecipitated with a monoclonal anti-*ras* antibody (A lanes b-e), or a monoclonal anti T-antigen antibody (B lanes b'-4'). Lanes a and a' are immunoprecipitates with nonimmune serum. (By permission from ref. 38).

Transformed granulosa cells provide a very useful model for the study of the mechanism of inducible steroidogenesis (for review see [30,31,42]). Cells transfected with SV40 alone show only traces of progesterone production. However, cells transfected with SV40 and Ha-*ras* show a similar low basal level of steroid hormone production, but upon stimulation with 8-Br-cAMP or substances elevating intracellular cAMP (i.e. forskolin, cholera toxin or BPAC) these cells produce progesterone at levels comparable to those produced by stimulated primary cells[41].

In normal cells obtained from preantral or antral follicles, the expression of steroidogenic enzymes is relatively high, even without stimulation. Unlike in normal cells, a lag period of 12-24 h is observed until elevated progesterone production can be detected[41] (Fig. 3). This lag period is due to the time required for the *de novo* synthesis of the steroidogenic enzymes of the cytochrome P450 side chain cleavage complex (P450$_{scc}$).

Upon examining the induction of steroidogenesis in transformed cells in greater detail, interesting phenomena are observed. There is no parallelism in the kinetics of induction of the individual steroidogenic enzymes, cytochrome P450$_{scc}$ and adrenodoxin (ADX)[39]. Although both contain cAMP response elements in their promoter regions, induction of ADX is faster than that of cytochrome P450$_{scc}$. The induction of the latter was parallel to the induction of progesterone production in these cells and therefore may be the limiting step in the induction of steroidogenesis[39]. The induction of the steroidogenic enzymes allows us to follow their distribution in individual mitochondria. We found that the enzymes are transported and appear in all mitochondria of a given cell[39] (Fig. 4). From these studies, we can conclude that there are no specialized mitochondria which carry the steroidogenic enzymes; all cellular mitochondria become steroidogenic.

We have found recently that the steroidogenic granulosa cells express also cholesterol carrier proteins which are essential for the transport of cholesterol into the mitochondria where the first step of conversion of cholesterol to a steroidogenic hormone takes place[43] (and J. Strauss and A. Amsterdam, in preparation). This was true both for sterol carrier protein 2 (SCP$_2$) and for the peripheral benzodaizapine receptor (PBR). Moreover the SCP$_2$ and PBR expression was enhanced by 8-Br cAMP. Interestingly, the SCP$_2$ modulation by the cyclic nucleotide was preserved also in a single transfected line (with SV40 alone) that showed only traces of steroidogenic activity[44]. This would suggest that unlike the cytochrome P450scc enzyme system which is probably associated with *ras* expression, the cholesterol carrier proteins can be modulated independently.

It was of interest to test whether the delayed response in the *ras* transformed cells is observed when cells are stimulated by the native stimulants, the gonadotropins. Unfortunately the transformed cells lose their gonadotropic receptor upon transformation. To overcome this loss, we succeeded recently in transfecting these cells with DNA coding for the luteinizing hormone (LH) receptor[45,46], which is now expressed in the transformed cell lines[47]. These cells show a restored steroidogenic response to the gonadotrophic hormone and therefore can be used to investigate the mechanism of gonadotropic action in greater detail.

The Cytoskeleton

Since primary granulosa cells undergo a dramatic change in organization and expression of the cytoskeleton during differentiation, it was of interest to analyze the cytoskeleton of SV40 compared to SV40 and Ha-*ras* transfected cells since they show considerably different morphologies[37,41]. In ultrathin sections, the SV40 transformed cells show a well developed network of thin filaments, very often parallel to the cell surface. In contrast, the SV40 and Ha-*ras* cotransfected cells show a poor network of thin filaments[41]. In 2D gels of metabolically labeled cell extracts, it was indeed found that actin expression was low in the cotransfected cells while tropomyosin isoforms 2 and 3, which show the highest affinities to actin among all isoforms of tropomyosin, were completely absent[38] (Fig. 5). Upon cAMP stimulation which induces rounding of the cells, no further down regulation of the actin cytoskeleton was observed. This would suggest that the lower expression of actin and actin binding proteins is a prerequisite for the development of steroidogenic capacity, not only in normal but also in transformed cells. However, since differential changes in modulation of tropomyosin isoforms were achieved in the

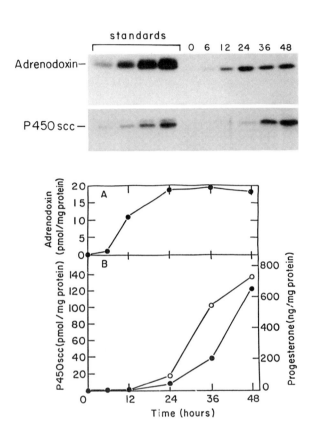

Fig. 3. **Correlation of progesterone synthesis and induction of cytochrome P450$_{scc}$ in a granulosa cell line cotransfected with SV40 DNA and the Ha-*ras* oncogene.** The top portion of the figure shows autoradiograms of Western blots of two different gels containing either purified adrenodoxin or P450$_{scc}$, and aliquots of cell protein isolated at indicated times after stimultion of the cells with 1 mM 8-Br-cAMP. In each gel the standards were 0.25, 0.5, 1, and 2 pmol of the indicated protein purifed from bovine adrenal cortex. The quantitation of the enzymes is based on densitometric scanning of the Western blots. (*Bottom*) Induction of adrenodoxin, cytochrome P450$_{scc}$ (•) and progesterone (o) synthesis in the cell line after stimulation with 8-Br-cAMP. (By permissio from ref. 39).

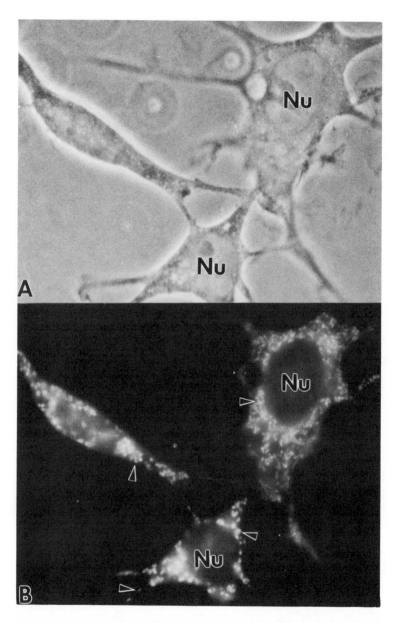

Fig. 4. **Localization of a steroidogenic enzyme in mitochondria of rat granulosa cells transformed with SV40 DNA and Ha-*ras* oncogene**. Phase contrast (A) and immunofluorescent (B) micrographs of the same PO-GRS1 cells reacted with antibodies to adrenodoxin subsequent to 48 h incubation at 37°C with 0.1 mM forskolin and permeabilization of the cells[39]. The enzyme is located in mitochondria dispersed throughout the cytoplasm. The arrowheads mark some regions where individual mitochondria can be distinctly visualized. Nu; nucleus. Magnification x 2000.

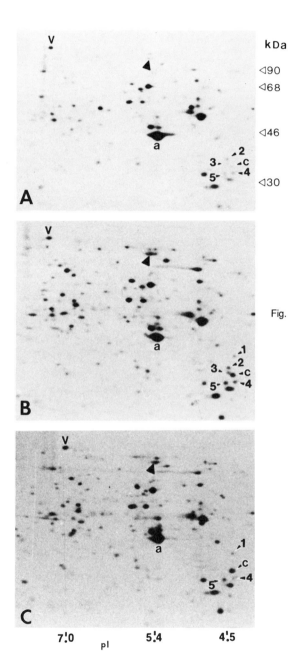

Fig. 5. **Downregulation of tropo-myosin expression in differentiating primary granulosa cells and in a SV-40 Ha-ras transfected granulosa cell line which maintains the ability to differentiate in response to cAMP.** A, primary granulosa cells stimulated for 48 h with cAMP; B, granulosa cell lines obtained by transfection with SV40, which lost the ability to differentiate; C, granulosa cell lines obtained by transfection with SV40 and Ha-*ras* oncogene a, actin; 1,2,3,4,5 - tropomyosin isoforms; v-vinculin, large arrowhead points to the position of α-actinin; c, cyclin. Note the very low expression of all tropomyosin isoforms in stimulated primary cells (A) and the absence of tropomyosin isoforms 2 and 3 in cells transformed by SV40 + Ha-*ras* oncogene (C). (By permission from ref. 38).

233

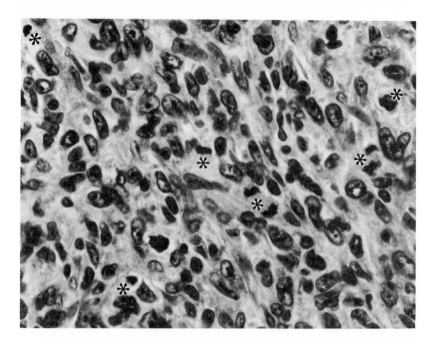

Fig. 6. **Granulosa cell tumor in a nude mouse.** Mice were injected subcutaneously with 2×10^6 cells and a piece of a developed tumor was fixed and sectioned (parafin section stained with hemotoxilin eosin) 4 weeks after injection. The tumor cells are tightly packed showing coffee bean shape nuclei. Very often mytotic figures are visible (astrisks). Magnification x 400.

transformed cells, it is concluded that the isoforms 2 and 3 are the most important isoforms to modulate the organization of the actin cytoskeleton in these cells. The exact role of the expression of the *ras* oncogene on cytoskeleton organization and expression is not yet understood. However, since expression of *ras* and related proteins could modulate cell shape in yeast[48] and other cell types, one cannot exclude the possibility that this effect is an important signal for differentiation implicated with *ras* protein expression.

Carcinogenesis

The oncogene transformed granulosa cells rapidly proliferate *in vitro*; however, when stimulated by cAMP they stop dividing[41]. When injected to nude mice they form large tumors which, within a couple of months, lead to the death of the animals. Lines obtained from the undifferentiated granulosa cells, those of preantral follicles, developed tumors more rapidly than cells obtained from preovulatory follicles which possess a higher steroidogenic capacity. When injected into the foot pad, or intravenously, the SV40 Ha-*ras* transformed cells produce metastases in ovaries, lung and kidney, but not in liver. The morphology of these cells is of a typical granulosa cell tumor which occasionally develops spontaneously in aged animals. They show an epithelioid shape with a typical coffee-bean shaped nucleus (Fig. 6). When cells are treated with the invasive BPAC *in vitro* prior to their injection to the animals, they are unable to produce metastases, probably because continuous elevated cAMP levels prevent their proliferation[49]. These observations suggest that the oncogene transformed cells can serve as an experimental model of ovarian cancer tumours, which in most cases are malignant and lethal. The study of drugs which suppress cell growth by forcing them to differentiate may be a useful approach against metastasis associated with the *ras* oncogene expression.

PROSPECTIVE RESEARCH

Ras-transformed granulosa cells can serve as a useful model for studying the control of steroidogenesis and tumorigenesis. Moreover, when cells were recently transfected with an LH-receptor expression plasmid, reconstitution of the hormonal response was achieved. It is hoped that these unique transformed cell lines, which are sensitive to cAMP and the gonadotropic hormone stimulation will serve as a useful model for studying the complex relationships between proliferation, differentiation and transformation, and hopefully lead to a better understanding of how *ras* proteins control these interconversions.

ACKNOWLEDGEMENTS

We thank all collaborators and especially Drs. A. Ben Ze'ev, M. Oren and I. Hanukoglu for their fruitful collaboration and Mrs. M. Kopelowitz for excellent secretarial assistance. This work was supported in part by the Minerva Foundation (Munich/Germany) and by a grant from the Joseph and Ceil Mazer Center for Structural Biology, and the Leo and Julia Forchheimer Center for Molecular Genetics at the Weizmann Institute of Science. A. Amsterdam is the Joyce and Ben B. Eisenberg Professor of Molecular Endocrinology and Cancer Research.

REFERENCES

1. R. A. Weinberg, The action of oncogenes in the cytoplasm and nucleus, *Science* 230:770 (1985).
2. R. A. Weinberg, Oncogenes, antioncogenes, and the molecular basis of multistep carcinogenesis, *Cancer Res.* 49:3714 (1989).
3. J. M. Bishop, The molecular genetics of cancer, *Science* 235:305 (1987).
4. D. A. Spandidos, and M. L. M. Anderson, Oncogenes and onco-suppressor genes: Their involvement in cancer, *J. Pathol.* 157:1 (1989).
5. D. Bar-Sagi, and J. R. Fermisco, Microinjection of the *ras* oncogene protein into PC12 cells induces morphological differentiation, *Cell* 42:841 (1985).
6. R. Muller, Proto-oncogene and differentiation, *Trends Biochem. Sci* 11:129 (1986).
7. R. Beug, P. A. Blandell, and T. Graf, Reversibility of differentiation and proliferation capacity in avian myelomonocytic cells transformed by ts E26 leukemia virus, *Genes & Develop.* 1:277 (1987).
8. C. Shih, L. C. Padley, M. Murray, and R. A. Weinberg, Transforming genes of carcinomas and neuroblastomas introduced into mouse fibroblasts, *Nature* 290:261 (1981).
9. T. G. Krontiris, and G. M. Cooper, Transforming activity of human tumor DNAs, *Proc. Natl. Acad. Sci. U.S.A* 78:1181 (1981).
10. J. L. Bos, *ras* oncogenes in human cancer: A review, *Cancer Res.* 49:4682 (1989).
11. A. Balmain, and I. B. Pragnell, Mouse skin carcinoma induced in vivo by chemical carcinogenes have atransforming Harvey-*ras* oncogene, *Nature* 303:72 (1983).
12. A. Eva, and S. Aaronson, Frequent activation of c-Kis a transforming gene in fibrosarcoma induced by methylchlanthrene, *Science* 220:506 (1983).
13. R. W. Ellis, D. Defeo, T. Y. Shih, M. A. Gonda, H. A. Young, N. Tsuchida, D. R. Lowy, and E. M. Scolnick, The p21 src genes of Harvey and Kirstein sarcoma viruses originate from divergent members of a family of normal vertebrate genes, *Nature* 292:506 (1981).
14. L. F. Parada, C. J. Tabing, C. Shih, and R. A. Weinberg, Human EJ bladder carcinoma oncogene is homologue of Harvey sarcoma virus *ras* gene., *Nature* 297:474 (1982).
15. A. Hall, C. J. Marshall, and R. A. Weiss, Identification of transforming gene in two human sarcoma cell lines as a member of the *ras* gene family located on chromosome 1., *Nature* 303:396 (1983).
16. E. M. Scolnick, A. G. Papagerorge, and T. Y. Shih, Guanine nucleotide-binding activity as an assay for src protein of rat-derived murine sarcoma viruses, *Proc. Natl. Acad. Sci. U.S.A.* 76:5355 (1979).
17. M. R. Hanley, and T. Jackson, The ras gene: transformer and transducer, Nature (Lond.) 328:668 (1987).
18. J. P. McGrath, D. J. Capon, D. V. Goeddel, and A. D. Levinson, Comparative biochemical properties of normal and activated human *ras* p21 protein, *Nature* 310:644 (1984).
19. M. C. Willingham, I. Pastan, T. Y. Shih, and E. M. Scolnick, Localization of src gene product of the Harvey strain of MSV to plasma membrane of transformed cells by electronmicroscopic immunocytochemistry., *Cell* 19:1005 (1980).
20. B. M. Williamsen, A. Christensen, N. L. Hubbert, A. G. Papageorge, and D. R. Lowry, The p21 *ras* C-terminus is required for transformation and membrane association., *Nature* 310:583 (1986).
21. J. B. Hurley, M. I. Simon, D. B. Teplow, J. D. Robishaw, and A. G. Gilman, Homologies between signal transducing G proteins and *ras* gene products, *Science* 226:860 (1984).

22. M. A. Lochrie, J. B. Hurley, and M. L. Simon, Sequence of the alpha subunit of phosphoreceptor G protein: Homologies between transducin, *ras*, and elongation factor, *Science* 228:96 (1985).

23. M. Barbacid, *ras* genes, *Ann. Rev. Biochem.* 56:779 (1987).

24. I. G. Macara, and A. Wolfman, Signal transduction and *ras* gene family: Molecular switches of unknown function, *Trends in Endocr.* 1:26 (1989).

25. G. P. Dotto, L. F. Parada, and R. A. Weinberg, Specific growth response of *ras* transformed embryo fibroblasts to tumor promoters, *Nature* 318:472 (1985).

26. I. Guerrero, H. Wong, A. Pellicer, and D. Burnstein, Activated N-*ras* gene induces neuronal differentiation of PC12 rat pheochromocytoma cells., *J. Cell. Physiol.* 129:71 (1986).

27. A. Amsterdam, A. Berkowitz, A. Nimrod, and F. Kohen, Aggregation of luteinizing hormone receptors in granulosa cells: A possible mechanism of desensitization to the hormone, *Proc. Natl. Acad. Sci. U.S.A.* 77:3440 (1980).

28. M. M. Sanders, and J. A. R. Midgley, Rat granulosa cell differentiation: An *in vitro* model., *Endocrinology* 111:614 (1982).

29. M. Knecht, T. Ranta, and K. J. Catt, Granulosa cell differentiation *in vitro*: Induction and maintenance of follicle-stimulating hormone receptors by adenosine 3',5'-monophosphate, *Endocrinology* 113:949 (1983).

30. A. J. W. Hsueh, E. Y. Adashi, P. B. C. Jones, and J. Welsh T.H. , Hormonal regulation of the differentiation of cultured ovarian granulosa cells, *Endocr. Rev.* 5:76 (1984).

31. A. Amsterdam, and S. Rotmensch, Structure-function relationships during granulosa cell differentiation., *Endocr. Rev.* 8:309 (1987).

32. A. Amsterdam, S. Rotmensch, A. Furman, E. A. Venter, and I. Vlodavsky, Synergistic effect of human chorionic gonadotropin and extracellular matrix on *in vitro* differentiation of human granulosa cells: progesterone production and gap junction formation, *Endocrinology* 124:1956 (1989).

33. J. S. Richards, Maturation of ovarian follicles: Action and interactions of pituitary and ovarian hormones on follicular cell differentiation, *Physiol. Rev.* 60:51 (1980).

34. M. Knecht, A. Amsterdam, and K. J. Catt, The regulatory role of cyclic AMP in hormone-induced granulosa cell differentiation, *J. Biol. Chem.* 256:10628 (1981).

35. J. S. Davis, L. L. Weakland, R. V. Farese, and L. A. West, Luteinizing hormone increases inositol trisphosphate and cytosolic free Ca^{2+} in isolated bovine luteal cells, *J. Biol. Chem.* 262:8515 (1987).

36. T. A. Fitz, R. M. Wah, W. A. Schmidt, and C. A. Winkle, Physiological characterization of transformed and cloned rat granulosa cells, Biol. Reprod. 40:250 (1989).

37. A. Amsterdam, A. Zauberman, G. Meir, O. Pinhasi-Kimhi, B. S. Suh, and M. . Oren, Cotransfection of granulosa cells with simian virus 40 and Ha-*ras* oncogene generates stable lines capable of induced steroidogenesis, *Proc. Natl. Acad. Sci. U.S.A.* 85:7582 (1988).

38. G. Baum, B. S. Suh, A. Amsterdam, and A. . Ben-Ze'ev, Regulation of tropomyosin expression in transformed granulosa cell lines with steroidogenic ability, *Dev. Biol.* 142:115 (1990).

39. I. Hanukoglu, B. S. Suh, S. Himmelhoch, and A. . Amsterdam, Induction and mitochondrial localization of cytochrome P450scc system enzymes in normal and transformed ovarian granulosa cells, *J. Cell Biol.* 111:1973 (1990).

40. D. Michalovitz, A. Amsterdam, and M. . Oren, Interactions between SV40 and cellular oncogenes in the transformation of primary rat cells, *in:* "Current Topics in Microbiology and Immunology," ed., Springer-Verlag, (1989).

41. B. S. Suh, and A. . Amsterdam, Establishment of highly steroidogenic granulosa cell lines by cotransfection with SV40 and Ha-*ras* oncogene: Induction of steroidogenesis by cAMP and its suppression by TPA, *Endocrinology* 127:2489 (1990).

42. I. Hanukoglu, Molecular biology of cytochrome P450 systems in steroidogenic tissues, in: "Follicular Development and the Ovulatory Response," A. Tsafriri, and N. Dekel, ed., Ares-Serono Symposia Review, 23:233-252 (1989).

43. A. Amsterdam, and B. S. Suh, An inducible functional peripheral benzodiazepine receptor in mitochondria of steroidogenic granulosa cells, Endocrinology in press (1991).

44. H. Rennert, A. Amsterdam, J. T. Billheimer, and J. T. Strauss, Regulated expression of sterol carrier protein2 in the ovary: A key role for cyclic AMP, Submitted (1991).

45. K. C. McFarland, R. Sprengel, H. . Phillips, M. Kohler, N. Rosemblit, K. Nikolics, D. L. Segaloff, and P. H. Seeburg, Lutropin-choriogonadotropin receptor: An unusual member of the G protein-coupled receptor family, *Science* 245:494 (1989).

46. H. Loosfelt, M. Misrahi, M. Atger, R. Salesse, M. T. V. Hai-Lui Thi, A. Jol'vet, A. Guiochon-Mantel, S. Sar, B. Jallai, J. Garnier, and E. . Milgrom, Cloning and sequencing of porcine LH-hCG receptor cDNA: Variants lacking transmembrane domain, Science 245:525 (1989).

47. B. S. Suh, R. Sprengel, P. H. Seeburg, and A. Amsterdam, Functional receptors to gonadotropins in oncogene-transformed steroidogenic granulosa cells, Proceedings of the 73rd Annual Meeting of the American Endocrine Society, 1991; Abstract 1851.

48. H. D. Schmitt, P. Wagner, E. Pfaff, and D. Gallwitz, The ras-related YPT1 gene product in yeast: a GTP-binding protein that might be involved in microtubule organization, Cell 47:401 (1986).

49. B. S. Suh, L. Eisenbach, and A. Amsterdam, Cyclic AMP suppresses metastatic spread in nude mice induced by steroidogenic rat granulosa cells transformed by SV40 and Ha-*ras* oncogene(s), Proceedings of the Annual Meeting of the Israeli Endocrine Society.(1991) ; Abstract 74.

236

BIOLOGICAL FUNCTION OF *Aplysia californica rho* GENE

Rafael P. Ballestero, Pilar Esteve, Rosario Perona, Benilde Jiménez
and Juan Carlos Lacal

Instituto de Investigaciones Biomédicas
Arturo Duperier 4, 28029 Madrid

SUMMARY

rho genes are a family of genes which are structurally related to the oncogenic *ras* family. The primary structure of *rho* genes has been elucidated for the marine snail *Aplysia californica,* two *S. cerevisiae* genes, and three human versions, *rho* A, B and C. They all codify for proteins of an approximate M.W. of 21 kDa (*rho*-p21) which show 35% homology to the *ras* proteins. It has been observed that *rho* proteins are ADP-ribosylated by the botulinum toxin C3 exoenzyme, suggesting that *rho* proteins could be involved in regulating neuronal function. However very little is known about their actual biological functions. While the human *rho* A and *rho* C products have been related to cytoeskeleton organization, the *rho* A product has a weak transforming activity. We have investigated the biological properties of the *Aplysia californica rho*-p21 protein when introduced into an heterologous system, and found that it does not induce foci in a regular NIH-3T3 transfection assay. However, the morphology of the cells was slightly altered and cells grew to higher cell densities. Moreover, transforming activity was detected when isolated cell lines were inoculated into nude mice. To further investigate the potential transforming activity of the *Aplysia* gene, we have also generated a Gly->Val mutation at position 14, equivalent to the activating mutation found in oncogenic *ras* genes. No apparent increase in the transforming activity was observed, indicating that the effects on growth behaviour are probably not the primary function of *rho* proteins.

INTRODUCTION

ras genes were first isolated and characterized as the oncogenes of two strains of acute rat retroviruses designated as Harvey- and Kirsten-MSV (1,2). Both Harvey- and Kirsten-*ras* genes encode 21 Kda proteins, designated as p21, responsible for their transforming activities (3). The retroviral *ras*-p21 molecules were found to be guanine-nucleotides binding proteins (4,5), located in the internal leaflet of the cytoplasmic membrane (6).

Analysis of a number of human tumors from a diversity of tissues revealed the presence of activated *ras* genes with variable incidence (reviewed in 7, 8). Most of these were found to be homologues of either Harvey- or Kirsten-*ras* genes. A new member of this family designated N-*ras* was detected in a human neuroblastoma cell line. Comparison of the sequence of the normal *ras* genes to those of their activated counterparts showed that single point mutations within the coding sequence were sufficient to confer the transforming phenotype. Most of the activating lesions from naturally occuring tumors have been localized at either residue 12 or 61 of the *ras* p21 molecule. Other mutated positions have been also found in tumors from human tissues and experimental animal systems (9,10).

ras proteins are GTP-binding proteins with a weak GTPase activity (11-13). The intrinsic GTPase activity *in vivo* is regulated by a factor, designated as GAP for $\underline{G}$TPase $\underline{A}$ctivating $\underline{P}$rotein, present in a large variety of tissues and established cell lines (14-15). Three factors regulating the GTP/GDP cycle are guanine-nucleotide interchange proteins that have been identified and designated as rGEF (16), *ras*-GRF (17), and REP (18).

The Superfamily of ras-Related Genes
Edited by D.A. Spandidos, Plenum Press, New York, 1991

The GDP/GTP binding activity of *ras* proteins implicated them as putative signal-transducing proteins involved in the transmission of information from the extracellular milieu into the cell (19). Moreover, such proteins as the G_S and G_i of the adenylate cyclase system (20), transducin (21), or the elongation factor of protein synthesis, EF-Tu (22), show a certain degree of homology to the *ras* p21 in regions known to be associated with the GTP-binding domain. Since mutated *ras* genes have a dominant effect on cell proliferation when transfected into NIH-3T3 cells, the *ras* p21 product likely plays an important role in the regulation of cell proliferation in this system.

Among the extensive superfamily of *ras* related genes, the *rho*-family comprise a group of genes highly conserved in evolution, with detected members in yeasts, *Drosophila*, *Aplysia* (marine snail), rat and humans (23). The *Aplysia* and the three human proteins show over 85% homology and the aminoacid homology with *ras* is about 30%. Furthermore, the homology with *ras*-p21 proteins comprise regions related to either GTP binding or hydrolysis, and membrane attachment (23). In fact, both recombinant and mammalian *rho* proteins bind and hydrolize GTP at similar rates to those of other *ras*-related proteins (24,25).

The three mammalian genes (rhoA, rhoB and rhoC) are ubiquitously expressed (23,26). The products of the genes probably have a preferential membrane localization, although the exact intracellular location awaits further investigation, since there are reports finding significant amounts of *rho* proteins in the cytoplasm (27), synaptosomal membranes (28), golgi (29) and membranes of the rod outer segments (30). It has been shown that the C3-ADP-ribosyl transferase, a exoenzyme which is a component of some botulinum toxins, ADP-ribosylates *rho* proteins (31,32). The site of ADP-ribosylation is asparagine 41, in the putative effector domain (33). *rho* proteins have been involved in cytoskeletal control since when recombinant *rho* A is microinjected into NIH-3T3 cells it disrupted the intermediate filaments and actin microfilaments network (34). Moreover, constitutive overexpression of *rho* A protein in fibroblasts reduces serum dependence for growth and cells become tumorigenic when inoculated into nude mice (35).

We sought to investigate whether the *Aplysia rho* gene shows any functional relationship to its human counterparts regarding the ability to induce morphological alterations as well as its tumorigenic activity in mice.

RESULTS

BIOLOGICAL ACTIVITY OF THE WILD-TYPE *Aplysia rho* GENE

The *Aplysia californica* wild-type gene (a gift from Dr. R. Axel) was subcloned into an eukaryotic expression vector under the control of the Abelson-MuLV Long Terminal Repeat (LTR) promoter as described in Fig.1. Transfection experiments into NIH-3T3 cells by the electrophoration method were performed to investigate the putative transforming activity of the gene. No obvious foci were observed in several experiments, but consistent areas of higher cell density and sligth morphological alterations were detected in all experiments (results not shown), suggesting the possibility of an altered growth behaviour of transfected cells. Thus, we decided to isolate clones with elevated expression of the p21-*rho* protein to further investigate for these putative growth alterations. Cotransfection experiments were carried out using the pSV$_2$-neo selection marker (50:1 ratio) and several independent cell lines resistant to G418 selection were isolated as mass-cultures.

Three independent isolates from transfected NIH-3T3 cells were inoculated into nude mice and tumor progression observed twice a week (Table I). One of these clones, WT-1, induced tumors in all inoculated animals within three weeks, while the other two clones, WT-2 and WT-3, induced tumors between thre and four weeks. Explants of the tumors from one representative animal from each cell line were isolated and corresponding cell lines established and designated as T-WT-1, T-WT-2, and T-WT-3. All isolated cell lines conserved morphological alterations and growth properties.

The presence of integrated *Aplysia rho* gene in both transfected cell lines as well as tumor-derived cell lines was determined by Southern blot analysis utilizing an *Aplysia rho* carrying probe. In addition, polymerase chain reaction amplification assay (PCR) analysis of total genomic DNA from each cell line was performed using specific synthetic olygonucleotides for the *Aplysia* gene. Both methods indicated that all transfected cell lines and their corresponding tumor-derived cell lines carriedat least one integrated copy of the exogenous *rho* gene that was efficiently transcribed (results not shown).

BIOLOGICAL ACTIVITY OF A GLY->VAL ^{14} MUTANT

The wild type gene was utilized as substrate for the generation of a Gly->Val^{14} mutant by PCR amplification method using a synthetic oligonucleotide carrying the point mutation GGT->GTT at codon 14, as described in Fig. 1. This mutant has been reported to reduce the intrinsic GTPase activity in the human *rho* A gene (34), as found also in activated *ras* oncogenic mutants. Several

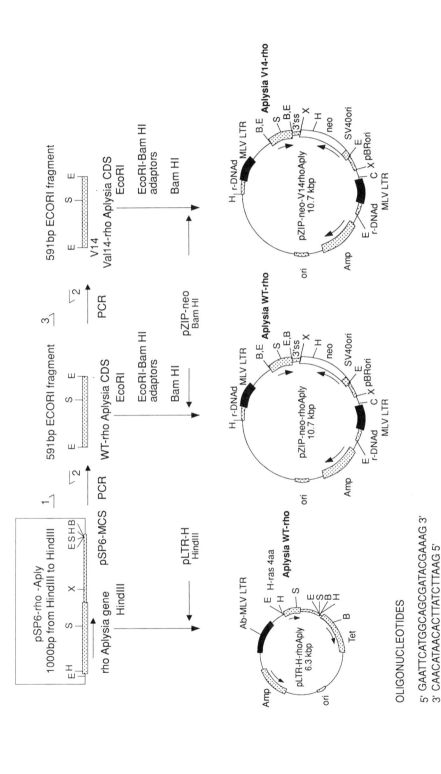

OLIGONUCLEOTIDES

5' GAATTCATGGCAGCGATACGAAAG 3'

3' CAACATAACACTTATCTTAAG 5'

5' GAATTCATGGCAGCGATACGAAAGAAGCTTGTTATAGTCGGAGATGTTGCGTGT 3'

Figure 1. Subcloning of wild-type and PCR generated VAL14 mutant *Aplysia californica rho* genes into eukaryotic expression vectors. pLTR-H is a pBR322 derived plasmid carrying Abelson-MLV LTR (Lacal, J.C. et al.; 1986. EMBO J. 5:679–687). Some restriction enzyme sites are represented. (E: EcoRI; H: HindIII; B: BamHI; S: SacI; X: XhoI; C: ClaI). r-DNAd means rat derived DNA sequences.

TABLE I.- Tumorigenicity of cell lines containing pLTR-rho constructions

NEO-RESISTANT CELL LINES	TUMORIGENICITY IN NUDE MICE	
	LATENCY	INCIDENCE
pSV$_2$-neo	-	0/3
WT-1	3 w	4/4
WT-2	4 w	3/3
WT-3	4 w	3/3
Val^{14}-2	3 w	0/2
Val^{14}-3	3 w	2/2
Val^{14}-5	3 w	3/3

Athymic nude mice were inoculated with 10^6 cells per injection. Tumor formation was followed up twice a week. Animals were sacrified when tumor formation reached at least 2 mm. and explants made as indicated in the text. Control cells were inoculated under identical conditions, and animals were followed up for 3 months after injections. Latency indicated in weeks (w).

clones were isolated and sequenced by the dideoxy-mediated chain-termination (Sanger) method (36) to select for one clone with only the desired point mutation. Both wild-type and the selected Val^{14} mutant were introduced into the pZip-neo eucaryotic vector. NIH-3T3 cells were transfected with both vectors. Although no clear foci were detected in two separate transfection experiments, using the calcium-phosphate precipitation technique, a similar effect to that found with the LTR-driven wild-type gene was observed in terms of areas of higher cell density and morphological alterations. Several clones were also isolated by selection with G418, and designated as Val^{14}-2, Val^{14}-3, and Val^{14}-5 . Inoculation of at least three independent clones carrying the Val^{14} mutation into nude mice did not reduce the lag period of tumor formation (Table I).

Cell behaviour of the various stablished cell lines WT-1, Val^{14}-2, Val^{14}-3, and Val^{14}-5 was studied by stablishing their respective growth curves in DMEM medium supplemented with 0.5% serum (low serum) (Fig. 2). While parental NIH-3T3 cells transfected with plasmid alone (NIH-3T3-neo cells) were serum-dependent, and showed contact inhibition at confluency, most rho-transfected clones showed a partial serum-independent growth and grew to higher cell densities at saturation (results not shown). Finally, a good correlation between mRNA expression levels and biological activity was also observed (results not shown).

DISCUSSION

There is an obvious conservation in the structural features of rho -p21 proteins and ras-p21 proteins due to their similar biochemical properties. In addition to these conserved structural features, there is also a low degree of conservation in regions of still unknown importance. Thus, it could be argued that these proteins share also a certain degree of conservation in their cellular functions. An easy way to study this possibility is to investigate whether expression of some members of the rho gene family into an heterologous system such as the NIH-3T3 cell line, where ras has a cell-growth promoting function, does affect cellular growth as the ras genes do. Therefore, the NIH-3T3 cell system could be an excellent assay to investigate how proximal are their biological functions. It has been shown that microinjection of the human p21-rho A, induced rapid changes in cell morphology and collapse of the intermediate filament network (34). On the other hand, expression of the same protein in Rat-1 and NIH-3T3 cells by transfection of appropriate expression plasmids, induced cell growth alterations, and showed a weak tumorigenic activity (35). These apparent contradictory results suggest a critical role of rho proteins in intracellular signalling.

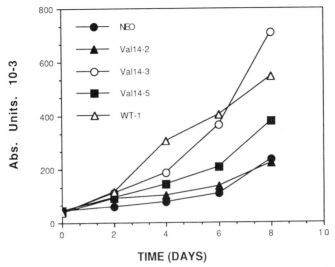

Figure 2. Growth of *Aplysia rho* transfected cell lines in low serum media. Cells were grown in 24-well miniwell plates in 0.5% foetal calf serum. At times indicated, cell number was estimated by cristal violet staining as described elsewhere (Gillies, R.J., Didier, N. & Denton, M. 1986 Anal. Biochem. 159:109–113).

We have observed that expression of the *Aplysia rho*-p21 protein into NIH-3T3 cells confers tumorigenic activity and morphological, as well as functional, alterations in close similarity to those induced by *ras* proteins. These results indicate that besides little homology between them, *rho* proteins are probably functionally related to *ras* proteins on regulating some important aspects of cell growth and /or the morphological alterations associated to cell growth. Since there is no increase in the biological function of a Val^{14} mutant, *rho* proteins are most likely not involved directly in regulating cell proliferation, but somehow are responsible for part of the associated changes observed in tansformed cells. Further investigations is needed to resolve wether there is a functional connection between *rho* and *ras* proteins.

REFERENCES

1. Harvey, J.J. (1964). Nature 204: 1104-1105.
2. Kirsten, W.H. and Mayer, L.A. (1967) J. Natl. Cancer Inst. 39:311-334.
3. Shih, T.Y., Weeks, M.O., Young, H.A., and Scolnick, E.M.(1979) Virology 96: 64-79.
4. Scolnick, E.M., Papageorge, A.G., Stokes, P.E., and Shih, T.Y.(1979) Proc. Natl. Acad. Sci. USA 76: 5355-5359.
5. Shih, T.Y., Papageorge, A.G., Stokes, P.E., Weeks, M.O., and Scolnick, E.M. (1980). Nature 287: 686-691.
6. Willingham, M.C., Pastanm, I., Shih, T.Y., and Scolnick, E.M.(1980) Cell 19: 1005-1014.
7. Barbacid, M. (1987) Ann. Rev. Bioch. 56: 779-827.
8. Lacal, J.C., and Tronick, S.E. (1988) The ras oncogene. In The Oncogene Handbook. Reddy, P., Curran, T, and Skalka,A. edts. Elsvier, Holland.
9. Bos, J.L. et al. (1985) Nature 315:726-730.
10. Reynolds, S. et al (1987) Science 237:1309-1316.
11. McGrath, J.P., Capon, D.J., Goeddel, D.V., and Levinson, A.D. (1984) Nature 310: 644-649.
12. Gibbs, J.B., Sigal, I.S., Poe, M., and Scolnick, E.M. (1984) Proc. Natl. Acad. Sci. USA 81: 5704-5708.
13. Sweet, R.W., Yokoyama, S., Kamata, T., Feramisco, J.R., Rosenberg, M., and Gross, M. (1984) Nature 311: 273-275.
14. Trahey, M., and McCormick, F. (1987) Science 238: 542-545.

15. Adari, H., Lowy, D.R., Willumsen, B.M., Der, C.J., and McCormick, F. (1988) Science 240: 518-521.
16. HUang, Y.K., Kung, H.-F., and Kamata, T. (1990) Proc. Natl. Acad. Sci. USA 87, 8008-8012
17. Wolfman, A., and Macara, I. (1990). Science 248, 67-70.
18. Downward, J., Riehl, R., Wu, L. and Weinberg, R.A. (1990) Proc. natl. Acad. Sci. 87, 5998-6002
19. Berridge, M.J., and Irvine, R.F. (1984) Nature 312: 315-319
20. Gilman, A.G. (1984) Cell 36: 577-579.
21. Stryer, L. (1986) Ann. Rev. Neurosc. 9: 787-819.
22. Ochoa, S. (1986) Arch. Biochem. Biophys. 223: 325-349.
23. Madaule, P. & Axel, R.. (1985). Cell. 41: 31-40.
24. Anderson, P.S., and Lacal, J.C. (1987) Mol. Cel. Biol. 7, 3620-3628.
25. Yamamoto, K.; Kondo, J.; Hishida, T.; Teranishi, Y. & Takai, Y.. (1988). J. Biol. Chem. 263: 9926-9932.
26. Olofsson, B.; Chardin, P.; Touchot, N.; Zahraoui, A. & Tavitian, A.. (1988). Oncogene. 3: 231-234.
27. Narumiya, S.; Sekine, A. & Fujiwara, M.. (1988). J. Biol. Chem. 263: 17255-17257.
28. Kim, S.; Kikuchi, A.; Mizoguchi, A. & Takai, Y.. (1989). Mol. Brain Res. 6: 167-176.
29. Toki, C.; Oda, K. & Ikehara, Y.. (1989). Biochem. Biophys. Res. Commun. 164: 333-338.
30. Wieland, T.; Ulibarri, I.; Gierschik, P.; Hall, A.; Aktories, K. & Jakobs, K. H.. (1990). FEBS Lett. 274: 111-114.
31. Kikuchi, A.; Yamamoto, K.; Fujita, T. & Takai, Y.. (1988) J. Biol. Chem. 263: 16303-16308.
32. Didsbury, J.; Weber, R. F.; Bokoch, G. M.; Evans, T. & Snyderman, R. (1989). J. Biol. Chem. 264: 16378-16382.
33. Sekine, A.; Fujiwara, M. & Narumiya, S. (1989). J. Biol. Chem. 264: 8602-8605.
34. Paterson, H. F.; Self, A. J.; Garrett, M. D.; Just, I.; Aktories, K. & Hall, A. (1990). J. Cell Biol. 111: 1001-1007.
35. Avraham, H. & Weinberg, R. A. (1989). Mol. Cell. Biol. 9: 2058-2066.
36. Sanger, F., Nicklen, S., and Coulson, A.R. (1977) Proc. natl. Acadm. Sci. 74:5463-5467.
37. Gillies, R.J., Didier, N., and Denton, M. (1986) Anal. Biochem. 159:109-113.

THE c-Ha-ras ONCOGENE INDUCES INCREASED EXPRESSION OF β-GALACTOSIDE α-2,6-SIALYLTRANSFERASE IN RAT FIBROBLAST (FR3T3) CELLS

Nadia Le Marer,[1+] Vincent Laudet,[2] Eric C. Svensson,[3] Haris Cazlaris,[1*]
Benoit van Hille,[2] Christian Lagrou,[2] Dominique Stehelin,[2]
Jean Montreuil,[1] André Verbert,[1] and Philippe Delannoy[1]

[1]Laboratoire de Chimie Biologique, Unité Mixte de Recherche du CNRS
N°111, Université des Sciences et Techniques de Lille-Flandres-Artois
59655 Villeneuve d'Ascq Cedex, France
[2]INSERM U 186/CNRS UA 041160, Institut Pasteur de Lille
1 rue Calmette, 59019 Lille Cedex, France
[3]Department of Biological Chemistry, UCLA School of Medicine
Los Angeles, CA 90024, USA

Key Words : α–2,6–sialyltransferase, c-Ha-ras, transformation, sialic acids.

INTRODUCTION

Elucidation of the molecular and cellular changes that accompany malignant conversion of normal cell populations is central to the understanding of cancer. Cell surface carbohydrates display structural alterations concomitant with malignant transformation (1). Transformation by chemical mutagens as well as by oncogenic viruses results in changes in the size of N-linked and O-linked glycans (1, 2). The increased size of these carbohydrate structures has been attributed to multi-antennarisation and to increased sialylation (3). Multi-antennarisation has been well studied and, in particular, has been associated with increased GlcNAc(β1-6)Man(α1-6) branching of complex type oligosaccharides. This phenomenon is directly associated with elevated N-Acetylglucosaminyltransferase V activity (4). Several recent observations have suggested that increased expression of β-1,6- branched oligosaccharides may be required for tumor cell metastasis (5). Multi-antennarisation of N-linked glycans synthesized by transformed cells is widely accepted. Increased sialylation in these cells is not a general phenomenon but has nevertheless been associated with metastatic potential (6).

Point mutations can convert cellular ras genes into transforming oncogenes (7). Moreover, in numerous types of human cancer, activated ras genes have been detected (8). In spite of the known biochemical properties of the ras gene product

---

+Corresponding author: present address in N°2 above.
*Present address: Hellenic Pasteur Institute, 127 Vas. Sofias Ave., 11521 Athens, Greece.

(p21^{ras}), namely GTPase activity (9), the molecular mechanisms of action of activated, as well as normal, p21^{ras} in the cell is poorly understood. However, p21^{ras} is found associated with a GTPase activating protein (GAP) which may serve as an effector molecule for ras action (10). Both proteins are implicated in the pathways governing signal transduction and cellular growth. *In vitro* transfection of cells with an activated ras gene can induce the same glycosylation modifications as those observed in tumor cells (11).

The aim of our work was to study the effect of various oncogenes such as c-Ha-ras and v-myc on the activity of β–galactoside sialyltransferases. We show that c-Ha-ras induces an increase of the activity of the β–galactoside α-2,6-sialyltransferase (Gal-α-2,6-ST) but not of the β–galactoside α-2,3-sialyltransferase (Gal-α-2,3-ST). In addition, we find that other oncogenes such as v-myc, v-src, polyoma virus middle T (mT) or the transforming Bovine Papilloma Virus 1 (BPV1) cannot enhance Gal-α-2,6-ST activity. This increased Gal-α-2,6-ST activity causes an increased α-2,6 linked sialic acid on cell surface glycoconjugates. Moreover, we demonstrate that this ras-mediated enhancement is caused by an increase of the enzyme level and of the mRNA encoding Gal-α-2,6-ST.

EXPERIMENTAL PROCEDURES

Cell lines

G418-resistant, transformed derivatives of FR3T3 cells were previously characterized. Transformation of FR3T3 with the v-myc and v-src oncogenes (FRmyc and FRsrc) were performed by infection with the murine retrovirus MMCV-neo, carrying both the avian v-myc^{OK10} (FRMC14 clone in ref. 12) or the avian v-src oncogene (FRSR13 clone in ref. 13) and the neo gene. Transformation of FR3T3 with ras (FRras, FREJ4 clone in ref. 13) and super-transformation of FRmyc with ras (FRmyc+ras) were performed by transfection with pSV2neoEJ. FRBPV1 was obtained by transfection of the Bovine Papilloma Virus 1 genome (FRBP8 clone in ref. 12), and FRmT by transfection with the Polyoma Virus middle T antigene (14). FR3T3 and G418-selected pSV2-neo-transfected FR3T3 cells (FRneo) were used as controls. The cells were cultured in DMEM-10% FCS and confluent cultures were harvested by scraping in PBS.

Sialyltransferase assays

The cell homogenates and the acceptors were prepared as in ref. 15 with minor modifications. Enzyme activity assays were performed according to the procedure described in the same reference.

Fluorescence microscopy

Visualization of Neu5Ac(α2-6)Gal sequence on the cell surface with SNA-FITC was obtained by the procedure described in ref. 16.

Expression of Gal-α-2,6-ST

Immunoprecipitation of the enzyme was carried out as in ref. 44. For Northern blot analysis, total RNA from rat liver was isolated as previously described (17), while polyA+ RNA from FR3T3, FRras and FRmyc was isolated using Fast Track (Invitrogen). Blotted RNA was hybridized with a gel-purified, radiolabelled *Eco*RI fragment of ST3 cDNA as reported (17).

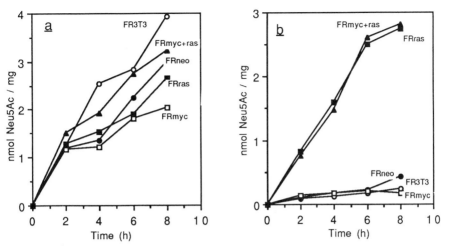

Figure 1. Kinetics of sialyltransferase activity in FR3T3 cells and its transformed derivatives using ASFET (a) and ASOR (b) as acceptors. O, FR3T3 ; ●, FRneo ; □, FRmyc ; ■, FRras; ▲, FRmyc+ras.

RESULTS

Sialyltransferase assays

The degree of transformation of FR3T3 cells obtained by the action of different oncogenes is a useful biological tool to evaluate the impact of oncogene expression on glycosyltransferases activity. Different sialyltransferases vary in their activity according to the type of glycan acceptors. Fetuin contains three tri-antennary complex type chains and three O-linked chains. Asialo-fetuin (ASFET) may therefore serve as an acceptor for the different sialyltransferases acting either on N- or O-linked chains of glycoproteins. In a first round of experiments, the global activity of sialyltransferases was determined in cell extracts. The kinetics of sialyltransferase activities on this acceptor (fig. 1a) did not reveal obvious modifications of activity between the different cells used. In contrast, kinetic assays on asialo-orosomucoid (ASOR) which contains only N-linked glycans, showed a 10-fold increase exclusively in ras transformed cells (fig. 1b). The capacity of transfer on ASFET is virtually identical in all the cell extracts, but there is a striking difference between normal and ras-transformed cells when ASOR was used. In order to determine whether the modulation of the specificity of sialyltransferase activity for N-linked glycans can explain this variation, we determined the apparent Km and Vm with variable concentration of glycopeptides of asialo-serotransferrin (ASSTFgp) containing exclusively bi-antennary glycans. As indicated in fig. 2, ras-transformed cells showed that both the apparent Km and the Vm values vary : the apparent Km was 3-fold lower and the Vm around 5-fold higher when compared with those of non-transformed cells. The apparent capacity of the different homogenates to transfer Neu5Ac on ASSTFgp was expressed as the ratio Vm/Km in Table I. Both FRras and FRmyc+ras exhibited similar changes in Km and Vm. The transfer of sialic acid to N-linked glycans is commonly catalyzed by two enzymes, β–galactoside α-2,6- and the β–galactoside α-2,3-sialyltransferases.

Table I. Kinetic parameters of sialyltransferase activity measured in cellular homogenates of FR3T3 cells and its transformed derivatives using ASSTFgp as an acceptor.

CELLS	Km (mM)	Vm (nmol/ mg/h)	Vm/km
FR3T3	0,89	0,16	0,18
FRneo	0,71	0,19	0,26
FRmyc	0,70	0,17	0,24
FRras	0,27	0,68	2,52
FRmyc+ras	0,27	0,56	2,09

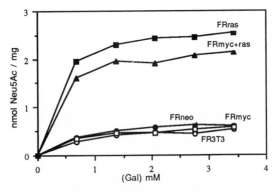

Figure 2 . Influence of asialo-serotransferrin glycopeptide (ASSTFgp) concentration on sialyltransferase activity of cellular homogenates of FR3T3 cells and its transformed derivatives. ○, FR3T3 ; ●, FRneo ; ◻, FRmyc ; ■, FRras; ▲, FRmyc+ras.

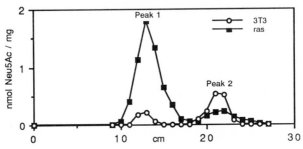

Figure 3 . Descending paper chromatography profiles of the [^{14}C]-Neu5Ac labeled mono-antennary oligosaccharides.
Peak 1 :
Neu5Ac(α2-6)Gal(β1-4)GlcNAc(β1-2)Man(α1-6)Man(β1-4)GlcNAc.
Peak 2 :
Neu5Ac(α2-3)Gal(β1-4)GlcNAc(β1-2)Man(α1-6)Man(β1-4)GlcNAc.
○, FR3T3 ; ■, FRras.

To compare both activities, we used a mono-antennary oligosaccharide as an acceptor in sialyltransferases assays of cell extracts. After a 6-hour incubation of cell extracts with this mono-antennary oligosaccharide, we separated the reaction products which differ in the linkage of sialic acid (Neu5Ac(α2-6)Gal-R and Neu5Ac(α2-3)Gal-R) by descending paper chromatography. Figure 3 shows the results obtained with FR3T3 and FRras. The profile obtained with FRmyc+ras was similar to that obtained with FRras (data not shown). This confirms that the capacity of transfer of Neu5Ac on N-acetyllactosaminic type glycans in an α-2,6 linkage is 7-fold increased in FRras and FRmyc+ras. Moreover, the ratio of activities of both sialyltransferases is clearly different.

Increase of the activity of Gal-α-2,6-ST in ras-transformed FR3T3 cells correlates with the increase of Neu5Ac(α2-6)Gal- sequence on membrane cell glycoproteins

In order to determine if the differential activity of Gal-α-2,6-ST and Gal-α-2,3-ST in ras-transformed FR3T3 cells induces modifications of membrane glycoprotein sialylation, *Sambucus nigra* agglutinin (SNA) was used. SNA recognizes the Neu5Ac(α2-6)Gal sequence with 50- to 100-fold higher affinity than the Neu5Ac(α2-3)Gal sequence (18) and proved to be a sharp tool to show differential transfer activity. SNA-FITC revealed a striking increase of fluorescence on the membrane of ras-transformed cells (FRras and FRmyc+ras) whereas this fluorescence was weak in FR3T3 (fig. 4) and in all other cell lines tested (data not shown). These results largely confirm our observations on the increased Gal-α-2,6-ST activity with cell extracts.

No correlation between the transformed phenotype and the Gal-α-2,6-ST activity

We addressed the question of whether this phenomenon could be correlated with the transformed phenotype of the various FR3T3 cells. We have tested the Gal-α-2,6-ST activity of FR3T3 transformed by other transforming oncogenes such as v-src, BPV1 or Polyoma middle T. Surprisingly, we did not find any differences in activity in these transformed cells using ASOR as acceptor suggesting that the Gal-α-2,6-ST activity is not enhanced in these cells. Nevertheless, the possibility remained that, in the particular clone of FRras used, the Gal-α-2,6-ST activity was elevated for a reason independant of the action of ras. Thus, we tested several independent clones (FREJ1, 2 and 3 in ref. 13), which displayed the same increased activity as the original FRras used (FREJ4 in ref. 13, data not shown). Furthermore, the FRmyc+ras has the same increased Gal-α-2,6-ST activity as our original FRras. Consequently, these results suggest that the increase of Gal-α-2,6-ST is a specific effect of the expression of the p21[ras] oncoprotein.

The increased Gal-α-2,6-ST activity correlates with a higher amount of Gal-α-2,6-ST protein resulting from an increase of the steady state level of the correspondent mRNA

We chose two strategies to determine if the level of expression of the Gal-α-2,6-ST enzyme in these cells correlated with the results obtained in enzymatic assays. In a first round of experiments we tested the amount of the enzyme using rabbit Gal-α-2,6-ST antibodies and secondly we studied the mRNAs encoding Gal-α-2,6-ST by Northern blot analysis. The level of expression of the enzyme was also determined by immunoprecipitation with L-[35S] methionine labeled FR3T3 and FRras cells. The results of the immunoprecipitation are shown in fig.5a. We can see a 44 kDa band which matches with the known size of Gal-α–2,6-ST (19). The signal is stronger in FRras than in FR3T3 cells. Thus, the increased Gal-α2,6-ST activity observed

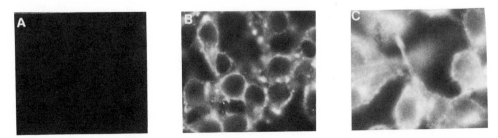

Figure 4. Detection of Neu5Ac(α2-6)Gal sequence on the cell surface of FR3T3 (A), FRras (B) and FRmyc+ras (C) cells.

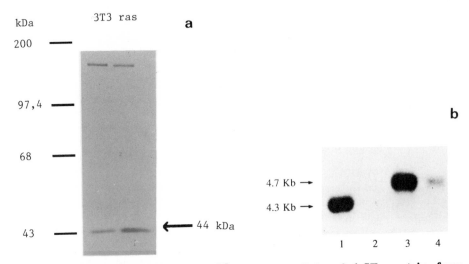

Figure 5. Immunoprecipitation of [^{35}S]-labelled Gal-α-2,6-ST protein from normal cells (FR3T3) and _ras_-transformed cells (FRras)(a). _Ras_ transformed fibroblasts express high levels of Gal-α-2,6-ST mRNA (b).

in FRras may be attributed, at least in part, to an increase of the number of enzyme molecules present in the cells.

The increased Gal-α-2,6-ST enzyme levels may be due to either a modification of Gal-α-2,6-ST protein synthesis or turnover rates, or to an increased steady state level of Gal-α-2,6-ST mRNA. To address this question, we performed Northern analysis on RNA isolated from FR3T3, FRras and FRmyc cells (fig. 5b). The results show that FRras produce dramatically higher levels of Gal-α-2,6-ST mRNA compared to FR3T3 and FRmyc. In addition, only the 4.7 kb mRNA level is increased, while the expression of the 4.3 kb "liver-restricted" species is not induced upon transformation by ras, even though both of these mRNAs are believed to be produced from the same gene (20, 21).

These results suggest that the increase in Gal-α-2,6-ST enzyme level upon ras transformation is due to an increase in Gal-α-2,6-ST mRNA synthesis, probably via an increased transcription of the Gal-α-2,6-ST gene .

DISCUSSION

Results obtained from the apparent capacity of transfer of sialic acid, using different concentration of bi-antennary glycans of ASSTFgp (Table I) indicate that sialyltransferases of N-linked glycans are weakly expressed in FR3T3 and suggest that the differences observed in the apparent Vm values, among all the cell types, are most likely due to an increase of the number of active molecules in ras transformed cells. Likewise, the variation of Km is probably due to a change in the quantity ratio of the enzymes, Gal-α-2,6-ST and Gal-α-2,3-ST. We know that ASSTFgp is a better acceptor for the Gal-α-2,6-ST than for the Gal-α-2,3-ST (22). The apparent Km decrease in ras-transformed cells could account for an increase of Gal-α-2,6-ST activity. This has been confirmed, first by using a mono-antennary glycan as acceptor (fig. 3, showing increased Gal-α-2,6-ST activity) and second by immunoprecipitation and Northern blot analysis (fig. 5a,b, demonstrating increased expression of the Gal-α-2,6-ST gene in ras-transformed cells). The modification of the quantity ratio of sialyltransferases in ras-transformed versus normal FR3T3 cells, particularly Gal-α-2,6-ST and Gal-α-2,3-ST, induces a modification of cell surface sialylation. These results correlate with recent studies (16, 20, 23) which show that the regulation of the biosynthesis of terminal glycosylation depends on the regulation of the expression of these enzymes and, more specifically, in our case, of those responsible for sialylation. Nevertheless, using SNA-FITC, we observe a large difference in cell surface α-2,6 linked sialylation between transformed and control cells. The difference observed with this technique seems higher than that observed with other biochemical parameters, i.e. activity, protein and mRNA levels. A precise rate of these differences is difficult since experimental conditions are very different and not directly comparable.The differences may be due to the fact that measurement of activity was carried out in vitro whereas cellular sialylation is an in vivo observation. In this latter case, the intracellular environment may regulate sialyltransferase activity which can be different from the in vitro activity tested in our assays with respect to competition for substrates and interactions between glycosyltransferases.

Surprisingly, the phenomenon we observed seems restricted to the ras oncogene. Indeed, with the other oncogenes tested (v-myc, v-src, polyoma middle T and BPV1) we did not see any enhancement of Gal-α-2,6-ST activity. Furthermore, we have detected no difference in the cell membrane sialylation pattern by SNA-FITC visualization. We show that the increase of Gal-α-2,6-ST level is not correlated with cell transformation but is induced specifically by the c-Ha-ras oncogene. This was not expected since numerous reports have associated modifications of cell surface glycoconjugates, such as multi-antennarisation, to the transformed phenotype of the cell but not to the biochemical action of the transforming oncogene. The c-Ha-ras oncogene can induce a modification of glycosyltransferase expression. Bolscher et al.

(24) have recently described glycosylation modifications in NIH3T3 cells transiently transfected by this oncogene. Moreover, these alterations were present before the manifestation of the transformed phenotype, suggesting that the p21^{ras} may induce transformation by modulation of glycosylation mechanisms. Bolscher *et al.* also propose that the modifications of glycosylation are not a direct consequence of transformation. This observation, although obtained in a different system, fits well with our results.

How does the ras oncogene regulate the Gal-α-2,6-ST level in rat fibroblast cells, such as FR3T3 ? The mechanism of action of the p21^{ras} is unclear but p21^{ras} is known to interfere with the intricate pathways governing cellular proliferation. Ras activity depends on a correct anchoring of the protein to the cell membrane via a lipidic anchor. Moreover, the normal p21^{ras} is known to interact with the GAP protein which increases its GTPase activity. The ways by which ras may modulate Gal-α-2,6-ST level remain elusive but it is highly probable that ras acts in an indirect manner (i.e. transcriptional activation). Recently, the Gal-α-2,6-ST gene has been cloned and its promoter partially characterized (21). However, although we can see a clear increase in the Gal-α-2,6-ST mRNA level, we have no evidence suggesting whether ras can regulate Gal-α-2,6-ST gene expression or its mRNA turnover. Work is in progress in our laboratory to test how ras may achieve this regulation which leads to important physiological consequences for the cells.

This phenomenon could be determinant for metastasis establishment, since we know that it has been observed recently in metastatic models (25, 26) where ras is often activated (see other communications in these proceedings).

REFERENCES

1- Smets, L.A., and Van Beek, W.P., Carbohydrates of the tumor cell surface, *Biochim. Biophys. Acta* **738**:237-249 (1984).

2- Santer, U.V., and Glick, M.C., Partial structure of a membrane glycopeptide from virus-transformed hamster cell, *Biochemistry* **18**:2533-2540 (1979).

3- Warren, L., Fuhrer, J.B., and Buck, C.A., Surface glycoproteins of normal and transformed cells : a difference determined by sialic acid and a growth-dependant sialyltransferase *Proc. Natl. Acad. Sci. USA* **69**:1838-1842 (1972).

4- Dennis, J.W., Kosh, K., Bryce, D.M., and Breitman, M.L., Oncogenes conferring metastatic potential induce increased branching of Asn-liked oligosaccharides in Rat2 fibroblasts, *Oncogene* **4**: 853-860 (1989).

5- Dennis, J.W., Laferté, S., Waghorne, C., Breitman, M.L., and Kerbel, R.S. β1-6 branching of Asn-linked oligosaccharides is directly associated with metastasis *Science* **236**:582-585 (1987).

6- Nicolson, G.L., Cancer metastasis - Organ colonization and the cell surface properties of malignant cells, *Biochim. Biophys. Acta* **695**:113-176 (1982).

7- Santos, E., Tronick, S.R., Aronson, S.A., Puciani, S., and Barbacid, M., T24 human bladder carcinoma oncogene is an activated form for the human homologous of BALB- and Harvey-MSV transforming genes, *Nature* **298**:343-347 (1982).

8- Forrester, K., Almoguera, C., Han, K., Grizzle, W.E., and Perucho, M., Detection of high incidence of K-ras oncogenes during human colon tumorigenesis, *Nature* **327**:298-303 (1987).

9- Mc Grath, J.P., Capon, D.J., Goeddel, D.V., and Levinson, A.D., Comparative biochemical properties of normal and activated human ras p21 protein, *Nature* **310**:644-649 (1984).

10- Mc Cormick, F., Ras GTPase activating protein : signal transmitter and signal terminator, *Cell* 56:5-8 (1989).

11- Collard, J.G., Van Beek, W.P., Janssen, JW.G., and Schijen, J.F., Transfection by human oncogenes : concomitant induction of tumorigenesis and tumor-assiociated membrane alterations, *Int. J.Cancer* **35**:207-214 (1985).

12- Salomé, N., van Hille, B., Duponchel, N., Menguizzi, G., Cuzin, F., Rommelaere, J., and Cornelis, J.J., Sensitization of transformed rat cells to parvovirus MVMp is restricted to specific oncogenes, *Oncogene* 5:123-130 (1990).

13- van Hille, B., Duponchel, N., Salomé N., Spruyt, N., Cotmore, S., Tattersall, P., Cornelis, J.J., and Rommelaere, J., Limitations to the expression of parvoviral nonstructural proteins may determine the extent of sensitization of EJ-ras-transformed rat cells to minute virus mice,*Virology* **171**, 89-97 (1989).

14- Mousset, S., Cornelis, J., Spruyt, N., and Rommelaere, J., Transformation of established murine fibroblasts with an activated cellular Harvey-ras or the polyoma virus middle T gene increases cell permissiveness to parvovirus minute-virus-of-mice, *Biochimie* 6:951-955 (1986).

15- Cazlaris, H., Le Marer, N., Laudet, V., Lagrou, C., Zhu, Q., Delannoy, P., and Montreuil, J., Modifications de la sialylation des cellules BHK 21/C13 après transfection in vitro par l'oncogène humain c-Ha-ras, *C. R. Acad.Sci. Paris*, **312**, Série III:293-300 (1991).

16- Lee, E.U., Roth, J., and Paulson, J.C., Alteration of terminal glycosylation on N-linked oligosaccharides of chinese hamster ovary cells by expression of β-galactoside α2,6-sialyltransferase, *J. Biol. Chem.* **264**:13848-13855 (1989).

17- Paulson, J.C., and Colley, K.J., Glycosyltransferases : Structure, localization and control of cell type-specific glycosylation, *J. Biol. Chem.* **264**:17615-17618 (1989).

18- Shibuya, N., Goldstein I.J., Brockaert, W.F., Nsimba-Lubaki, M., Peeters B., and Peumans, W.J., The elderberry (*Sambuccus nigra* L) bark lectin recognizes the Neu5Ac(α2-6)Gal/GalNAc sequence *Arch. Biochem. Biophys.* **254**:1-8 (1987).

19- Weinstein J., Lee E.U., Mc Entee K., Lai P.H., and Paulson, J.C., Primary structure of β-galactoside α2,6-sialyltransferase, *J. Biol. Chem.* **262**:17735-17743 (1987).

20- Paulson, J.C., Weinstein, J., and Schauer, A., Tissue-specific expression of sialyltransferases, *J. Biol. Chem.* **264**:10931-10934 (1989).

21- Svensson, E.C., Soreghan, B., and Paulson, J.C., Organization of the β-galactoside α2,6-sialyltransferase gene, *J. Biol. Chem.* **265**:20863-20868 (1990).

22- Joziasse, D.H., Schiphorst, W.E.C.M., Van den Eijnden, D.H., Van Kuik, J.A., Van Halbeek, H., and Vliegenthart, J.F.G., Branch specificity of bovine colostrum CMP-sialic acid : Gal(β1-4)GlcNAc-R α2,6-sialyltransferase, *J.Biol. Chem.* **262**:2025-2033 (1987).

23- Colley, K.J., Ujita, L., Beverly, A., Browne, J.K., and Paulson, J.C., Conversion of a Golgi apparatus sialyltransferase to a secretory protein by replacement of NH2-terminal signal anchor with a signal peptide, *J. Biol. Chem.* **264**:17619-17622 (1989).

24- Bolscher, J.G.M., vander Bijl, M.M.W., Neefjes, J.J., Hall, A., Smets, L.A., and Ploegh, H.L., Ras (proto-oncogene) induces β-linked carbohydrate modification : temporal relationship with invasive potential, *EMBO J.* 7:3361-3368 (1988).

25- Dall'Olio, F., Malagolini, N., Di Stefano, G., Ciambella, M., and Serafini-Cessi, F., α2,6 sialylation of N-acetyllactosaminic sequences in human colorectal cancer cell lines. Relationship with non-adherent growth, *Int. J. Cancer* **47**:291-297 (1991).

26- Bresalier, R.S., Rockwell, R.W., Dahiya, R., Duh, Q., and Kim, Y.S., Cell surface sialoprotein alterations in metastatic murine colon cancer cell lines selected in an animal model for colon cancer metastasis, *Cancer Res.* **50**:1299-1307 (1990).

LOCALIZATION OF RAB PROTEINS

Philippe Chavrier, Jean-Pierre Gorvel, Kai Simons,
Jean Gruenberg and Marino Zerial

European Molecular Biology Laboratory
Postfach 10.2209, D-6900 Heidelberg, Germany

INTRODUCTION

A number of *in vitro* and *in vivo* studies have shown that the
large superfamily of Ras-related low molecular weight GTP-
binding proteins includes proteins which are involved in the
control of membrane traffic. In the yeast Saccharomyces
cerevisiae, the SEC4 gene encodes a 23 Kd ras-related GTP-
binding protein involved in the regulation of vesicular
traffic from the Golgi apparatus to the plasma membrane. The
protein is found associated with both the cytoplasmic surface
of the plasma membrane and secretory vesicles (Salminen and
Novick, 1987; Goud et al., 1988). Temperature-sensitive and
dominant SEC4 mutants lead to a block in transport from the
Golgi apparatus and accumulation of post-Golgi secretory
vesicles (Salminen and Novick, 1987; Walworth et al., 1989).
The YPT1 gene product is a 23 Kd GTP-binding protein (Gallwitz
et al., 1983) which functions at an earlier step of the
secretory pathway, from the ER to or within the Golgi
apparatus (Schmitt et al., 1986; 1988; Segev and Botstein,
1987; Segev et al., 1988; Baker et al, 1990). Two other low
molecular weight GTP-binding proteins belonging to a subfamily
of ras-related proteins distinct from that of Ypt1p and Sec4p
have been shown to be functionally associated with the yeast
secretory pathway. Sar1p regulates vesicular traffic between
ER and Golgi membranes (Nakano and Muramatsu, 1989). Arf1p is
associated with the Golgi apparatus in mammalian cells and
expression of mutants of this protein in yeast leads to a
defect in secretion (Sewell and Kahn, 1988; Stearns et al,
1990).
The general requirement for GTP-ases function in different
transport reactions has been indirectly revealed by the strong
inhibitory effect of a non-hydrolyzable GTP analogue, GTPγS,
in several different *in vitro* reconstituted transport
reactions. On the exocytic pathway, GTPγS caused inhibition of
ER to Golgi transport (Baker et al., 1988; Ruohola et al.,
1988; Beckers and Balch., 1989) and *in vitro* transport between
Golgi cisternae (Melancon et al , 1987; Orci et al., 1989) An
early (Mayorga et al., 1989; Tuomikoski et al., 1989) as well

The Superfamily of ras-Related Genes
Edited by D.A. Spandidos, Plenum Press, New York, 1991

as a late (Bomsel et al., 1990) endocytic fusion event have also been shown to be inhibited by GTPγS, indicating that GTP-binding proteins may also play a role in the transport of molecules along the endocytic pathway. Recycling of the mannose 6-phosphate/IGF II receptor to the trans Golgi network (TGN) *in vitro* also requires GTP-hydrolysis (Goda and Pfeffer, 1988). Finally, budding from the TGN of both constitutive secretory vesicles and immature secretory granules is sensitive to GTPγS (Tooze and Huttner, 1990).

The general involvment of GTP-binding proteins in different transport reactions *in vivo* and *in vitro* suggest that similar biochemical mechanisms regulate both exocytosis and endocytosis. How could they function in the regulation of membrane traffic? As proposed by Bourne (1988) they could act in the same way as elongation factor Tu in protein synthesis and direct the energy dependent, unidirectional, specific delivery of vesicles to the target organelle. GTP hydrolysis would be required to trigger a conformational change of the GTP-binding protein, as shown for p21ras (deVos et al., 1988; Pai et al., 1989), and promote fusion of the vesicle membrane with the acceptor membrane.

LOW MOLECULAR WEIGHT GTP-BINDING PROTEINS IN MAMMALIAN CELLS

Mammalian cells express four different sub-families of low molecular weight GTP-binding proteins: RAS/RAP, RAB, RHO and SAR/ARF (see Bourne et al., 1990 for a review). Of these, Rab proteins are the most closely related to the yeast Ypt1 and Sec4 proteins. All rab proteins display the four highly conserved regions participating in the formation of the GTP-binding site (Figure 1 and 4) and one or two cysteine residues at their C-termini required for membrane association. A number of studies on H-ras, K-ras, Ras2, Ypt1, Sec4 and rab proteins have established that these cysteines are required for membrane association via post-translational addition of aliphatic chains (Willumsen et al., 1984; Deschenes and Broach, 1987; Molenaar et al., 1988; Walworth, et al., 1989). In the case of p21[Hras] three post-translational modifications of the C-terminal cysteine motif, a CAAX box (C=cysteine, A=aliphatic, X=any amino acid), have been shown to occur (Gutierrez et al., 1989; Hancock et al., 1989). While rho proteins have a typical CAAX box, rab proteins are much more heterogeneous in their cysteine motifs. Sequence homology is

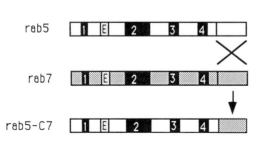

Fig.1. Diagram of rab5, rab7 and rab5C7 hybrid protein. The four conserved regions participating in the formation of the GTP-binding site are indicated as black boxes, the effector region "E" as a striped box.

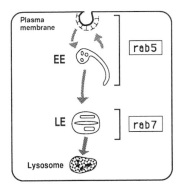

Fig.2. Scheme of the localization of rab proteins on the endocytic pathway in mammalian cells. Rab5 was localized to the plasma membrane and early endosomes (EE), rab7 to late endosomes (LE).

observed also for the so called "effector loop" which in the case of p21^{ras} is known to interact with the GTP-ase activating protein GAP (Adari et al., 1988). Rab proteins exhibit a much higher sequence homology in the "effector loop" with Ypt1p and Sec4p compared to rho proteins.

Arf and Sar proteins clearly belong to another subfamily of low molecular weight GTP-binding proteins and they lack the cysteines present at the carboxyl terminus of rab and rho proteins (Sewell and Kahn, 1988; Nakano and Muramatsu, 1989; Stearns et al, 1990).

Mammalian cells express more than thirty low molecular weight GTP-binding proteins related to YPT1 and SEC4 products (Touchot et al; 1987; Bucci et al., 1988; Matsui et al., 1988; Zahraoui et al., 1989; Didsbury et al., 1989; Polakis et al., 1989; Chavrier et al., 1990a,b). This large number was extrapolated from the frequency of number of cDNAs cloned over the total number of screened (Touchot et al., 1987; Chavrier et al., 1990a;b). Experimental evidence supporting this complexity has been recently obtained with the isolation of fifteen additional cDNAs using a PCR-based screening approach (Chavrier et al., submitted).

LOCALIZATION OF RAB PROTEINS

The intracellular localization of a few rab proteins has also been determined. By using affinity purified antibodies in immunofluorescence and electron microscopic studies, rab2 was localized to an "intermediate compartment" between the Endoplasmic Reticulum (ER) and Golgi apparatus (Chavrier et al., 1990a). This vesiculo-tubular compartment is likely to correspond to the so called "salvage compartment" postulated to recycle ER-resident proteins to the ER. Rab2 co-localized with the 53 Kd protein (Schweizer et al., 1988) a marker of the "intermediate compartment". Using similar techniques, rab6 was localized to later subcompartments of the exocytic pathway, the middle- and trans-Golgi cisternae (Goud et al., 1990). Rab3a (called also smg25A) was found associated with specialized organelles of the regulated scretory pathway, synaptic vesicles in neurons and chromaffin granules in adrenal medulla (Fischer von Mollard et al., 1990; Darchen et al., 1990; Mizoguchi et al., 1990).

Two other rab proteins were found associated with compartments along the endocytic pathway (Figure 2): rab5 was detected at the cytoplasmic surface of both the plasma membrane and early endosomes, whereas rab7 was associated with late endosomes (Chavrier et al., 1990a).
Both the large number of these proteins and their localization to distinct subcompartments along the exocytic and endocytic pathway support their proposed role in the regulation of membrane traffic at specific steps (Bourne, 1988). Accordingly, we recently demonstrated that rab 5 is involved in early endosome fusion *in vitro* (Gorvel et al, 1991).

THE TARGETTING SIGNAL ON RAB PROTEINS

The mechanism responsible for the specific localization of rab proteins is at present unknown. A first step towards the elucidation of this process is to investigate which region of rab proteins serves as targetting signal responsible for their specific localization. The C-terminal cysteine motif required for membrane association is variable among rab proteins. However, it seems unlikely that it could function alone as a specific localization signal since for instance, rab6 and rab7 have the same C-terminal CSC sequence, but are localized to different compartments: rab6 to the Golgi apparatus and rab7 to late endosomes (Chavrier et al., 1990a; Goud et al., 1990).

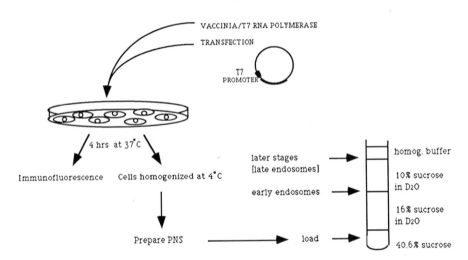

Fig.3. Distribution of rab proteins to early or late endosome fractions. Rab proteins are transiently expressed using a system based on the recombinant T7 RNA polymerase vaccinia virus (Fuerst et al., 1986). The cells are first infected with the vaccinia hybrid virus and then transfected with a plasmid containing the DNA to transcribe under the control of the T7 RNA polymerase promoter. After four hours incubation the cells can be analysed by confocal immunofluorescence microscopy or, alternatively by cell fractionation. After homogeneization a post nuclear supernatant is prepared and fractions enriched in early and late endosomes are separated by flotation using a sucrose/D_2O gradient (Gorvel et al, 1991).

The C-terminal hypervariable domain (downstream from the fourth conserved region among ras-related proteins; Figure 1) represents an interesting candidate for this function. It is located adjacent to the cysteine motif and must lie close to the membrane and it is highly variable both in sequence and length. To investigate the possible role of this C-terminal region in targetting rab proteins to their corresponding acceptor membranes, we constructed chimeric proteins between rab5 and rab7 (Figure 1) and transiently expressed them in BHK cells using the T7 RNA polymerase recombinant vaccinia virus system (Fuerst et al., 1986; Figure 3). The distribution of these proteins was analysed by immunofluorescence confocal microscopy and sub-cellular fractionation using affinity purified antibodies at concentration selected to detect only the overexpressed proteins. BHK cells were cotransfected with the wt rab5 and rab7 protein constructs and a plasmid encoding the human transferrin receptor (hTR; Zerial et al., 1986), a marker for plasma membrane and early endosomes (Hopkins, 1983; Schmid et al., 1988). By immunofluorescence most of the rab5 and hTR staining overlapped as expected since the two proteins are localized to the same compartment. In contrast, segregation of hTR and rab7 was clearly observed consistent with the localization of this rab protein to late endosomes (Chavrier et al., 1990a). Further evidence for the localization of rab5 and rab7 was obtained by sub-cellular fractionation. Fractions enriched in early and late endosomes were separated by flotation using a sucrose/D_2O gradient (Gorvel et al., 1991; Figure 2). The over-expressed rab5 and rab7 proteins were detected in the early and late endosome fractions, respectively, thus confirming our previous observation that high expression levels do not lead to mis-localization.(Chavrier et al., 1990a).
We progressively replaced C-terminal sequences of rab5 with those from rab7. When we exchanged the CCSN motif of rab 5 with the CSC sequence of rab7, the hybrid protein displayed an immunofluorescence staining pattern very similar to that of wt rab5 and was detected in the early endosome fraction. Replacing the last 8 and 13 C-terminal residues of rab5 with

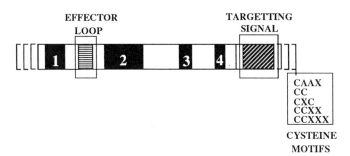

Fig.4. Scheme of a typical rab protein showing the variable lengths of the N- and C-termini, the conserved regions participating in the formation of the GTP-binding site (black boxes), the "effector region" (striped box) and the C-terminal targetting signal containing the diverse cysteine motifs.

the corresponding sequences from rab7 did not change this localization In contrast, an N-terminal 181 residue long rab5 reporter fragment fused to the C-terminal 34 amino acids of the rab 7 protein (rab5C7; Figure 1) displayed a similar localization as rab7 in the perinuclear region of the cells and was detected exclusively in the late endosome fraction. These results indicate that the C-terminal 34 residues of rab7 can target an N-terminal 181 residue-long rab5 fragment to late endosomes. Neither the cysteine motif of rab7 nor adjacent sequences up to 13 amino acids) can mediate this process. Therefore, these results suggest that the targetting signal involves residues between position 174 and 195 of the rab7 protein (Chavrier et al., submitted).

We next examined whether a rab protein from the exocytic pathway could be targetted to an endocytic compartment by exchanging its C-terminal region. As a reporter molecule we used rab2 which is localized to the intermediate compartment between the ER and the Golgi complex.(Chavrier et al., 1990a). The over-expressed rab2 protein was hardly detected in the endosomal fractions. A hybrid protein consisting of 166 N-terminal rab2 residues fused with the 35 C-terminal aminoacids of rab 5 was detected only in the early endosome fraction (Fig.3a), while the same reporter fragment fused to the C-terminal 35 aminoacids of rab7 was found in the late endosome fraction. These data indeed suggest that the C-terminal region of rab5 and rab7 are sufficient to target a rab2 reporter fragment to early and late endosomes, respectively.

CONCLUSION

Our data indicate that the localization signal for rab proteins resides in the highly variable C-terminal sequence. This region is both necessary and sufficient for correct targetting of the rab proteins we have tested. While the C-terminal cysteine motif is necessary for membrane association (Willumsen et al., 1984; Deschenes and Broach, 1987; Molenaar et al., 1988; Walworth, et al., 1989) it does not alone specify the targetting. Nor substitution of the last 8 and 13 C-terminal residues of rab5 with the corresponding from rab7 did affect the localization of the hybrid protein to early endosomes. Sequences next to the CAAX box have also been shown to be required for plasma membrane association of p21ras (Hancock et al., 1990). Therefore, membrane localization is a multi-step process involving structural determinants required for post-translational modifications and others for the specificity of targetting.

It is possible that rab proteins interact via their C-terminal hypervariable domain with specific receptors found on the target membrane. Alternatively, the docking event might be mediated by a cytosolic factor interacting with the targetting signal on the rab protein. The identification of the targetting signal on the rab proteins should facilitate the identification of the factor(s) playing a role in this process.

REFERENCES

Adari, H., Lowy, D.R., Willumsen, B.M., Der, C.J. and McCormick, F. ,1988, Guanosine triphosphatase activating protein (GAP) interacts with the p21ras effector binding domain, Science, 240:518.

Baker, D., Wuestehube, L., Schekman, R., Botstein, D., and Segev, N. , 1990, GTP-binding Ypt1 protein and Ca^{2+} function independently in a cell-free protein transport reaction. Proc. Natl. Acad. Sci. USA, 87:355.

Barbacid, M. ,1987, Ras Genes, Ann. Rev. Biochem., 56:779.

Becker, J., Tan, T.J., Trepte, H.-H and Gallwitz, D. , 1991, Mutational analysis of the putative effector domain of the GTP-binding Ypt1 protein in yeast suggests specific regulation by a novel GAP activity. EMBO J., 10:785.

Beckers, C.J.M. and Balch, W.E. , 1989, Calcium and GTP: essential components in vesicular trafficking between the endoplasmic reticulum and the Golgi apparatus, J. Cell Biol., 108:1245.

Bomsel, M., Parton, R., Kuznetsov, S.A., Schroer, T.A.,and Gruenberg, J., 1990, Microtubule- and motor-dependent fusion in vitro between apical and basolateral endocytic vesicles from MDCK cells, Cell, 62:719.

Bourne, H.R. , 1988, Do GTPases direct membrane traffic in secretion? Cell, 53:669.

Bourne, H.R., Sanders, D.A., and McCormick, F., 1991, The GTPase superfamily: conserved structure and molecular mechanism, Nature, 349:117.

Bucci, C., Frunzio, R., Chiariotti, L., Brown, A.L., Rechler, M.M. and Bruni, C.B., 1988, A new member of the ras gene superfamily identified in a rat liver cell line, Nucl. Acids Res., 16:9979.

Cales, C., Hancock, J.F., Marshall, C.J. and Hall, A., 1988, The cytoplasmic protein GAP is implicated as a target for regulation by the ras gene product, Nature, 332:548.

Chardin, P., and Tavitian, A., 1986, The ral gene: a new ras related gene isolated by the use of a synthetic probe, EMBO J., 5:2203.

Chavrier, P., Parton, R.G., Hauri, H.P., Simons, K. and Zerial, M., 1990a, Localization of low molecular weight GTP-binding proteins to exocytic and endocytic compartments, Cell, 62:317.

Chavrier, P., Vingron, M., Sander, C., Simons, K. and Zerial, M., 1990b, Molecular cloning of YPTq/SEC4-related cDNAs from an epithelial cell line, Mol. Cell. Biol., 10:6578.

Darchen, F., Zahraoui, A., Hammel, F., Monteils, M.-P., Tavitian, A., and Scherman, D., 1990, Association of the GTP-binding protein Rab3A with bovine adrenal chromaffin granules, Proc. Natl. Acad. Sci. USA, 87:5692.

Deschenes, R.J.,and Broach, J.R., 1987, Fatty acylation is important but not essential for Saccharomyces cerevisiae RAS function, Mol. Cell. Biol., 7:2344.

deVos, A., Tong, L., Milburn, M.V., Matias, P.M., Jancarik, J., Noguchi, S., Nishimura, S., Miura, K., Ohtsuka, E., and Kim, S.-H. , 1988, Three-dimensional structure of an oncogene protein: catalytic domain of human c-H-ras p21, Science, 239:888.

Didsbury, J., Weber, R.F., Bokoch, G.M., Evans, T., and Snyderman, R., 1989, Rac, a novel ras-related family of proteins that are Botulinum Toxin substrates., J. Biol. Chem., 264:16378.

Fawell, E., Hook, S., Sweet, D. and Armstrong, J., 1990, Novel YPT1-related genes from Schizosaccharomyces pombe, Nucl. Acids Res. 18:4264.

Fisher v.Mollard, G.F., Mignery, G.A., Baumert, M., Perin, M.S., Hanson, T.J., Burger, P.M., Jahn, R., and Sudhof, T., 1990, Rab3 is a small GTP-binding protein exclusively localized to synaptic vesicles, <u>Proc. Natl. Acad. Sci. USA</u>, 87:1988.

Fuerst, T.R., Niles, E.G., Studier, F.W. and Moss, B., 1986, Eukaryotic transient-expression system based on recombinant vaccinia virus that synthesizes bacteriophage T7 RNA polymerase, <u>Proc. Natl. Acad. Sci USA</u>, 83:8122.

Gallwitz, D., Donath, C., and Sander, C., 1983, A yeast gene encoding a protein homologous to the human c-has/bas proto-oncogene product, <u>Nature</u>, 306:704.

Goda, Y., and Pfeffer, S.R. 1988, Selective recycling of the mannose 6-phosphate/IGFII receptor to the trans Golgi network <i>in vitro</i>, <u>Cell</u>, 55:309.

Gorvel, J.P., Chavrier, P., Zerial, M. and Gruenberg, J., 1991, Rab5 controls early endosome fusion <i>in vitro</i>, <u>Cell</u>, 64:915.

Goud, B., Salminen, A., Walworth, N.C., and Novick, P.J., 1988, A GTP-binding protein required for secretion rapidly associates with secretory vesicles and the plasma membrane in yeast, <u>Cell</u>, 53:753.

Goud, B., Zahraoui, A., Tavitian, A., and Saraste, J., 1990, Small GTP-binding protein associated with Golgi cisternae, <u>Nature</u>, 345:553.

Gutierrez, L., Magee, A.I., Marshall, C.J. and Hancock, J.F., 1989, Post-translational processing of p21ras is two-step and involves carboxy-methylation and carboxy-terminal proteolysis <u>EMBO J.</u>, 8:1093.

Hancock, J.F., Magee, A.I., Childs, J.E. and Marshall, C.J., 1989, All ras proteins are polyisoprenilated but only some are palmitoylated, <u>Cell</u>, 57:1167.

Hancock, J.F., Paterson, H. and Marshall, C.J., 1990, A polybasic domain or palmitoylation is required in addition to the CAAX motif to localize p21ras to the plasma membrane, <u>Cell</u> 63:133.

Haubruck, H., Disela, C., Wagner, P., and Gallwitz, D., 1987, The ras-related ypt protein is an ubiquitous eukaryotic protein: isolation and sequence analysis of mouse cDNA clones highly homologous to the yeast YPT1 gene, <u>EMBO J.</u>, 6:4049.

Haubruck, H., Prange, R., Vorgias, C., and Gallwitz, D., 1989, The ras-related mouse ypt1 protein can functionally replace the YPT1 gene product in yeast, <u>EMBO J.</u>, 8:1427.

Hopkins, C.R., 1983, Intracellular Routing of transferrin and transferrin receptors in epidermoid carcinoma A431 cells, <u>Cell</u>, 35:321.

Kawata, M., Farnsworth, C.C., Yoshida, Y., Gelb, M.H., Glomset, J.A., and Takai, Y., 1990, Posttranslationally processed structure of the human platelet protein smg p21B: Evidence for geranylgeranylation and carboxyl methylation of the C-terminal cysteine, <u>Proc. Natl. Acad. Sci. USA</u>, 87:8960.

Matsui, Y., Kikuchi, A., Kondo, J., Hishida, T., Teranishi, Y., and Takai, Y., 1988, Nucleotide and deduced amino acid sequences of a GTP-binding protein family with molecular weight of 25,000 from bovine brain, <u>J. Biol. Chem.</u>, 263:11071.

Mayorga, L.S., Diaz, R., and Stahl, P.D., 1989, Regulatory role for GTP-binding proteins in endocytosis, _Science_, 244:1475.

Melançon, P., Glick, B.S., Malhotra, V., Weidman, P.J., Serafini, T., Gleason, M.L., Orci, L., and Rothman, J.E., 1987, Involvement of GTP-binding "G" proteins in transport through the Golgi stack, _Cell_, 51:1053.

Miyake, S. and Yamamoto, M., 1990, Identification of ras-related, YPT family genes in Scizosaccharomyces pombe, _EMBO J._, 9:1417.

Mizoguchi, A., Kim., S., Ueda, T., Kikuchi, A., Yorifuji, H., Hirokawa, N., and Takai, Y., 1990, Localization and subcellular distribution of smg p25A, a ras p21-like GTP-binding protein, in rat brain, _J. Biol. Chem._, 265:11872.

Molenaar, C.M.T., Prange, R., and Gallwitz, D., 1988, A carboxyl-terminal cysteine residue is required for palmitic acid binding and biological activity of the ras-related yeast Ypt1 protein, _EMBO J._, 7:971.

Nakano, A., and Muramatsu, M., 1989, A novel GTP-binding protein, Sar1p, is involved in transport from the endoplasmic reticulum to the Golgi apparatus, _J. Cell Biol._, 109:2677.

Orci, L., Malhotra, V., Amherdt, M., Serafini, T., and Rothman, J.E., 1989, Dissection of a single round of vesicular transport: sequential intermediates for intercisternal movements in the Golgi stack, _Cell_, 56:357.

Pai, E.F., Kabsch, W., Krengel, U., Holmes, K.C., John, J., and Wittinghofer, A., 1989, Structure of the guanine-nucleotide-binding domain of the Ha-ras oncogene product p21 in the triphosphate conformation, _Nature_, 341:209.

Polakis, P.G., Weber, R.F., Nevins, B., Didsbury, J.R., Evans, T., and Snyderman, R., 1989, Identification of the ral and rac1 gene products, low molecular mass GTP-binding proteins from human platelets, _J. Biol. Chem._, 264:16383.

Rothman, J.E., and Orci, L., 1990, Movement of proteins through the Golgi stack: a molecular dissection of vesicular transport, _FASEB J._, 4:1460.

Ruohola, H., Kastan Kabcenell, A., and Ferro-Novick, S., 1988, Reconstitution of protein transport from the endoplasmic reticulum to the Golgi complex in yeast: the acceptor Golgi compartment is defective in the _sec23_ mutant, _J. Cell Biol._, 107:1465.

Salminen, A., and Novick, P.J., 1987, A ras-like protein is required for a post-Golgi event in yeast scretion, _Cell_, 49:527.

Schmid, S.L., Fuchs, R., Male, P. and Mellman, I., Two distinct subpopulations of endosomes involved in membrane recycling and transport to lysosomes, _Cell_ **52**, 73-83 (1988).

Schmitt, H.D., Wagner, P., Pfaff, E. and Gallwitz, D., 1986, The ras-related YPT1 gene product in yeast: a GTP-binding protein that might be involved in microtubule organization, _Cell_, 47:401.

Schmitt, H.D., Puzicha, M., and Gallwitz, D., 1988, Study of a temperature-sensitive mutant of the ras-related YPT1 gene product in yeast suggests a role in the regulation of intracellular calcium, _Cell_, 53:635.

Schweizer, A., Fransen, J.A.M., Bachi, T., Ginsel, L., and Hauri, H.-P., 1988, Identification, by a monoclonal antibody, of a 53kD protein associated with a tubular-vesicular compartment at the cis-side of the Golgi apparatus, J. Cell Biol., 107:1643.

Segev, N. and Botstein, D., 1987, The ras-like Yeast YPT1 gene is itself essential for growth, sporulation, and starvation response, Mol. Cell. Biol., 7:2367.

Segev, N., Mulholland, J., and Botstein, D., 1988, The yeast GTP-binding YPT1 protein and a mammalian counterpart are associated with the secretion machinery, Cell, 52:915.

Sewell, J.L., and Kahn, R.A., 1988, Sequences of the bovine and yeast ADP-ribosylation factor and comparison to other GTP-binding proteins, Proc. Natl. Acad. Sci. USA, 85:4620.

Stearns, T., Willingham, M.C., Botstein, D. and Kahn, R.A., 1990, ADP-ribosylation factor is functionally and physically associated with the Golgi complex, Proc. Natl. Acad. Sci. USA, 87:1234.

Tooze, S.A., Weiss, U. and Huttner, W.B. , 1990, Requirement for GTP hydrolysis in the formation of secretory vesicles, Nature, 347:207.

Touchot, N., Chardin, P., and Tavitian, A., 1987, Four additional members of the ras gene superfamily isolated by an oligonucleotide strategy: molecular cloning of YPT-related cDNAs from a rat brain library, Proc. Natl. Acad. Sci. USA, 84:8210.

Tuomikoski, T., Felix, M-A., Dorée, M., and Gruenberg, J., 1989, The cell-cycle control protein kinase cdc2 inhibits endocytic vesicle fusion in vitro, Nature, 342:942.

Walworth, N.C., Goud, B. Kastan Kabcenell, A., and Novick, P.J., 1989, Mutational analysis of SEC4 suggests a cyclical mechanism for the regulation of vesicular traffic, EMBO J., 8:1685.

Willumsen, B.M., Norris, K., Papageorge, A.G., Hubbert, N.L., and Lowy, D.R., 1984, Harvey murine sarcoma virus p21 ras protein: Biological and biochemical significance of the cysteine nearest the carboxy terminus, EMBO J., 3:2585.

Yeramian, P., Chardin, P., Madaule, P. and Tavitian, A., 1987, Nucleotide sequence of human rho cDNA clone 12., Nucl. Acids Res., 15:1869.

Zahraoui, A., Touchot, N., Chardin, P., and Tavitian, A., 1989, The human Rab genes encode a family of GTP-binding proteins related to yeast YPT1 and SEC4 products involved in secretion, J. Biol. Cell., 264:12394.

Zerial, M., Melancon, P., Schneider, C. and Garoff, H. , 1986, The transmembrane segment of the human transferrin receptor functions as a signal peptide, EMBO J., 5:1543.

THE DROSOPHILA ras2/cs1 BIDIRECTIONAL PROMOTER SHARES CONSENSUS SEQUENCES WITH THE HUMAN c-Ha-ras1 PROMOTER

O. Segev and K. Lightfoot

Department of Zoology, University of the Witwatersrand, P O Wits 2050 Johannesburg, South Africa

INTRODUCTION

A very high proportion - up to 30% of a variety of human tumours including carcinomas, sarcomas and neuroblastomas contain activated ras oncogenes[1]. Several mechanisms which activate the transforming potential of normal ras cellular genes have been described[2]. In some cases alteration of the regulatory sequences of the gene may also induce its transforming activity[3]. Evidence exist for the role of enhanced expression of the normal ras onco-protein in ras activation in human tumours. Increase amounts of the normal ras protein were observed in breast carcinomas, lung, colon, bladder and gastric cancers[4] and in bone marrow cells from patients with Myelo dysplastic syndrome[5].

Cellular oncogenes are highly conserved through evolution in vertebrate and invertebrate species[6,7] as well as in yeast[8]. In Drosophila melanogaster, cellular genes related to several viral oncogenes have been detected[9]. Three ras genes have been isolated and termed ras1, ras2, and ras3. These genes were mapped to loci 85D, 64B and 62B respectively, on the polytene chromosomes of larval salivary glands[10]. They share on average 80% similarity with vertebrate ras p21 proteins at their N-terminus[10,11]. The structural and functional homologies between the Drosophila and vertebrate ras proteins were demonstrated by precipitating 21 and 27 - 28Kd proteins from Drosophila cell extract using monoclonal antibodies raised against the v-Ha-ras p21 protein[12]. Furthermore, it was previously shown that ras genes derived from fruitflies can promote neoplastic transformation of mammalian cells[13].

Interestingly, the promoter of the Drosophila ras2 gene exhibits bidirectional activity as was shown for the human c-Ha-ras1 and the mouse c-Ki-ras[14]. From this observation and because over-expression of the ras oncogene can cause cell transformation and tumour induction[2], we decided to study the mechanism controlling the Drosophila ras2 expression. The reported evidence that alteration of the ras regulatory region may induce transformation[3], led us to assume that a fine analysis of the ras2 promoter elements and the bound trans-factors should contribute to the understanding of ras regulation.

We have previously completed the isolation and characterization of the Drosophila ras2 promoter region[15]. Our studies revealed that another transcription unit, which we termed cs1 is located in the opposite orientation only 94 bases upstream to the ras2 gene. Apparently, the ras2/cs1 promoter is one of the shortest bidirectional promoters reported so far. From the results currently available, it seems likely that this bidirectional promoter is supporting transcription of both genes probably by one main mechanism and not by two separate transcription complexes regulating each gene.

Mobility-shift assays indicate the binding of a single major protein complex to the ras2/cs1 promoter. Evidence for the interaction of two different factors (components of this major protein complex), with specific promoter elements has been obtained by competition experiments, partial purification of nuclear extract and DNaseI footprinting analysis. The regions demarcated by these sequence classes have been found necessary for maximal transcription in both directions.

RESULTS

Sequences Required for Bidirectional Activity of the ras2/cs1 Promoter

To demonstrate bidirectional RNA transcription activity, the putative ras2 promoter and 5'-end were fused to the bacterial chloramphenicol acetyl transferase (CAT) gene in the promoterless vector p106 (a derivative of pSVOCAT)[16]. The same was done for the cs1 gene flanking the other side of the promoter. Both constructs were tested by transfecting Drosophila Schneider 2 culture cells supplemented with two different sera. While no CAT activity was detected in cells transfected with the p106 vector, significant activity was obtained with the ras2-promoter/CAT fusion and with the cs1-promoter/CAT fusion.

By introducing progressive deletions into sequences upstream of the ras2 or cs1 we were able to identify the regulatory promoter elements. The location of the deletions and the relative CAT activity obtained with each deletion are shown in Fig. 1. In cells which were supplemented with fetal bovine serum (Fig. 1C), only 22 bases 5' to the cap site of the ras2 gene were shown to be essential for obtaining up to 60% transcription efficiency in both directions (nt. 264-286 in particular).

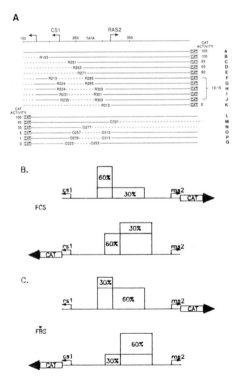

Fig. 1. (A). Deletion analysis of the bidirectional ras2/cs1 promoter. Schneider 2 (S2) cells were transfected with deleted ras2-CAT and cs1-CAT fusions and percentage of relative CAT activity was monitored. (---): deleted sequences. (B) and (C) Schematical presentation of relative promoter strength as detected by the CAT assay. S2 cells were supplemented with FCS or FBS respectively. Arrows represent transcription start sites. Large arrow represents the linked CAT gene. The promoter region spanning the nucleotides 250-290 is necessary for obtaining up to 85% - 90% expression of both genes.

264

However all upstream sequences up to the cs1 cap site seem to be required for maximal transcription efficiency of the ras2 gene and necessary for cs1 transcription to occur. By transfecting cells that were enriched with fetal calf serum (Fig. 1B), we obtained the opposite results. Most of the bidirectional expression was achieved by transfecting a plasmid which carries 56bp 5' to the cs1 transcription start site, (nt. 257-277 in particular) while upstream sequences contributed to 100% promoter activity.

In this context it should be mentioned that by injecting growing Drosophila embryos with these constructs the former results were obtained[16]. Thus is might be that different sera conditions can alter DNA binding properties and as a result affect gene regulation.

Distinct Factors can Interact with the ras2/cs1 Bidirectional Promoter

To identify trans-acting factors which interact with the cis-acting DNA sequences, appropriate DNA fragments were incubated with nuclear extracts and assayed as described below by gel retardation experiments. The extracts were obtained from nuclei of tissue culture cells and growing embryos, where the promoter is similarly active in both directions.

Labelled restriction fragments taken from the promoter region, were incubated with nuclear extracts in the presence of poly[dI-dC] to prevent 'non-specific' protein-DNA interactions. The resulting complexes were separated by electrophoresis on non-denaturating high ionic strength polyacrylamide gels (Fig. 2). Incubation of the entire promoter region (Fig. 2A), or 62 nt upstream of the ras2 cap site, with Drosophila nuclear extract generated one main retarded complex.

By using smaller synthetic oligonucleotides spanning the promoter region one could detect 2 distinct d.s. specific factors. Three DNA-protein complexes were identified (Fig. 3)

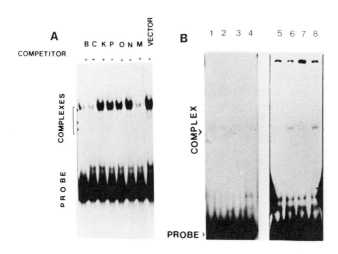

Fig 2.   In vitro protein binding assay profile. Gel electrophoresis mobility shift essays were performed with 10^4 cpm of the following probes, 8ug of Drosophila Schneider 2 cells extract, (supplemented with FBS) and 2ug poly (dI-dC). Products were analyzed by electrophoresis in a 6% polyacrylamide gel. Probes: panel (A) : the whole promoter region - a 120bp fragment (193-313). (B) a 90bp fragment (167-257). Unlabelled DNA added: (A) 1ug (+) of the indicated plasmid vectors presented in Fig. 1. (B) 1 - no competitor added. 2 - 1ug control vector sequence. [3,4 plasmid E; 5,6 - plasmid O; 7,8 plasmid P; -1ug and 2ug respectively].

using two 32bp oligomers (fragment I and II Fig. 8) overlapping part of the 22 bp region mentioned above. Given the results obtained in Fig. 3, the upper DNA-protein complex seem to be a specific double-strand binding protein, as addition of identical unlabelled oligomers at a 10-300 fold excess successfully competes for the formation of the complex proteins.

The other DNA-protein complexes represent non-specific proteins unable to interact specifically with fragment I. Using the two 14 bp repeats included in fragment I as probes, we could identify two out of the three factors that were described previously - a single strand binding protein and the former specific double strand binding protein (Fig. 4).

Another distinct factor was also identified using a 46bp fragment representing the cs1 5' end region (Fig. 2B, fragment IV-Fig. 8). From the results obtained it seems likely that this factor binds this specific promoter element with low affinity and is consequently not as important (as a single protein) for gene activation.

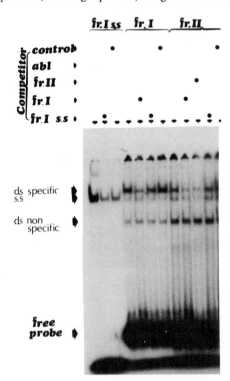

Fig 3. Gel retardation experiment using end labelled oligomers. Lanes 1-3 : s.s. fragment I (268-300), Lanes 4-7: d.s. fragment I, Lanes 8-12 : fragment II (287-325). Each lane contains a constant amount of end labelled DNA (10^4 cpm), 15ug Schneider 2 nuclear extract, 0.25ug synthetic poly (dI-dC) and 1pmol competitor DNA. Competitor DNAs are represented.

Serum Specificity

When various deletions of the ras2/cs1 promoter were tested as competitors in a mobility shift assay, two regions were found to be required for binding (Table 1) in cells supplemented with different sera. When S2 cells were induced with fetal bovine serum, the right central region of the promoter (nt. 264-286 Fig. 8) was shown to bind a large protein complex. The same region of DNA gave the highest levels of induction in those cells and in growing embryos, as was demonstrated by the CAT assays. By treating the cells with fetal calf serum, the same large protein complex binds the left central region of the promoter (nt. 240-277 Fig. 8). As was indicated before, this region was shown to be the potent activator of bidirectional transcription in these conditions.

The results indicate the presence of one main protein complex which can interact with the promoter central region and regulate both genes expression. Apparently by altering serum components which could probably affect growth factor balance within the cell, the relative DNA binding specificity of this protein complex can be modified (Table 1B).

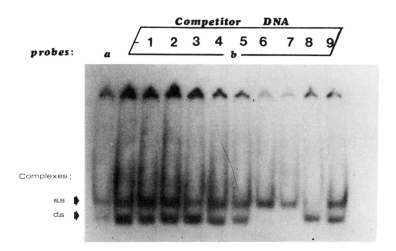

Fig 4. Similar in vitro protein binding assay profile using 20bp oligomers carrying the two repeats found in fragment I (see fig.8). Assays were performed with 10^4 cpm labelled oligomers, 15ug Drosophila Schneider 2 nuclear extract, 0,05ug poly (dI-dC)and 1 pmol competitor DNA. Probes: Lane 1 left inverted repeat within fragment I (265-282), Lanes 2-9 right inverted repeat within fragment I (285-301). Complexes are blocked by the following competitors: 6 (fragment I); 7 (fragment II); 8(S.S control DNA); 9(d.s. non specific oligonucleotide).

Table 1.   (A). Identification of specific protein binding sites in the ras2/cs1 promoter. S2 cells were grown in the presence of FBS or FCS, extracted and used for gel retardation assays as was described previously. 16h embryo extract was also included. The labelled DNA-protein complex was formed in vitro by the whole labelled 120 bp ras2/cs1 promoter region. Unlabelled 1ug competitor DNA   plasmids are indicated and illustrated in Fig 1. (+) - indicates competition for complex formation with the labelled DNA. (B) Model of protein complex binding sites in 16h embryos and S2 cells supplemented with different sera types.   In FCS induced cells the large protein complex binds the A region (plasmids N and O) but not the B region (plasmids D and E). Region B competes very efficiently when using FBS induced cell extract or embryo extract, for the binding of the same protein complex, while region A does not.

Main Protein Complex binding sites.

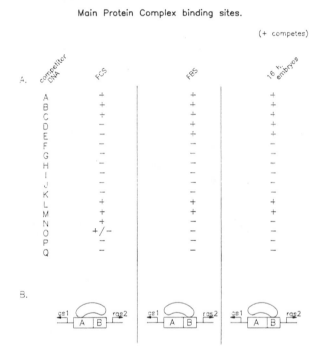

Partial Purification of ras2/cs1 Transcription Factors, and DNase I Footprinting Assays Identifying the Exact Binding Sites

The   major protein complex was partially purified by chromatography on a Heparin-Ultragel or phosphocellulose columns. Table 2 (A,B) represents the results of a mobility shift assay of the various fractions eluted from the Heparin-Ultragel and Phosphocellulose columns respectively. When the whole labelled promoter region was used as a probe, the majority of the major complex was eluted in the 300mM Nacl step elute in both columns. This represented approximately 40-fold purification with respect to the crude NCE (note that equal amounts of protein were used in all the reactions). When gel retardation assays were carried out using aliquots from the Heparin-Ultragel column fractions and a [^{32}p]-labelled fragment I (or either left or right inverted repeats of fragment I), the d.s specific protein-DNA complex (mentioned before) was eluted also in the 300mM Nacl elute. A larger DNA-protein complex was also observed using 300mM Nacl and 400mM Nacl fractions.

Addition of an aliquot from the 200mM Nacl fraction to the 400mM fraction aliquot, resulted in enhancement of the larger complex compared with reduced intensity of the d.s. specific complex (Fig. 5). Surprisingly, using the Phosphocellulose column fractions, the identical d.s. specific complex (which binds fragment I) was eluted in the 500mM Nacl elute.

These results raise the possibility of a hetero-dimer formation by direct interaction between the d.s. specific complex and another protein factor which cannot interact directly with fragment I but can interact very poorly with region A. Thus the latter factor presumably is present in the 200mM and 300mM samples and contributes to the formation of the larger complexes seen when fragment I or the whole promoter region are used as probes. All these complexes were specific for the ras2/cs1 promoter region (fragment I + region A) since they were greatly reduced when the same unlabelled promoter elements were added to the reaction.

In several occasions incubation of the whole labelled promoter fragment with Heparin-Ultragel samples or nuclear-cell extract, resulted in one smaller retarded complex (Fig. 6) which was shown to interact with fragment I. As was indicated above, a large main protein complex can be seen interacting with the ras2/cs1 promoter prior to the purification step. This data strongly support disassociation of the protein complex sub-units. If this smaller retarded complex represents one protein component which is capable of interacting with fragment I, why can we not identify the other sub-unit which was supposed to interact with region A? The answer to this question is not known yet, but it seems likely from the results obtained to date that in the absence of a hetero-dimer formation, this partially purified factor cannot bind efficiently to its cognate promoter region in vitro.

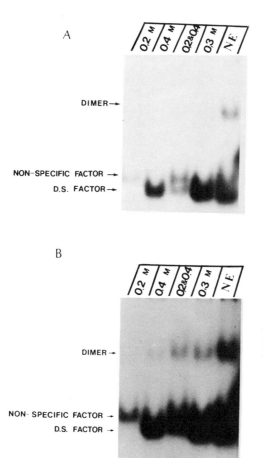

Fig. 5. Hetero-dimer formation. Heparin-Ultralgel chromatographic fractions of S2 cells nuclear extract were incubated with labelled 20 b.p. oligomer spanning the right inverted repeat of fragment I. Three DNA-protein complexes can be identified. D.s. specific complex, s.s. non specific complex and a larger d.s. specific complex (dimer). Binding reactions contained 1ug aliquots of the fractions indicated above the autoradiogram. The molarity of Nacl of every fraction is indicated. Only the relevant gradient fractions are shown. The dimer and d.s. DNA complexes reproducibily eluted at similar ionic strengths in several experiments using independently prepared nuclear extracts. A -24h exposure; B- longer exposure. N.E. nuclear extract.

Table 2.   Representation of DNA - binding assays of Heparin - Ultragel and Phosphocellulose column fractions. Binding reaction mixtures (12ul) included 0.05ug poly (dI-dC), 10^4 cpm labelled oligomer (indicated in this figure) and 1ug of the column fractions. Salt concentration (Nacl) is indicated. (+) represents specific DNA-protein complex formation ( ) low affinity complex formation or traces amounts of detected complex.

	A. Heparin — ultragel	B. Phospho — cellulose.
whole promoter region.	+ +	+
fragment I	+ +	+
right repeat of fragment. I	+ +	+
region A	+ +	+ +
	100 200 300 400 500 600 700 800 900 1000	100 200 300 400 500 600 700 800 900 1000 (mM Nacl)

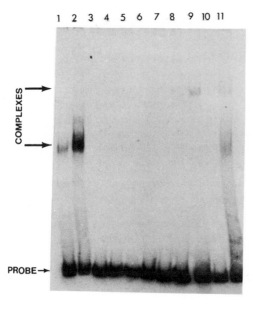

Fig. 6.   Disassociation of the large protein complex sub-units. Chromatographic fractionation of S2 cells nuclear extract on Heparin Ultragel column. Complexes formation was followed using the whole labelled 120 bp ras2/cs1 promoter region. Reaction conditions described above. The arrows represent two DNA-protein complexes: the large DNA-protein complex (formed with fetal bovine serum or fetal calf serum induced cell extracts), lanes 9, 10 and 11 respectively. The smaller DNA-protein complex represents a sub unit of the former large complex. Lanes 1-8 represent salt molarity fractions of 0.2M - 0.9M respectively

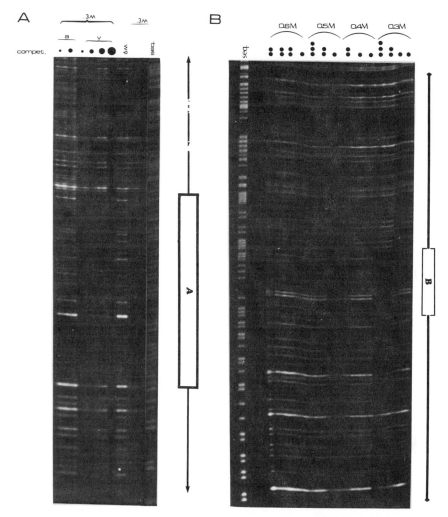

Fig. 7.   DNase I footprint of full-length ras2/cs1 promoter region. A 340-bp fragment was end-labelled with ^{32}P, incubated with Heparin-Ultragel column fractions aliquotes as indicated and subjected to DNase cleavage.   (A) 1ug protein elute was added to every reaction. Fragments used as competitors: (a) 0.5 pmol and 2 pmol of a 120 bp promoter region. (V) 0.5, 2, 5 and 25 pmoles of a non-specific vector sequence. Box A on right margin represents the position of protein-complex binding sequences. (B) 1ug, 2ug. 3ug and 5ug protein aliquotes were added to the binding reaction. The nuclear extract used for the purification was derived from cells grown in FBS containing medium, whereas FCS was added in panel A. A G+A sequencing reaction was run in adjacent lanes. The strongly protected region is indicated by the B box.

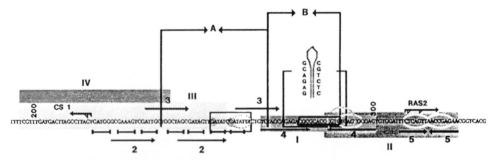

Fig. 8.   Consensus sequences and protein binding sites of the ras2/cs1 promoter. Black and open triangles: transcription start sites determined by T4 pol. external primer extension and by RNase protection, respectively. The arrows above the triangles show the direction of the transcription; 4 large boxes: fragments used as probes for gel retardation assays; numbered arrows: direct and inverted repeats with more than 80% similarity; open bars - DSE and GC-like sequence; open circle - AP-I like sequence; elipse-TATA like sequence and direct and inverted insect "cap box", large open squares - protein binding sites A and B.

DNase footprinting assays confirmed that these sequences (region A and fragment I Fig. 8) bound selectively to the sub-units which forms the large protein complex (Fig. 7). Inducing the cells with 2 different sera caused the formation of the same large protein complex which, however, confers different DNA-binding specificities (dependent) on the type of serum used. Closer examination of the two DNase I protected regions within the promoter, revealed different sequence motifs which exhibit homology to known consensus sequences (Fig. 8).

Region A carries a hexamer GATA(TT/GC) which is repeated seven times overlapping an element showing significant homology to the dyad-symmetry element [(DSE)-part of the serum response element]. A "TATA"- like sequence is also present in this region. A hair-pin structure and 2 motifs which resemble an AP-1/TRE (TPA responsive element) and a sp1-binding site have been noted in the B region which overlaps fragment I. The same DNase I footprints were identified using the coding and non-coding strands. Taken together, these observations imply that the context of the cell can clearly affect the nature of the signal response by altering DNA-binding specificity.

DISCUSSION

There is intense interest in elucidating the biological function of the ras oncogene product and the components interacting with it or regulating its expression in the pathway leading to normal or transformed growth promotion.   Integrated into the functional network of oncoproteins must be mechanisms for the mutual regulation of oncogene expression, since the dose and activity of each member of a network needs to be tightly controlled. Since none of the cellular factors is constitutively active, there is always the possibility that oncogenicity results from altered response to regulation.

In this article we have presented structural and functional evidence suggesting that the two genes flanking the bidirectional promoter (ras2 and cs1), are regulated by one main mechanism and not by two separate transcription complexes regulating each gene. Sharing regulatory elements between two divergent genes was shown for the α1(IV) and α2(IV) collagen genes. These genes share a bidirectional promoter that utilizes the same enhancer within the first intron of the α1(IV) gene for efficient transcriptional activity [17]. One of the examples for separate transcriptional mechanisms was described for the bidirectional promoter of the his3 and pet56 genes in yeast. Among the trans-factors interacting with this promoter, the GCN4 trans activator can induce the his3 expression but not the pet56 expression to the opposite direction[18]. Another aspect concerned with bidirectional promoters is the transcriptional efficiency of each gene. The Drosophila ras2 gene is not a stage- specific gene [19,20] and the spatial and temporal specificities of the cs1 gene have not yet been studied. We do know that these two genes are co-regulated during development[21]. In this context it should be noted that several gene pairs regulated by bidirectional promoters are also ubiquitously expressed (the mouse DHFR gene and small nuclear RNAs; the Drosophila tl and t2 genes[22]; and the mouse surf-1 and surf2-genes[23]). Recently Z.Lev and his colleagues have shown that both genes are expressed simultaneously in all developmental stages. Studies are now in progress to establish whether they are expressed in the same tissues (personal communication). Interestingly, the promoters of the human c-Ha-ras1 and the mouse c-Ki-ras also exhibit bidirectional activity[14] but in these cases no specific divergent transcription unit have yet been identified.

Vertebrate ras promoters which are generally GC rich, do not contain the consensus sequences directing polymerase II transcription such as the TATA box, but do contain a number of GC boxes[14,24,25,26]. These features are very similar to those of the human EGF receptor promoter[27] and to those of a number of cellular house keeping gene promoters (e.g. HMG coenzyme A reductase[28]; mouse and human DHFR[29]; human superoxide dismutase[30]). In the human c-Ha-ras1 promoter region, multiple elements for transcriptional control are dispersed. Among them are the Sp1-binding sites, a CCAAT-box element and a CACCC-box element (also found in the globin gene promoter), which are necessary for maximal promoter activity[31]. Furthermore, four motifs which resemble TPA-inducible and AP-1-binding sites were also noted conferring TPA inducibility[32]. Previous findings show the most active form of AP-1 to consist of a hetero-dimer between JUN and FOS, the products of the c-jun and c-fos proto-oncogenes respectively[33]. As Ha-ras was shown to trigger fos expression[34] it might be that Ha-ras can be autoregulated by activating the c-fos gene. Evidence for positive autoregulation of the H-ras1 gene, came from studying the effects of insulin on H-ras1 gene expression through mutant T24 ras p21[35]. The Drosophila ras2 promoter region shares with the human Ha-ras1 promoter an sp-1 like sequence, a CACCC-box element and an Ap-1 like sequence. In this paper we identified a specific transcription factor which binds these sites on the DNA and can form a complex with another specific DNA-binding protein. We are currently in the process of finding out which of these sites is the main control region.

The binding sites (A+B - see Fig. 8)for the two distinct protein factors (see Fig. 8) that were identified, were found to be essential for up to 90% expression of both genes. We are considering the possibility that in the absence of one of the protein factors the level of expression or tissue specificities of these two genes would be modified. Studies are underway to test this hypothesis. Recent results indicate binding of a large protein complex to the promoter region in embryos, whereas only one component: the B-region binding factor (factor B), was identified in pupae and larvae. Another component of the large complex: the A-region binding factor (factor A), was shown to interact very poorly with the promoter region. By forming a hetero-dimer with the B factor, binding efficiency to the A region can be enhanced. Many transcription factors bind to DNA as dimers, either of homomeric or heteromeric nature, and these hetero-dimers may have sequence specificities different from either parent homodimer[36]. Whether the "TATA"-like sequence within region A serves to identify the sense strand, as well as the precise transcription start point for the cs1 gene, remains to be established. If this is the case, deletion of the TATA box will diminish cs1 expression and possibly affect ras2 expression. Fine structural and functional analysis of the regulatory elements together with the purification and characterization of the trans- factors controlling the bidirectional expression are clearly required to elucidate their role in the transcriptional machinery.

In addition it is of interest to test the possibility that other promoters carrying the same cis-elements are regulated by the same transcription factors, and whether these DNA elements can either act as bidirectional promoter elements or enhancer motifs.

As was represented in this article, altering DNA-protein binding specificities can be achieved by supplementing the growth media with different sera. Changing serum conditions might cause modifications in enzymes that act upon these factors, and this in turn can influence the regulatory machinery. Neoplastic transformation caused by transcriptional alteration or by other means of oncogene activation, might be triggered by the existence of a balance of oncogenic and suppressive proteins, alteration of which can switch a cell from a state of restrained into unrestrained growth.

REFERENCES

1. Bos, J.L., E.R. Fearon, S.R. Hamilton, M. Verlaan-de Vries, J.H. Van Boom, A.J. Van der Eb & B. Vogelstein. 1987. Nature 327: 293. Forrester, K., C. Almoguera, K. Han, W.E. Grizzle, & M. Perucho. 1987. Nature 327: 298.
2. Chang, E.H. & M.E. Furth. 1982. Nature 297: 479. Alitalo, K. 1984. Med. Biol. 62: 304. Fasano, O., T. Aldrich, F. Tamanoi, E. Taparowsky, M. Furth & M. Wigler. 1989. Proc. Natl. Acad. Sci. USA. 81: 4008.
3. Theillet, C. 1986. Cancer Res. 46: 4776.
4. Biasi, F.D., G.D. Sal, P.H. Hand. 1989. Int. J. Cancer 43:431.
5 Srivastava, A., H. Scott Boswell, N.A. Heerema, P. Nahreini, R.C. Laner, A.C. Antony, R. Hoffman & G.J. Tricot. 1988. Inter Society for Experimental Hematology meeting, Buffalo.
6. Shilo, B.-Z. 1989. Adv. Viral. Oncol. 4: 29.
7. Shilo, B.-Z. & R.A. Weinberg. 1981. Proc. Natl. Sci. USA. 78: 6789.
8. Dhar, R., A. Nigtu, R. Koller, D. Defeo-Jones, E.M. Scolnick. 1984. Nucleic Acids Res. 12: 3611.
9. Liven, E., L. Glazer, D. Segal, J. Schlessinger & B.-Z. Shilo. 1985. Cell 40: 599. Wadsworth, S.C., W.S. Vincent, D. Bilodean-Wentworth. 1985. Nature 314: 178. Shilo, B.-Z. 1987. Trands Genet. 3: 69.
10. Neuman-Silberberg, F.S, E. Schejter, F.M. Hoffmann & B.-Z. Shilo. 1984. Cell 37: 1027.
11. Mozer, B., R. Marlor, S. Parkhurst, V. Corces. 1985. Mol. Cell. Biol. 5, 885. H.W. Brock. 1987. Gene 51: 129.
12. Papageorge, A.G., D. De Feo-Jones, P. Robinson, G. Tememles & E.M. Scolnick. 1984. Mol. Cell. Biol. 4: 23.
13. Shilo, B.-Z. 1985. EMBO J. 4: 407.
14. Spandidos, D.A. & M. Biggio. 1986. FEBS lett. 203: 169. Hoffman, E.K., S.P. Trusko, N. Freeman & D.L. George. 1987. Mol. Cell. Biol. 7: 2592.
15. Cohen, N., A. Salzberg & Z. Lev. 1988. Oncogene 3: 137.
16. Lev, Z., O. Segev, N. Cohen, A. Salzberg & T. Shemer. 1989. NATO ASI SERIES on the "ras oncogenes", ATHENS.
17. Burbelo, P.D., G.R. Martin, & Y. Yamada. 1988. Proc. Natl. Acad. Sci. USA. 85: 9679.
18. Sruhl, K. 1989. Presented at the First Natl. Conference on Gene Regulation/Oncogenesis/Aids, Lutraki.
19. Lev, Z., Z. Kimchie, R. Hessle & O. Segev. 1985. Mol. Cell. Biol. 5: 1540.
20. Segal, D. & B.-Z. Shilo. 1986. Mol. Cell. Biol. 6: 2241.
21. Farnham, P.J., J.M. Abrams & R.T. Schike. 1985. Proc. Natl. Acad. Sci. USA. 82: 3978.
22. Swaroop, A., J-W. Sun, M.L. Paco-Larson & A. Garen. 1986. Mol. Cell. Biol. 6: 833.
23. Williams, T.J. & M. Fried. 1986. Mol. Cell. Biol. 6: 4558.
24. Ishii, S., G.T. Merlino & I. Pastan. 1985. Science 230: 1378.
25. McGrath, J.P., D.J. Capon, D.V. Goeddel & A.D. Levinson. 1984. Nature 310: 644.
26. Hall, A. & R. Brown. 1985. Nucl. Acids Res. 13: 5255.

27. Ishii, S., Y.-H. Xu, R.H. Stratton, G. Roe, T. Merlino & I. Pastan. 1985. Proc. Natl. Acad. Sci. USA. 82: 4920.
28. Osborne, T.F., J.L. Goldstein & M.S. Brown. 1985. Cell 42: 203.
29. Masters, J.N. & G. Attardi. 1985. Mol. Cell. Biol. 5: 493. McGrogan, M., C.C. Simonsen, O.T. Smouse, P.J. Farnham & R.T. Schimke. 1985. J. Biol. Chem. 260: 2307.
30. Levanon, D., J. Lieman-Hurwits, N. Dafni, M. Wigderson, L. Sherman, Y. Bernstein, Z. Laver-Rudich, E. Danciger, O. Stein & Y. Groner. 1985. EMBO J. 4: 77.
31. Jones, K.A.,J. Paul, J. Win & M. Allan. 1989. Mol. Cell Biol. 9: 3758. Takahiro, N., U. Yoshio & I. Shonsuke. 1990. Gene 94: 249.
32. Spandidos, D.A., R.A.B. Nichols, N.M. Wilkie & A. Pintzas. 1988. FEBS lett. 240: 191.
33. Nakabeppu, Y., K. Ryder & D. Nathans. 1988. Cell 55: 907. Halazonetis, T.D., K. Georgopoulos, M.E. Greenberg & P. Leder. 1988. Cell 55: 917.
34. Stacey, D.W. 1987. Mol. Cell. Biol. 7: 523. Wyllie, A.H. 1987. Br. J. Cancer 56: 251. Schonthal, A., P Herrlich, H. Jobst Rahmsdorf & H. Ponta. 1988. Cell 54: 325. .
35. Pintzas, A. & D.A. Spandidos. 1989. Gene Anal. Tech. 6: 125.
36. Jones, N. 1990. Cell 61: 9. Hunter, T. 1991. Cell 64: 249. Lewin, B. 1991. Cell, 64: 303.

ONCOGENE EXPRESSION AND CERVICAL CANCER

R.P. Symonds[1], T. Habeshaw[1], J. Paul[1],
D.J. Kerr[1], A. Darling[2], R.A. Burnett[3],
F. Sotsiou[4], S. Linardopoulos[5] and
D.A. Spandidos[5]

1-Beatson Oncology Centre, Western Infirmary
Glasgow, G11 6NT, U.K.
2-Beatson Institute for Cancer Research
Glasgow, U.K.
3-Department of Pathology, Western Infirmary
Glasgow, U.K.
4-Department of Pathology, Evangelismos
Hospital, Athens, Greece
5-National Hellenic Foundation, Institute of
Biological Research and Biotechnology, Athens
Greece

INTRODUCTION

Evidence strongly suggests that oncogene expression is an important factor in the initiation and development of a range of malignant tumours including carcinoma of cervix (Slamon, 1987). Some patients with histologically similar tumours at the same clinical stage treated the same way do better or worse than others. Prognosis, radiosensitivity and chemotherapy resistance may be affected by oncogene expression. Riou and colleagues (1987) have shown that surgically treated patients with early stage (I and IIa) disease whose tumours show c-myc overexpression have an eight-fold increase in the incidence of relapse compared to other patients.

It has been suggested that Cis-platin resistance in human carcinomas may be associated with oncogene amplification. The oncogenes c-fos and c-H-ras have been shown to be amplified in patients failing treatment with Cis-platin and 5-Flurouracil (Scanlon et al., 1988). Cultured cell lines of Cis-platin resistant breast and ovarian tumours have shown a 3-5 fold increase in c-myc messenger RNA (Scanlon et al., 1989).

We have shown that long term disease free survival in patients with advanced carcinoma of cervix treated by neo-adjuvant chemotherapy is strongly associated with an objective responce to chemotherapy (Symonds et al., 1989). Differences in oncogene expression may account for the relatively good prognosis of patients associated with chemo-sensitive lesions. In order to detect clinically significant differences in oncogene expression we have retrospectively examined paraffin fixed biopsy material immunocytochemically stained for the protein products of the ras, myc and jun proto-oncogenes. Previous studies suggest there is an interaction between ras and myc in the progression of cervical cancer. C-myc amplification and overexpression is more frequent in advanced tumours (stage III and IV) than earlier stage disease (Riou, 1988). Similarly, c-H-ras mutations are seen more frequently in advanced tumours. Mutations at codon 12 of the c-H-ras gene were detected in 2% of stage I and II tumours and 24% of stage III and IV (Riou, 1988). Loss of one c-H-ras allele is relatively common in cervical cancer (36%) and is not associated with increasing tumour stage. However, tumours with loss of one c-H-ras allele often contain an activated c-myc gene. Both c-H-ras mutations and allele loss are associated with c-myc gene activation which was found in 100% and 70% of tumours containing mutation and deletion respectively.

In contrast to myc and ras, the role of c-jun in carcinoma of cervix has not been explored. C-jun interacts with c-fos to produce the transcriptional protein AP-1 which controls the transcription of other genes and the resultant synthesis of cellular proteins. Genes activated by AP-1 include those binding for metallothionine IIa, collagenase and stromelysin. Metallothionine binds to Cis-platin and overexpression of metallothionine confers Cis-platin resistance upon cells in culture (Andrews and Howell, 1990).

MATERIALS AND METHODS

Fifty-five patients were treated between January 1984 and June 1987. The median age was 49 (range 29-69) and all patients had a WHO performance status of 2 or more. All were fit for radical chemotherapy and radiotherapy with a creatinine clearance of greater than 50mls/min.

Staging was according to FIGO criteria after an examination under anaesthesia which included cystoscopy, vaginal and rectal examination, repeat biopsy and estimation of tumour size. All patients had a chest radiograph, an intravenous urogram and abdominal ultrasound. Forty patients were assinged to Stage III and 15 to Stage IVa.

Two pulses of chemotherapy consisting of Cis-platin 50mg/m^2, Bleomycin 30mg and Vincristine 2mg were given 14 days apart before radical radiotherapy. External beam x-ray treatment (4Mev) began 28 days after starting chemotherapy. A dose of 42.5Gy was given in 20 fractions over 28 days to the true pelvis using a 4-field technique (average volume 15x15x11 cm). This was followed by a single Cs^{137} insertion using "Manchester" type applicators. The dose to point A was

33.5Gy at a dose rate of 0.55Gy/hr. When the Selector afterloading machine was used (a point dose rate 1.2-1.7Gy/hr) the dose to point A was reduced according to the CRE formula. Further details including side-effects are published elsewhere (Symonds et al., 1989).

Response to chemotherapy prior to radiotherapy was assessed by clinical examination (in borderline cases by two examiners) and by pelvic ultrasound.

Paraffin-fixed histological material was immunohistochemically stained to try to correlate prognosis with expression of ras, c-myc and jun oncoproteins. The following antibodies were used: for ras p21, the rat monoclonal antibody Y13 259 (Furth et al., 1982); for c-myc p62, the mouse monoclonal antibody Mycl - 9E10 (Evan et al., 1985) and c-jun AP-1, the rabbit anti-peptide polyclonal antibody 890. This antibody was made against the peptide GSLKPHLRAKNSD which corresponds to amino acid 47-59 of human c-jun sequence (Hattori et al., 1988). Both Y13259 and mycl - 9E10 antibodies were prepared from hybridomas and previously described (Spandidos et al., 1987; Papadimitriou et al., 1988). Immunohistochemistry was carried out using the streptavidin-peroxidase DAB system (Spandidos et al., 1987; Papadimitriou et al., 1988).

The antibody staining was scored as positive, negative or equivocal. All of the sections were examined by two experienced pathologists who at the time of the examination were unaware of the clinical outcome.

RESULTS

Fifty-one percent (26/51) of patients had a partial response to chemotherapy according to WHO criteria. Twenty-five patients had no response to chemotherapy and in 4 cases response was not assessed. All patients who demonstrated a partial response to chemotherapy had a complete response to radiotherapy when assessed 3 months after the end of treatment. The actuarial 5-year survival of all patients is 43% (95% C.L. 29-56%). However, an initial response to chemotherapy is associated with significantly better longterm survival (see Fig.1). The 3-year survival of chemotherapy responders is 62% against 21% for non-responders (p=0.009 log-rank test).

Neither clinical factors, histological grade or the expression of oncogene proteins correlates with tumour shrinkage after chemotherapy and a subsequent good outlook. Factors such as patient's age, tumour stage and pathological grade are similarly distributed between responders and non-responders.

The frequency of expression of oncogene protein products is high in these advanced tumours. The percentage of biopsy speciments showing positive immunocytochemical staining was 90.4% for ras, 45.1% for c-myc and 39.2% for c-jun. In the 49 specimens in which an assay for all 3 oncoproteins was performed only 16.3% (8 cases) had no evidence of expression of at least one oncogene and 28.6% (14) exhibited all 3 oncogenic protein products.

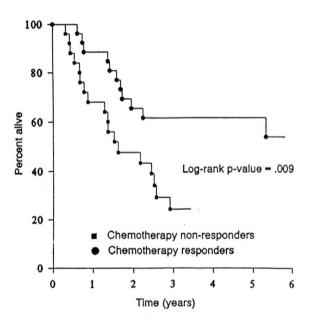

FIGURE 1. Survival in advanced Ca cervix by chemotherapy response.

Table 1. Association of oncoprotein expression and response to chemotherapy

Oncogene		PR		NR		P-value
Ras	Present	80.0%	(20)	77.3%	(17)	1.00*
	Absent	20.0%	(5)	27.7%	(5)	
		100%	(25)	100%	(22)	
C-myc	Present	50.2%	(13)	33.3%	(7)	.251
	Absent	50.0%	(13)	66.7%	(14)	
		100%	(26)	100%	(21)	
C-jun	Present	40.0%	(10)	31.8%	(7)	.560
	Absent	60.0%	(15)	68.2%	(15)	
		100%	(25)	100%	(22)	

    * This P-value from Fisher's exact test.
    PR = Partial response
    NR = no response

There is no statistically significant association with positive staining for oncogene products and response to chemotherapy (Table 1), survival, time to distant metastases or local recurrence (Table 2). Similarly pathological grade shows no statistically significant association with individual oncogene expression or the number of oncogenes detected in each biopsy specimen.

DISCUSSION

It has been suggested that Cis-platin resistance in human carcinomas may be associated with oncogene amplification. We can find no statistically significant association with oncogene expression and response to chemotherapy. In particular ras p21 protein was found in 80% of chemoresponsive tumours and 77.3% of non-responding lesions, a virtually even split.

One reason for the lack of prognostic significance of the gene products is high level of detection. Previously, activation of c-myc has been shown to be common in advanced tumours. Amplification was present in 49% and overexpression (increased c-myc RNA) was present in 75% (30/40) in biopsies taken from Stage III and IV patients (Riou,1988). Our frequency of expression of myc p62 (45%) is consistent with these findings. The frequency of detection of ras protein p21 (80.4%) is high but perhaps not surprising in these aggressive advanced tumours. The relatively high detection rate of all three oncogene protein products

Table 2. Log-rank p values for the association of oncoprotein expression with patient outcome

	Survival	Time to develop distant metastases	Time to local recurrence
ras present v absent	.295	.061	.974
c-myc present v absent	.183	.164	.601
c-jun present v absent	.923	.521	.995
Number of oncoproteins present in each biopsy 0v1v2v3	.709,.476*	208,.404	.607,.636

* First P-value compares 0v1v2v3 assuming no ordering for patients for whom 3 oncoproteins were assayed. The second P-value tests for linear trend across these categories. The results in this row are only for patients who had three oncoproteins assayed.

reflects the high cellular activity in these tumours. Our data suggests that the detection of protein oncogene products of c-myc, ras and c-jun by immynocytochemistry does not give useful prognostic information in advanced carcinoma of cervix and does not predict for response to chemotherapy or ultimate outcome.

ACKNOWLEDGEMENTS

We thank Mrs. Moira Stewart for data collection.

REFERENCES

Andrews, P.A., Howell, S.B., 1990. Cellular pharmacology of cis-patin. Perspectives on mechanisms of acquired resistance. Cancer Cells, 2:35-43.

Evan, G.I., Lewis, G.K., Ramsay, G., Bishop, J.M., 1985. Isolation of monoclonal antibodies specific for human c-myc proto-oncogene product. Mol.Cell.Biol., 5:3610-3616.

Furth, M.E., Davis, L.J., Fleurdelys, B., Sconlick, E.M., 1982. Monocolonal antibodies to the p51 products of the transforming gene of Harvey murine sarcoma virus of the cellular ras gene family. J.Virol., 43:294-304.

Hattori, K., Angel, P., Le Beau, M., Karin, M.M., 1988. Structure and chromosomal localisation of the functional intronless human jun proto-oncogene. Proc.Natl.Acd.Sci. USA., 85:9148-9152.

Papadimitriou, K., Yiagnisis, M., Tolis, G., Spandidos, D.A., 1988. Immunohistochemical analysis of the ras oncogene protein product in human thyroid neoplasmas. Anticancer Res., 8:1223-1228.

Riou, G., Le, M.G., Le Doussal, V., Barvois, M., Martine, G., Haire, C., 1987. C-myc proto-oncogene expression and prognosis in early carcinoma of the uterine cervix. Lancet., ii:761-763.

Riou, G.F., 1988. Proto-oncogene and prognosis in early carcinoma of the uterine cervix. Cancer Surveys., 7:441-455.

Riou, G., Barvois, M., Sheng, Z., Duvillard, P., Lhomme, C., 1988. Somatic deletions and mutations of c-Ha-ras gene in human cervical cancers. Oncogene., 3:329-333.

Scanlon, K.J., Lu, Y., Kashani-Sabet, M., Ma, J., Newman. E., 1988. Mechanisms for Cis-platin-FUra synergism and Cis-platin resistance in human ovarian carcinoma cells both in vitro and in vivo, in: Rustum Y.& McCuire J. (Eds). The expanding role of Folates and Fruoro-pyrimides in Cancer Chemotherapy. New York, Plenum Press, p131-139.

Slamon, D.J. 1987. Proto-oncogenes and human cancers. N.Engl.J. Med., 317:955-957.

Spandidos, D.A., Pintzas, A., Kakkana, A., 1987. Elevated expression the myc gene in human benign and malignant breast lesions compared to normal tissue. Anticancer Res., 7:1299-1304.

Symonds, R.P., Burnett, R.A., Habeshaw, T., Kaye, S.B., Snee, M.P., Watson, E.R. 1989. The prognostic value of a response to chemotherapy given before radiotherapy in advanced cancer of cervix. Br.J.Cancer., 59:473-475.

283

THE EFFECT OF H-RAS AND C-MYC ONCOGENE TRANSFECTION ON THE
RESPONSE OF LUNG EPITHELIAL CELLS TO GROWTH FACTORS AND
CYTOTOXIC DRUGS

D.J. Kerr[1,3], J.A. Plumb[1], G.C. Wishart[1],
M.Z. Khan[1], R.I. Freshney[1] and D.A. Spandidos[2]

1-CRC Dept of Medical Oncology, Alexander Stone
Building, Garscube Estate, Bearsden
Glasgow G61 1BD, U.K.
2-Institute of Biological Research and
Biotechnology, National Hellenic Research
Foundation, 48 Vas Constantinou Ave, Athens
11635, Greece
3-To whom requests for reprints should be
addressed.

INTRODUCTION

Clinical applications consequent upon elucidation of the
role of oncogenes in contributing to the malignant cellular
phenotype have mainly focussed on refinement of existing
prognostic models for patient survival. Current research on
breast cancer has shown a relationship between amplification
of the c-erbB-2 or Her-2/neu oncogenes and disease progression
and patient survival (Slamon et al, 1987). Similarly,
amplification of the N-myc oncogene was found consistently in
patients with advanced state neuroblastoma which was
relatively resistant to chemotherapy (Schwab et al, 1984).
The association between oncogene expression and prognosis
could be due to two related variables: oncogene activation
could be associated with a particularly aggressive malignant
phenotype, e.g. they could confer metastability; or the
oncogenes could confer cellular drug resistance to conventional
chemotherapy resulting in a tumour refractory to standard
therapy. A number of in vitro studies have indicated that
oncogene transfection of immortalised cells can result in
relative resistance to a number of antineoplastic drugs,
including cisplatin (Sklar 1988).
    In addition, the rate of proliferation of tumour cells
may be governed by the pattern of oncogene activation and the
cocktail of growth factors operant in their microenvironment.

The Superfamily of ras-Related Genes
Edited by D.A. Spandidos, Plenum Press, New York, 1991

The chemoresponsiveness of tumour cells to cytotoxic drugs has been related to proliferative capacity in that rapidly dividing cells are usually more sensitive to the cycle specific agents used in common clinical practice. In this study, we report the effect of exogenous growth factors, EGF and TGF-β, on the chemosensitivity of mink lung epithelial cells and sublines derived by transfection with c-myc and activated H-ras oncogenes.

MATERIALS AND METHODS

Cell lines

The parent mink cell line MV1Lu (ATCC-CCL64) was obtained from the European Collection of Animal Cell cultures, Porton Down, Salisbury, Wilts, England. It was maintained in a 1:1 mixture of Ham's F10 and Dulbecco's modification of Eagle's basal medium (F10/DMEM), supplemented with 10% foetal bovine serum and under a gas phase of 2% CO_2 in equilibrium with 8mM $NaHCO_3$. The cell lines were characterised in vitro and in vivo, grown as xenografts in nude mice. (Khan et al, 1991).

Transfections

Three other cell lines were obtained from the MV1Lu cell line by insertion of mutated (T24) and non-mutated H-ras1 and human c-myc genes in high expression vectors, by the calcium phosphate technique (Graham and Van der Eb, 1973) as modified by Spandidos and Wilkie (1984a). The plasmids used for transfection were pH06T1, pH06N1 (Spandidos and Wilkie 1984b) and pMCGM1 (Spandidos, 1985) containing the entire T24 H-ras1, the entire normal human H-ras1 proto-oncogene and the entire human c-myc gene respectively. In pH06T1 and pH06N1 the ras genes are surrounded by both SV40 and Moloney virus long terminal repeats (LTR) enhancers, whilst in pMCGM1, the Moloney virus LTR sequence is linked to the human c-myc gene. All the plasmids contain the aminoglycoside phosphotransferase gene (aph), conferring resistance to geneticin, which was used in the selection of the lines. Each derivative line was expanded from a single clone, and their identity confirmed as mink by chromosomal analysis. They were designated MLMC (myc), H06N1 (normal H-ras), and and H06T1 (mutated H-ras).

Cytotoxicity Assay

Drug sensitivity was determined by a tetrazolium based microtitration assay (Plumb et al 1989). Cells were plated out at a density of 5×10^2 per well in 96 well plates (Linbro; from Flow Labs, High Wycombe, Bucks) and allowed to attach and grow for 48 hours. Cells were then exposed to drug for 24 hours after which the drug containing medium was removed and replaced with drug free medium. The medium was changed on each of the following 3 days and on the third day 50μl of 3-(4,5-dimethylthiazol-2-yl) -2,5-diphenyltetrazolium bromide

(MTT, 5mg/ml) was added to the wells. Plates were incubated in the dark for 4 hours after which the medium was removed from the wells and the MTT-formazan crystals dissolved in dimethyl sulphoxide (200μl/well). Sorensen's glycine buffer (25μl, pH 10.5) was added to each well and the absorbance read immediately in a 96 well plate reader (Model 2550 EIA reader, Bio-Rad Laboratories, Watford). Cytotoxicity was expressed as the ID_{50}, which is the drug concentration required to give an absorbance per well equal to half that of the control, untreated wells.

^{3}H-Thymidine incorporation assay

Cells were seeded at initial density of 5×10^4 cells per well in 24 well plates. After 1 hour, TGF-β (purified in house from human expired platelets) or EGF (purchased from Rand D Inc., Minnesota, USA) was added singly or in combinations to give the desired final concentration and the cells incubated at 37°C, in a humid 5% CO_2 atmosphere for 22 hours. ^{3}H-thymidine is then added (0.5μCi per well) and the cells cells incubated at 37°C for a further 2 hours. Following fixation for 1 hour (methanol/acetic acid, 3:1 vol/vol) the cells were washed twice with 80% methanol. Trypsin 0.25% (0.5ml) was added to each well and left at room temperature for 1 hour. 0.5ml SDS (1%) was added and the contents of each well transferred 5 minutes later for liquid scintillation counting in Ecoscint. The results were expressed as % incorporation of ^{3}H-thymidine for treated cells relative to control untreated cells. Each experiment was repeated 4 times with 3 wells per point for each separate experiment.

RESULTS

Cell line characteristics

Although the parental line is epithelioid, the mutated H-ras-transfected line lost its epithelial morphology resulting in a bipolar cell of fibroblastoid shape with evidence of piling up on culture plates.

Chromosomal analysis gave a diploid number of 30, the diploid number for mink cells, therefore there was no indication of a major ploidy change, or gross chromosomal abnormalities in the transfected cell lines (data not shown).

All the lines grew rapidly with doubling times under 20 hr following a short lag period around 5 hr. The population doubling times decreased from 19.5 hr+0.5 (mean + standard deviation) in the parental MV1Lu and normal (N1) H-ras transfected lines to 14.1 hr+0.9 in the mutant H-ras line (T1) with the c-myc-transfected line (M1) intermediate at 18.2+0.4 hr.

Southern blot analyses revealed that the sublines contained the expected transfected oncogenes and immuno-cytochemical staining with monoclonal antibodies directed against the protein products of the H-ras and c-myc oncogenes demonstrated that the cell lines stained positively for the appropriate oncogene product (Khan et al 1991). As a control, colonies were picked from cells transfected with the donor plasmids containing the transcriptional enhancers and the aph gene. The derived cell line was morphologically similar to the parent line with a comparable population doubling time (19.2±0.4 hrs). The plasmid transfected line had no evidence of human oncogene expression and responded to the growth factors TGF-β and EGF in an identical manner to the parent cell line.

Cytotoxicity studies

The drug concentration which kills 50% of cells (ID_{50}) is summarised for each cell line in Table 1. Transfection with activated H-ras (H06T1 line) conferred significant cellular resistance ($p < 0.05$, Students t-test) to doxorubicin and vincristine relative to the parent cell line ($P < 0.05$).

Table 1 The cytotoxic response of parent and transfected cell lines following 24 hour exposure to a range of antineoplastic drugs. (mean ± 1 standard error of mean). ID_{50} is the drug concentration which kills 50% of cells. n = 6

Cell line   ID50

	Cisplatin (μM)	Cytosine Arabinoside (μM)	Doxorubicin (nM)	Vincristine (nM)	Etoposide (μM)
MV1Lu	0.41±0.05	3.3±0.28	2.7±0.3	2.4±0.3	1.73±0.05
MLMC	0.46±0.04	3.1±0.2	3.4±0.3	3.7±0.9	0.81±0.05
H06N1	0.34±0.03	5.6±0.5	2.6±0.5	4.9±0.6	0.31±0.02
H06T1	0.50±0.04	2.9±0.3	4.4±0.3	4.9±0.2	0.56±0.02

Pretreatment with growth factors (EGF and TGF-β) for 24 hours and during 24 hour exposure to doxorubicin had no effect on the cellular sensitivity of any cell line. (Table 2).

The effect of oncogene transfection on growth factor responsiveness

The dose-response curve for TGF-β and inhibition of ^{3}H-thymidine incorporation is shown in Figure 1, for each of the cell lines.

Table 2   The effect of TGF-β (1 ng/ml) and EGF (1 ng/ml) treatment for 24 hours prior to and during 24 hour exposure to doxorubicin.

Cell lines	ID_{50}(nm) MV1Lu	MLMC	H06N1	H06T1
Doxorubicin	4.8±0.56	4.1±1.0	4.3±0.8	6.3±0.66
Doxorubicin + TGF-β	5.0±0.18	3.6±0.1	4.4±0.56	6.5±0.78
Doxorubicin + EGF	5.5±0.32	3.2±0.83	3.9±1	7.3±0.85

Results are expressed as mean ± SEM.

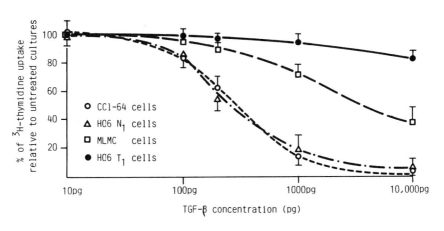

FIGURE 1

The effect of TGF-β on ^{3}H-thymidine incorporation by the cell lines. The symbols represent the mean value (n = 6) and vertical bar = 1 standard deviation.

The dose dependent inhibition of thymidine incorporation was similar for the parent MV1Lu and H06N1 (non-mutated H-ras1 proto-oncogene) lines with ID_{50} values of 310pg/ml and 280 pg/ml respectively. TGF-β had a less marked effect on the c-myc (ID_{50}= 3,800pg/ml) transfected cell line and virtually no effect on the H06T1 line (activated H-ras1 oncogene).

EGF significantly stimulates incorporation of ³H-thymidine into the MV1Lu and H06N1 lines with an increase of approximately 6-fold relative to control at an EGF concentration of 1ng/ml (Figure 2). The response of the MLMC (c-myc transfected cell line) line to EGF (1ng/ml) was exaggerated with a significantly higher response compared to the parent line (p<0.05). The H06T1 line did not respond to EGF.

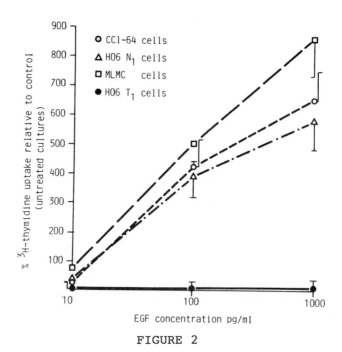

FIGURE 2

The effect of EGF on ³H-thymidine incorporation by the cell lines. The symbols represent the mean value (n = 6) and vertical bar = 1 standard deviation.

DISCUSSION

We have shown in this study that transfection of mink lung epithelial cells with c-myc and activated H-ras oncogenes alters the cellular response to exogenous growth factors and cytotoxic drugs. TGF-β inhibits incorporation of [³H]-thymidine by mink lung epithelial cells in a dose dependent manner with an ID50 of 310pg/ml. EGF has the opposite effect with stimulation of [^{3}H]-thymidine incorporation.

A different pattern of growth factor responsiveness is apparent in the oncogene transfected cells. C-myc transfected cells have a reduced response to TGF-β and an exaggerated response to EGF. Transfection of the mink lung epithelial cells with the normal ras proto-oncogene did not materially alter their response to TGF-β and EGF. However, activated H-ras transfection virtually abolished the effect of both TGF-β and EGF. Ras and c-myc transfection have been shown previously to effect the growth factor responsiveness of a range of cell types (Stern et al, 1986). One attractive hypothesis would be that the ras transfectants are synthesising and secreting large amounts of TGF-α and TGF-β and therefore additional exposure to these factors exogenously will have no effect due to receptor saturation. Alternatively, it is possible that ras transfection could downregulate the appropriate cell surface receptor for these growth factors and therefore reduce response in this way. Both alternatives are currently being explored in our laboratory.

Interestingly, although TGF-β and EGF have profound effects on DNA synthesis in MV1Lu, myc and normal ras transfected cells, exposure to these growth factors did not alter their response to doxorubicin and vincristine. It is possible that the duration of treatment with the growth factors was too short (i.e. 24 hours before and during drug exposure) to influence the cytotoxic effect, although there were DNA synthetic effects by 24 hours.

There are few published studies relating oncogene expression to cellular sensitivity to antineoplastic agents. There are clinical data suggesting that certain oncogenes, particularly c-fos and activated H-ras are amplified in ovarian cancer patients failing treatment with cisplatin and 5-fluorouracil (Scanlon et al 1988). These studies are supported by in vitro studies in which derived cisplatin resistant ovarian cancer cell have been shown to express elevated (2-4 fold) levels of mRNA transcripts for c-fos and activated H-ras (Lu et al 1988). Sklar has taken these observations a stage further by transfecting NIH 3T3 cells with activated H-ras. The subsequent transfectants were 4-8 fold more resistant to cisplatin although the mechanism underlying this resistance remained obscure (Sklar, 1988). We were unable to repeat these observations, however the main difference between Sklar's study and our own is that we transfected epithelial rather than mesenchymal cells. It is possible that some of the phenotypic alterations which occur on oncogene transfection of immortalised cells could be tissue specific and this could relate to differential chemosensitivity.

C-myc transfection was not found to alter chemosensitivity in this report however cisplatin resistant variants of ovarian cancer and oestrogen receptor positive human breast carcinoma cell lines have been found to have increased steady state levels of c-myc mRNA (3-5 fold increase) compared to wild type cells (Scanlon et al 1989). Interestingly this alteration in c-myc mRNA is not found in cisplatin resistant colorectal carcinoma cells, which again underlines the tissue specific nature of drug resistance.

The mechanisms underlying cytotoxic drug resistance have been reviewed for a number of different antineoplastic agents (Gottesman et al 1988). We chose a range of cytotoxic drugs with different mechanisms of action in an attempt to discover whether there were patterns of drug cross resistance in the transfected cell lines which could give clues as to the potential mechanism of resistance.

For example, the H-ras transfected cells were found to be relatively resistant to both doxorubicin and vincristine, but not etoposide. This pattern of cross resistance would be consistent with the the H-ras transfected cells expressing the multi-drug resistant P-glycoprotein which acts as an efflux pump. However, immunocytochemical studies in our laboratory using the anti P-glycoprotein monoclonal antibodies failed to demonstrate expression of the membrane glycoprotein in the transfected cell lines (Wishart et al, 1991).

Etoposide is thought to exert its cytotoxic effects on cancer cells by interacting with and stabilising the complex of the enzyme topoisomerase II with DNA, resulting in double strand breaks (Beck 1989). Recently, Woessner et al (1990) used a functional assay to show that total topoisomerase II activity was higher in ras transformed than normal cells, and was higher in exponentially growing than in plateau phase cells for both cell types. They also found that the ras transformed NIH-3T3 cells were more sensitive to etoposide than the normal cells. These results are consistent with ours, but it would appear that transfection with c-myc and the ras proto-oncogene confers a similar sensitivity to etoposide.

It is clear, from our experiments, that activated H-ras transfection has significant effects on growth control of the lung epithelial cells and on cellular sensitivity to cytotoxic agents (Kerr et al, 1991). The mechanisms underlying these effects are as yet unknown but continuing work in in our laboratory is aimed at their resolution.

ACKNOWLEDGEMENTS

The authors gratefully acknowledge the financial support of the Cancer Research Campaign and the Medical Research Council and wish to thank Ms F Conway for typing the manuscript.

REFERENCES

Beck, W.T. (1989). Unnkotting the complexities of multidrug resistance: the involvement of DNA topoisomerases in drug action and resistance. Journal. Natl. Can. Inst., 81:83.

Gottesman, M.M. & Pastan, I. (1988). Resistance to multiple chemotherapeutic agents in human cancer cells. TIPS Reviews, 9:54-58.

Graham, F.L. & Van der Eb, A.J. (1973). A new technique for the assay of infectivity of human adenovirus 5 DNA. Virology., 52:456.

Kerr, D.J., Plumb, J.A., Freshney, R.I., Khan, M.Z. & Spandidos, D.A. (1991). The effect of H-ras and c-myc oncogene transfection on response of mink lung epithelial cells to growth factors and cytotoxic drugs. <u>Anticancer Res.</u>,11:1349-1352

    Khan, M.Z., Spandidos, D.A., McNicol, A.M., Lang, J.C., Kerr, D.J., DeRidder, L. & Freshney, R.I. (1991). Oncogene transfection of mink lung cells: effect on growth characteristics in vitro and in vivo. <u>Anticancer Res.</u>,11:1343-1348

    Lu, Y., Han, J. & Scanlon, K.J. (1988). Biochemical and molecular properties of cisplatin resistance A2780 cells growth in folinic acid. <u>J. Biol. Chem.</u>, 263:4891-4894.

    Plumb, J.A., Milroy, R. & Kaye, S.B. (1989). Effects of the pH dependence of 3-(4,5-dimethylthiazol-2-yl) -2,5-diphenyltetrazoiium bromide-formazan absorption on chemosensitivity determined by a novel tetrazolium-based assay. <u>Cancer Res.</u>, 49:4435-3330.

    Scanlon, K.J., Lu, Y., Kashani-Sabet, M., Ma, J. & Newman, E. (1988). Mechanisms for cisplatin-FUra synergism and cisplatin resistance in human ovarian carcinoma cells both in vitro and in vivo. In: <u>Y. Rustum and J. McGuire, Eds., 244, Plenum Press, Inc., New York, pp131-139.</u>

    Scanlon, K.J. & Kashani-Sabet, M. (1988). Elevated expression of thymidine synthase cycle genes in cisplatin-resistant A2780 cells. <u>Proc. Natl. Acad. Sci. USA., 85:650-653</u>.

    Scanlon, K.J., Kashani-Sabet, M., Miyachi, H., Sowers, L.C. & Ross, J. (1989). Molecular baiss baiss of cisplatin resistance in human human carcinoma: model systems and patients. <u>Anticancer Res.</u>, 9:B:1301-1312

    Schwab, M., Ellison, J., Busch, M., Roseran, W., Varmus, H.E. & Bishop J.M. Enhanced expression of the human gene N-myc consequent to amplification of DNA may contrib ute to malignant progression of neuroblastoma.<u>Proc. Natl. Acad. Sci. 81:4940-4944</u>.

    Sklar, M.D. (1988). Increased resistance to cis-diamminedichloroplatinum (II) in NIH 3T3 cells transformed by <u>ras</u> oncogenes.<u>Cancer Res., 48:793-797</u>.

    Slamon, D.J., Clark, G.M., Wong, S.G., Levin, W.J., Ullrich, A. & McGuire, W.K.L. (1987). Human breast cancer: correlation of relapse and survival with amplification of the HER-2/neu oncogene. <u>Science, 235:177-182</u>.

    Spandidos, D.A. (1985). Mechanism of carcinogenesis: the role of oncogenes, transcriptional enhancers and growth factors. <u>Anticancer Res., 5:485</u>.

    Spandidos, D.A. & Wikie, N.M. (1984a). Expression of exogenous DNA in mammalian cells. In: <u>Hames, B.D., Higgings, S.J., Eds, IRL Press Oxford, pp1</u>.

    Spandidos, D.A. & Wilkie, N.M. (1984b). Malignant transformation of early passage rodent cells by a single single mutated human oncogene. <u>Nature, 310:469</u>.

Stern, D.F., Roberts, A.B., Roche, N.S., Sporn, M.B. & Weinberg, R.A. (1986). Differential responsiveness of myc and ras-transfected cells by epidermal growth factor. Mol. Cell. Biol. 6:870-877.

Woessner, R.D., Chung, T.D.Y., Hofman, G.A., Mattern, M.R., Mirabelli, C.K., Drake, F.H. & Johnson, R.K. (1990). Differences between normal and ras-transformed NIH-373 cells in expression of the 170kD and 180kD forms of topoisomerase II. Cancer Res., 50:2901-2908.

Wishart, G.W., Plumb, J.A., Spandidos, D.A. & Kerr, D.J. (1991). H-ras transfection in mink lung epithelial cells may induce "atypical" multi drug resistance. Eur. J Cancer, in press.

THE RAP PROTEIN FAMILY: RAP1A, RAP1B, RAP2A, AND RAP2B

Eduardo G. Lapetina, Michael J. Campa, Deborah A. Winegar
and Francis X. Farrell

Department of Cell Biology
Burroughs Wellcome Co.
Research Triangle Park, NC 27709

INTRODUCTION

Rap proteins were discovered and identified within the last three years by screening cDNA libraries with probes homologous to regions of ras (1, 2). Rap proteins, like ras, are low molecular weight GTP-binding proteins. They share 50% homology to ras and are expressed in a wide variety of tissues. Two rap familes, designated rap1 and rap2, have been identified to date. Within each family are two members denoted a and b, i.e., rap1a, rap1b, rap2a, and rap2b (1, 4). The two members within each family are 90% homologous. The rap proteins are members of a larger family of proteins homologous to ras including ral, rho, rac, and rab, which has been termed the ras-related protein family (5, 6).

More important than rap's 50% homology to ras is the fact that the greatest homology between rap and ras is in the regions important for ras transformation, including the nucleotide binding region, the putative effector region, and a membrane attachment site. As shown in Figure 1, rap1 and rap2 are almost identical to Kirsten ras in these regions. The membrane attachment site is the much studied CAAX motif (Cysteine-aliphatic-aliphatic-last amino acid) found at the C-terminal of ras and rap (7). This sequence motif has also been identified in other ras-related proteins, nuclear lamins, the gamma subunits of the heterotrimeric G-proteins, and several yeast mating factors. An important difference between ras and rap is that all rap proteins contain a threonine at position 61 instead of glutamine.

Rap1a reverts a K-ras transformed phenotype

Probably the most significant finding to date is that rap1a is able to revert a ras-transformed cell line to a normal or "flat" morphology. Kitayama et al. had observed that when a K-ras transformed NIH 3T3 cell line was transfected with a human fibroblast cDNA library, some cells exhibited a normal morphology (8). This experiment suggested that a gene (or genes) contained in this library was able to "revert" the ras transformed phenotype. The gene responsible for this revertant-inducing activity was found to be rap1a, also called Krev-1. This result has been repeated by transfecting the transformed cell line with rap1a contained on an expression vector. This confirmed that the transformation revertant

```
  1 MTEYKLVVVGAGGVGKSALTIQLIQNHFVDEYDPTIEDSYRKQVVIDGET  K-ras
    MREYKLVVIGSGGVGKSALTVQFVQGIFVEKYDPTIEDSYRKQVEVDCQQ  rap1a
    MREYKLVVIGSGGVGKSALTVQFVQGIFVEKYDPTIEDSYRKQVEVDAQQ  rap1b
    MREYKVVVIGSGGVGKSALTVQFVTGTFIEKYDPTIEDFYRKEIEVDSSP  rap2a
    MREYKVVVIGSGGVGKSALTVQFVTGSFIEKYDPTIEDFYRKEIEVDSSP  rap2b
          phosphate-binding              effector region
```

```
 51 CLLDIIDTAGQEEYSAMRDQYMRTGEGFLCVFAINNTKSFEDIHHYREQI  K-ras
    CMLEIIDTAGTEQFTAMRDLYMKNGQGFALVYSITAQSTFNDLQDLREQI  rap1a
    CMLEIIDTAGTEQFTAMRDLYMKNGQGFALVYSITAQSTFNDLQDLREQI  rap1b
    SVLEIIDTAGTEQFASMRDLYIKNGQGFILVYSLVNQQSFQDIKPMRDQI  rap2a
    SVLEIIDTAGTEQFASMRDLYIKNGQGFILVYSLVNQQSSQDIKPMRDQI  rap2b
          phosphate-binding
```

```
101 KRVKDSEDVPMVLVGNKCDLPS-RTVDTKQAQDLARSY-GIPFIETSAKT  K-ras
    LRVKDTEDVPMILVGNKCDLEDERVVGKEQGQNLARQWCNCAFIESSAKS  rap1a
    LRVKDTDDVPMILVGNKCDLEDERVVGKEQGONLARQWNNCAFIESSAKS  rap1b
    IRVKRYEKVPVILVGNKVDLESEREVSSSEGRALAEEW-GCPFMETSAKS  rap2a
    IRVKRYERVPMILVGNKVDLEGEREVSYGEGKALAEEW-SCPFMETSAKN  rap2b
          guanine-binding                    guanine-binding
```

```
151 RQGVDDAFYTLVREIRKHKEKMSKDGKKKKKKSKTKCVIM   K-ras
    KINVNEIFYDLVRQINRKTPVEKKKPKKKS------CLIL   rap1a
    KINVNEIFYDLVRQINRKTPVPGKARKSS------CQLL   rap1b
    KTMVDELFAEIVRQMNYAAQPDKDDPCCSA------CNIQ   rap2a
    KASVDELFAEIVRQMNYAAQSNGDEGCCSA------CVIL   rap2b
                                    CAAX motif
```

Figure 1. Alignment of rap protein sequences with human K-ras.
Phosphate and guanine binding domains are shown by the boxed regions. The
effector region is denoted by the heavy black line. The CAAX motif is shown
by the shaded box.

activity was the result of the expression of a single gene product, rap1a
(9). Given that rap proteins share strong homology, we are testing for
transformation revertant activity with rap1b.

Phosphorylation of rap1b by protein kinase A

Concurrent with finding that rap1a was responsible for reversion of a
ras transformed phenotype, our laboratory was investigating the role of
G-proteins in platelet function and signal transduction. Platelets contain
several G-proteins between 21-41 kDa including the well characterized
G-proteins, G_i and G_S which regulate adenylate cyclase (10, 11). We
observed that platelets contain a membrane associated 22 kDa protein, which
binds [^{32}P] GTP and is recognized by M90, an antibody which recognizes many
ras-related proteins. M90 is a monoclonal antibody directed against amino
acid residues 107-130 of ras (12-18). When platelets were treated with the
prostacyclin analog iloprost, an agonist that increases cAMP levels, we
observed a shift of the protein to an apparent molecular weight of 24 kDa
(12, 14, 16). This was also observed after platelets were treated with
PGE_1(19, 20). This shift in molecular weight was accompanied by the

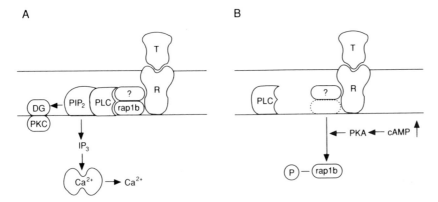

Figure 2.Receptor coupled activation of PLC in stimulated platelets.
Thrombin (T) activates a specific membrane-bound receptor (R) which causes
activation of phospholipase C (PLC). This activity hydrolyzes
phosphatidylinositol 4,5-bisphosphate to produce diacylglycerol stimulating
protein kinase C (PKC) and inositol 1,4,5-trisphosphate causing the release
of intracellular calcium. The coupling of PLC to the receptor seems to be
related to a "classical G-protein".which might be G_{i-2} in platelets. In
addition, rap1b seems to be involved in this pathway. The phosporylation
of rap1b by a cAMP-dependent protein kinase produces a translocation of
rap1b from the membrane to the cytosol possibly causing an uncoupling of
the receptor from PLC.

protein's translocation from the particulate to the cytosolic fraction
(12). When platelets were labeled with ^{32}P and then treated with iloprost
the 24 kDa protein was found to be phosphorylated (12, 14). In addition.
this phosphorylated protein was specifically immunoprecipitated by M90
(12). We concluded that this shift in molecular weight was the result of
phosphorylation by a cAMP dependent protein kinase and we sought to
identify the protein.
    To rule out the possibility that the protein was ras, we demonstrated
that it was not recognized by the ras-specific antibody Y13-259 (12). To
unambiguously identify the protein undergoing phosphorylation, the 24 kDa
phosphorylated protein was isolated and subsequently treated with the
proteases trypsin and V8 (14). The phosphopeptides were separated by HPLC
and subjected to amino acid analysis. The analysis revealed that the
phophopeptides were derived from rap1b. To further confirm our finding,
synthetic peptides corresponding to the C-terminal regions of rap1b and
rap1a were used as substrates for phosphorylation by protein kinase A (14).
Our results showed that only the rap1b peptide is phosphorylated by protein
kinase A (14). In addition, we determined using recombinant rap1b and
protein kinase A that the ratio of phosphate to protein is 1:1 (21).
Further studies using mutational analysis identified serine 179 as the site
of phosphorylation. (D. Altschuler and E. G. Lapetina, manuscript in
preparation).

**The relationship between the inhibitory action of cAMP on phospholipase C
and phophorylation of rap1b**

    The 22 kDa protein (rap1b) that shifts to an apparent molecular
weight of 24 kDa upon increasing cAMP levels was also seen in human
erythroleukemia (HEL) cells (13, 22). HEL cells possess a number of
platelet-like features including granules that morphologically resemble the
platelet alpha granule. The shift was observed when HEL cells were treated

Table 1. Xenopus laevis oocytes were isolated by manuel defolliculation and microinjected with the indicated protein solutions. BSA (bovine serum albumin), rap1b, and rap2b were microinjected at 100 ng per oocyte while ras [H-ras (Val-12)] was microinjected at 75 ng per oocyte. All proteins were microinjected in buffer containing 20 mM Tris-HCl, pH 7.5; 10 mM MgCl2; 7 mM 2-mercaptoethanol; and 10% (v/v) glycerol. Incubations were carried out at 19° C in modified Barth's medium. GVBD (germinal vesicle breakdown) was scored by the appearance of an unpigmented spot at the oocyte animal pole.

Table 1

Injection	% GVBD
ras	80
ras + rap2B	60
ras + BSA	70
ras + rap1B	0
rap1B	0

with prostacyclin, or theophylline and dibutyryl cAMP. On the other hand, no change in mobility was observed when HEL cells were treated with thrombin, phorbol dibutyrate, or the Ca^{2+} ionophore A23187 (13). This result suggests that rap1b is not a substrate of protein kinase C.

One of the first suggestions of a possible function of rap1b was the observation that in HEL cells the increase in phosphorylation of rap1b by the addition of iloprost occurred with a decrease in the phopholipase C-induced formation of inositol phospholipids (22, 23). This result suggests that rap1b has a functional role in the regulation of phospholipase C and the metabolism of inositol phospholipids. This model has been postulated on the basis that thrombin leads to the activation of phospholipase C, which hydrolizes phosphatidylinositol 4,5-bisphosphate to produce two second messengers, 1,2-diacylglycerol and 1,4,5-trisphosphate (23). Phosphorylation of rap1b by the action of the cyclic AMP-dependent protein kinase seems to uncouple the thrombin receptor from phospholipase. This possibility was further discussed elsewhere (23) and is indicated in Figure 2.

Rap1b can also be phosphorylated by a Ca^{2+}/calmodulin dependent protein kinase

In addition to the phosphorylation of rap1b by a cAMP dependent protein kinase, we recently demonstrated that rap1b is also phosphorylated by a neuronal Ca^{2+}/calmodulin dependent protein kinase isolated from cerebellum granule cells (21). As a control, we showed that the kinase does not phosphorylate rap1a, rap2b, or rab3a. This result suggests a connection between a neuronal Ca^{2+} signaling pathway and a specific G-protein signal transduction pathway.

Rap1b inhibits the effect of ras on GVBD in Xenopus oocytes

As mentioned earlier, rap1a will revert a ras-transformed cell line to a normal or "flat" morphology (8, 9). To investigate the potential role of rap1b in ras-mediated signal transduction, we employed the Xenopus laevis oocyte system. Oocytes obtained by ovaryectomy from Xenopus laevis are

```
  1  MREYKVVVLGSGGVGKSALTVQFVTGSFIEKYDPTIEDFYRKEIEVDSSP rap2b
     MREYKVVVLGSGGVGKSALTVQFVTGIFIEKYDPTIEDFYRKEIEVDSSP rap2a


 51  SVLEILDTAGTEQFASMRDLYIKNGQGFILVYSLVNQQSSQDIKPMRDQI rap2b
     SVLEILDTAGTEQFASMRDLYIKNGQGFILVYSLVNQQSEQDIKPMRDQI rap2a


101  IRVKRYERVPMILVGNKVDLEGEREVSYGEGKALAEEWSCPFMETSAKNK rap2b
     IRVKRYEKVPMILVGNKVDLESEREVSSSEGRALAEEWCPFMETSAKSK rap2a


                                      CLLL rap1a
151  ASVDELFAEIVRQMNYAAQSNGDEGCCSACVIL rap2b    183
     TMVDELFAEIVRQMNYAAQPDKDDPCCSACNIQ rap2a
                                      CQLL rap1b
                                      CAAX motif
```

Figure 3. Amino acid alignment of rap2a and rap2b protein sequence. Boxed amino acids indicate where rap2a and rap2b differ in sequence. The CAAX motif sequence of rap1a and rap 1b is also shown.

arrested in prophase of meiosis I. Such oocytes are incapable of being fertilized and hence, are often referred to as "immature". The maturation process can be induced in vitro by progesterone, insulin, insulin-like growth factor I (IGF-I), or by the microinjection of oncogenic forms of ras. Although the precise role of ras in oocyte maturation has yet to be elucidated, data from other laboratories suggest it is involved in maturation induced by insulin or IGF-I. Progesterone, on the other hand, functions via a G-protein mediated inhibition of oocyte adenylate cyclase. Experiments involving the microinjection of ras proteins mutated in such a way as to bind ras GTPase activating protein (ras-GAP) tightly while preventing its movement to the plasma membrane implicate ras-GAP in the mediation of ras-induced oocyte maturation (24).

We investigated the action of purified recombinant rap1b on ras-induced maturation in oocytes. When co-injected with the valine 12 mutant form of H-ras, rap1b blocked the maturation response (25). As shown in Table 1, rap1b was effective at blocking maturation when microinjected at approximately equimolar ratios to the ras protein. Although we have yet to decipher the means by which rap1b interferes with ras-mediated maturation, work carried out elsewhere utilizing the highly similar rap1a suggests that the interference may be at the level of ras-GAP (26). In other words, if interaction between ras and ras-GAP is required for the maturation response, rap1b could block maturation by competing with ras for binding to ras-GAP. Data indicate that rap1a can bind to ras-GAP with approximately 2 orders of magnitude greater affinity than can ras and with no subsequent increase in rap1a GTPase activity (26). Given the 100% identity between rap1a and rap1b in the regions believed to be directly involved with ras-GAP binding, it is conceivable that rap1b would bind ras-GAP in a similar manner.

The isolation and characterization of a new rap protein, rap2b

In addition to our studies on the role of rap1b, we initiated studies to identify other ras-related proteins by screening cDNA libraries with the monoclonal antibody M90. When a platelet cDNA library was screened with M90 we isolated a partial clone with strong sequence homology to the previously described rap2 (3, 4, 15). Upon rescreening to isolate a full length clone, we isolated a full-length clone that was different from rap2 (4). This new gene, which shares 90% homology with rap2 has been designated rap2b (see figure 3). Interestingly, the clone has the greatest divergence in sequence to rap2 in the C-terminal portion of the protein. In addition, the two proteins contain different CAAX motifs, the site of membrane attachment via isoprenylation for ras, nuclear lamins, and the gamma subunits of the heterotrimeric G-proteins. Biochemical studies of rap2b were initiated by expressing the protein in E. coli (15). The recombinant protein bound GTP and was not phosphorylated by a cAMP-dependent kinase. We are curently investigating the role of rap2b in cellular function.

The CAAX motif is the signal for prenylation of rap proteins

As stated earlier, the CAAX sequence at the C-terminal region of some proteins was found to be important for membrane binding (7, 27). For example, it has been shown that a ras protein containing mutations in this area loses its transforming ability by virtue of its inability to bind to the membrane.

The attachment of the isoprenoid group to the protein is only one step in a series of posttranslational modifications in this region. It has been shown that proteins incorporate an isoprenoid group at the cysteine residue by a thioether linkage. This is followed by a proteolysis of the last three amino acids (-AAX). The newly exposed cysteine is then methylesterfied in most cases. We have shown that rap1b undergoes these modifications by demonstrating that an antibody directed against this region does not recognize the processed protein, i.e., missing the last three amino acids in an insect/baculovirus system (18). Furthermore, we have shown that the protein is methylesterfied and that the incorporation of S-[methyl-3H]adenosylmethionine is stimulated by Guanosine 5'-(3-0-Thio)triphosphate (17). Finally, others have shown that rap1b incorporates the isoprenoid geranylgeranyl by virtue of its terminal leucine (28).

Rap2a incorporates both farnesyl and geranylgeranyl

We have recently demonstrated that rap2 is isoprenylated in platelets and HEL cells (11). In fact, we have show that both rap2a and rap2b are isoprenylated (F. X. Farrell, K, Yamamoto, and E.G. Lapetina, manuscript submitted). CAAX motif proteins have been shown to incorporate the isoprenoids farnesyl and geranylgeranyl. Proteins terminating in leucine are geranylgeranylated while those terminating in methionine or serine are farnesylated. Interestingly, as shown in Figure 3 rap2a contains a glutamine at its C-terminal. We sought to determine the isoprenoid attached to this protein; to our knowledge only one other CAAX motif protein, rod cGMP phosphodiesterase, terminates with this amino acid. Plasmid constructs containing the cDNA sequence of rap2a and rap2b were transfected into COS cells and incubated in the presence of [3H]mevalonolactone. Immunoprecipitated rap2 proteins were treated with methyl iodide and the released isoprenoid groups were analyzed by HPLC. We observed that rap2b, which terminates in CVIL, was geranylgeranylated while rap2a, which terminates in CNIQ, incorporated both geranylgeranyl and farnesyl. This result suggests that glutamine may not confer a strict isoprenoid specificity to rap2a.

That rap 2a and rap2b are highly homologous yet incorporate different lipid moieties is intriguing. All rap proteins discovered to date possess leucine at their C-terminal with the exception of rap2a. The relative abundance of rap2a versus rap2b in tissue is currently being studied. Although it appears that rap2a incorporates both farnesyl and geranylgeranyl in our tissue culture system, the relative amounts of each with respect to the native protein in tissue is unknown. If both modifications can be found, it seems probable that the two molecules may partition into different membrane compartments given their isoprenoid attachment.

CONCLUSION

As is true for other low molecular weight GTP-binding proteins, the function of the rap proteins is largely unknown. Several functions have been proposed, including a role in inositol metabolism, secretion, calcium signaling, second messenger transduction, and antioncogenesis. In addition, the molecules which act upon rap proteins including the rap-specific GTPases, nucleotide exchange proteins, and nucleotide exchange inhibitors are being characterized. The study of rap proteins will continue to be an area of intense study over the next few years.

REFERENCES

1. V. Pizon, P. Chardin, I. Lerosey, B. Olofsson, and A. Tavitian, Oncogene 3:201-204 (1988).
2. V. Pizon, I. Lerosey, P. Chardin, and A. Tavitian, Nucleic Acids Res. 16:7719 (1988).
3. F.X. Farrell, C.-A. Ohmstede, B.R. Reep, and E.G. Lapetina, Nucleic Acids Res. 18:4281 (1990).
4. C.-A. Ohmstede, F.X. Farrell, B.R. Reep, K.J. Clemetson, and E. G. Lapetina, Proc. Natl. Acad. Sci USA 87:6527-6531 (1990).
5. A. Hall, A. Science 249:635-640 (1990).
6. J. Downward, Trends Biochem. Sci. 15:469-472 (1990).
7. J.F. Hancock, A.I. Magee, J.E. Childs, and C.J. Marshall, Cell 57:1167-1177 (1989).
8. H. Kitayama, Y. Sugimoto, T. Matsuzaki, Y. Ikawa, and M. Noda, M. Cell 56:77-84 (1989).
9. K. Zhang, M. Noda, W.C. Vass, A.G. Papageorge, and D.R. Lowy, Science 249:162-165 (1990).
10. E.C. Lapetina, B. Reep, and K.-J. Chang, Proc. Natl. Acad. Sci USA 83:5880-5883 (1986).
11. M.F. Crouch, and E.G. Lapetina, J. Biol. Chem. 263:3363-3371 (1988).
12. E.G. Lapetina, J.C. Lacal, B.R. Reep, and L. Molina y Vedia, Proc. Natl. Acad. Sci USA 86:3131-3134 (1989).
13. E.R. Lazarowski, J.C. Lacal, and E.G. Lapetina, Biochem. Biophys. Res. Comm. 161:972-978 (1989).
14. W. Siess, D.A. Winegar, D. and E.G. Lapetina, Biochem. Biophys. Res. Comm. 170:944-950 (1990).
15. L. Molina y Vedia, C.-A. Ohmstede, and E.G. Lapetina, Biochem. Biophys. Res. Comm. 171:319-324 (1990).
16. D.A. Winegar, L. Molina y Vedia, and E.G. Lapetina, J. Biol. Chem. 266:4381-4386 (1991).
17. H. Akbar, D.A. Winegar, and E.G. Lapetina, J . Biol. Chem. 266:4387-4391 (1991).
18. D.A. Winegar, C.-A. Ohmstede, L. Chu, B. Reep, and E.G. Lapetina, E.G. J. Biol. Chem. 266:4375-4380 (1991).
19. M. Kawata, A. Kikuchi, M. Hoshijima, K. Yamamoto, E. Hashimoto, H. Yamamura, and Y. Takai, J. Biol. Chem. 264:15688-15695 (1989).
20. M. Hoshijima, A. Kikuchi, A.M. Kawata, T. Ohmori, E. Hashimoto, H. Yamamura, and Y. Takai, Biochem. Biophys. Res. Commun. 157:851-860 (1988).
21. N. Sahyoun, O.B. McDonald, F. Farrell, and E.G. Lapetina, Proc. Natl. Acad. Sci USA 88:2643-2647 (1991).
22 E.R. Lazarowski, D.A. Winegar, R.D. Nolan, E. Oberdisse, and E.G. Lapetina, J. Biol. Chem. 265:13118-13123 (1990).
23. E.G. Lapetina, FEBS Letters 268,400-404 (1990).
24. J.B. Gibbs, M.D. Schaber, T.L. Schofield, E.M. Scolnick, and I.S. Segal, Proc. Natl. Acad. Sci USA 86:6630-6634 (1989).
25. M.J. Campa, K.-J.Chang, L. Molina y Vedia, B.R. Reep, and E.G. Lapetina, Biochem. Biophys. Res. Comm. 174:1-5 (1991).

26. M. Frech, J. John, V. Pizon, P. Chardin, A. Tavitian, R. Clark, F. McCormick, and A. Wittenghofer, <u>Science</u> 249, 169-171 (1990).
27. L. Gutierrez, A.I. Magee, C.J. Marshall, and J.F. Hancock, <u>EMBO J.</u> 8:1093-1098 (1989).
28. M. Kawata, C.C. Farnsworth, Y. Yoshida, M.H. Gelb, J.A. Glomset, and Y. Takai, <u>Proc. Natl. Acad. Sci. U.S.A</u> 87:8960-8964 (1990).

EXPRESSION OF THE DROSOPHILA RAS2/CS1 GENE PAIR DURING DEVELOPMENT

Zeev Lev, Noa Cohen, Adi Salzberg, Ziva Kimchie,
Naomi Halachmi, and Orit Segev

Department of Biology, Technion - Israel Institute of Technology
Haifa 32000, Israel

INTRODUCTION

The Drosophila Ras Genes

Several homologs of vertebrate proto-oncogenes have been detected in D.melanogaster - abl, src, ras, myb, raf, erb-B (DER), fps (reviewed in Shilo, 1987), int-1 (wingless, Rijsewijk et al, 1987), rel (dorsal, Steward, 1987), TGF-B (decapentaplegic, Padgett et al, 1987) and jun and fos (Djra and Dfra, Perkins et al, 1988). Three homologs of the viral Ha-ras gene were also isolated. They were termed Ras1, Ras2, and Ras3 (Lindsley and Zimm, 1990) and mapped to regions 85D, 64B, and 62B, respectively, on the 3rd chromosome (Neuman-Silberberg et al, 1984). Synonymous names are Dras1, Dras2, and Dras3 (Neuman-Silberberg et al, 1984), or Dmras85D and Dmras64B (Mozer et al, 1985; Brock, 1987). Analysis of the nucleotide sequence of cloned genomic and cDNA sequences suggests that among the three Drosophila ras genes Ras1 is the genuine homolog of the human Ha-, Ki-, and N-ras genes. The Ras2 gene is more similar to the human R-ras gene family and the Drosophila ras3 is similar to the human rap gene family (Chardin et al, 1989). For example, in the 3-79 amino acids the human Ha-Ras protein has 100 percent homology with the Drosophila Ras1 protein and only 79 percent with the Drosophila Ras2 protein which, in the same region, has 87 percent homoloy with the R-ras protein. The homology between the Drosophila Ras1 and Ras2 proteins is significantly lower (78 percent) than the homology between them and their human counterparts.

The structural and functional homologies between the vertebrate ras and the Drosophila ras proteins were demonstrated by precipitating 21 kD protein from Drosophila cell extract using monoclonal antibodies raised against the v-Ha-ras p21 protein (D.

The Superfamily of ras-Related Genes
Edited by D.A. Spandidos, Plenum Press, New York, 1991

303

Sahar, unpublished results) and by transforming rat cells with a chimeric <u>ras</u> comprising the N-terminus of human activated Ha-<u>ras</u> gene and the C-terminus of the <u>Drosophila</u> <u>Ras3</u> gene (Schejter and Shilo, 1985). In vertebrate systems, expression of activated <u>ras</u> oncogenes causes growth abnormalities such as cell transformation and tumor induction (Reviewed by Barbacid, 1987). In <u>Drosophila</u> activated <u>Ras2</u> gene (Gly14 - Val14, which is analogous to Gly12 - Val12 in Ha-<u>ras</u>) driven by the hsp70 promoter caused developmental abnormalities and lethality (Bishop and Corces, 1988). The transgenic flies had low fertility, laying many eggs that failed to hatch, and showed developmental disturbances in their wing and eye structures.

The Ras2/CS1 Gene Pair

We have isolated and characterized the <u>Drosophila</u> <u>Ras2</u> promoter region. Our results show that this promoter is apparently a bidirectional promoter since it regulates another gene, temporarily termed CS1, in the opposite polarity relative to the <u>Ras2</u> gene (Cohen et al, 1988). Nucleotide sequence determination followed by external primer extension and RNase protection assays revealed that the proximal transcription initiation sites of the <u>Ras2</u> and <u>CS1</u> transcripts are merely 94 nucleotides apart. This is one of the shortest bidirectional promoters reported in metazoan so far, second only to the bidirectional promoter found in the mouse <u>surf</u> locus (Lennard and Fried, 1991). <u>Ras2</u>-CAT and <u>CS1</u>-CAT transcription fusions expressed transient CAT activity in transfected tissue culture cells and in injected embryos. The same constructs were introduced into wild-type flies by P-element-mediated transduction and the transgenic flies expressed CAT activity in embryonic, larval and adult tissues. Deletion analysis has proved that certain <u>cis</u>-acting elements within the promoter region are required for full   transcriptional activity of both genes simultaneously (Lev et al, 1989; Segev et al, in preparation). Thus, the <u>Drosophila</u> <u>Ras2</u> promoter is not composed of two closely located separated promoters but it is an authentic bidirectional promoter.

RESULTS AND DISCUSSION

Expression of the Drosophila Ras2 gene during development

The transcription pattern of the <u>Drosophila</u> <u>Ras2</u> gene during development was studied first by RNA blots of poly(A) RNA extracted from different stages throughout the life cycle of the fruit fly (Lev et al, 1985). The 1.8 Kb major Ras2 transcript was expressed in all the developmental stages tested: early and late embryos, the three larval stages, prepupae, pupae, and adult flies. A minor 1.4 Kb transcript was also detected, but only in oocytes and early embryos. Quantitative analysis suggests that the amount of Ras2 transcripts expressed in the different stages is rather constant comprising 0.05 percent of embryonic poly(A) RNA and about 0.02 percent of poly(A) RNA in later stages. Interestingly the Ras1 and Ras3 genes were expressed similarly. They code for two or three distinct transcripts, respectively. The larger transcript of each gene was expressed at a similar abundance during the entire life cycle of the fruit fly, and the shorter transcripts was maternal/embryonic specific.

The spatial distribution of the RNA transcripts expressed by the three Drosophila ras genes was examined in situ (Segal and Shilo, 1986). In the embryo the transcripts were uniformly distributed, and in the larva they were confined to growing cells in the imaginal discs, brain and gonads. In the adult most transcripts were found in the oocytes, but also in differentiated nervous tissues - the brain cortex and the ventral ganglion.

Tissue Specificity of Ras2

To obtain high resolution of the spatial distribution of Ras2 transcripts during development we constructed lines of transgenic flies containing transcription fusions of the gene and the bacterial lacZ gene. The activity of the promoter was monitored by X-Gal staining (N. Cohen, unpublished results). The Ras2 promoter was found to be homogeneously active in oocytes and in embryos. However, this ubiquitous pattern of expression is characteristic of the early stages only. In the larvae the X-Gal staining was confined to specific locations within certain organs e.g. the larval brain in which the promoter activity was limited to specific sites within the ventral ganglion and the brain hemispheres.

The CS1 Gene

To determine the nucleotide sequence of the CS1 coding region, an embryonic cDNA library was screened with a CS1-specific DNA probe and several CS1 cDNA clones were isolated. In addition genomic DNA clones were isolated from Drosophila genomic library using the same CS1 probes for screening. Sequence comparison between these clones indicated that the cDNA and the genomic DNA sequences are identical, suggesting that at least one CS1 mRNA is intronless. From the nucleotide sequence of several cloned cDNAs the sequence of the putative CS1 protein, 597 amino acid long, was deduced. The major databases were searched for nucleotide or protein sequences similar to CS1, but no significant similarity with any known gene has been found (Salzberg et al, in preparation).

The close proximity and coregulation of the Ras2/CS1 gene pair suggest that the two genes may interact with each other. Consequently the patterns of their expression during development should overlap, at least partially. Indeed, RNA blot analysis using a genomic DNA probe containing both genes showed that they were coexpressed, at about the same abundance, during embryonic, larval, pupal, and adult stages (Mozer et al, 1985). To find out if the two genes are expressed simultaneously in the same tissues, the spatial distribution of the CS1 transcripts during development is being determined by constructing lines of transgenic flies containing transcription fusions of the CS1 gene promoter and the bacterial lacZ gene and the activity of the promoter is monitored by X-Gal staining as described above. Preliminary results indicated that in general the tissue specificity of the CS1 transcripts is similar to that observed for the Ras2 transcript distribution. For example, the CS1 promoter is also homogeneously active in embryos, but in larval brains its activity is limited to specific sites within the ventral ganglion and the brain hemispheres (N. Cohen, unpublished results).

Discussion

Developmental Regulation: The regulation of the Drosophila Ras2 gene expression during development has been studied independently by three groups. RNA blot analysis revealed constitutive expression during the complete life cycle of the fruit fly, including the embryonic, larval, pupal, and adult stages (Lev et al, 1985; Mozer et al, 1985). The abundance of the Ras2 transcripts in all stages is moderate, in the range of 0.2 to 0.5 percent of poly(A) RNA (Lev et al, 1985). The tissue specificty of this gene, at the RNA level, was first studied by in situ hybridization using antisense RNA as probe (Segal and Shilo, 1986). In the embryo the transcripts were uniformly distributed, but in the larva they were detected only in imaginal discs, brain and gonads. In adult flies most transcripts were found in the oocytes, but also in differentiated nervous tissues - the brain cortex and the ventral ganglion. Even higher resolution is being obtained using Ras2/lacZ transcription fusions in which X-Gal staining enables detection of the Ras2 gene promoter activity. These experiments generally support the results obtained by in situ hybridization and in addition extend the resolution to the single-cell level (N. Cohen, unpublished results). Cellular transformation by activated ras genes and the acquisition of new growth properties by the transformed cells suggest a role for ras genes in transmitting growth signals. Therefore the presence of Ras2 transcripts in growing embryos and dividing cells within imaginal discs may be assigned to a role of this gene in proliferative tissues. However, most larval and adult tissues are differentiated and their cells do not divide anymore. The role of Ras2 in these tissues is not known, but it is conceivable to assume that the "ras casset" participates in the transmittance of signals which are not necessarily of the growth-signal type.

The CS1 gene: An interesting feature of this system is the CS1 gene located only 94 bases upstream to Ras2 in the opposite orientation (Cohen et al, 1988). CS1 is a novel gene and codes for a 65 Kd protein (A. Salzberg, in preparation). The Ras2 and CS1 genes are regulated by a common bidirectional promoter (Lev et al, 1989) and the tissue distribution of their expression is similar (N. Cohen, unpublished results). This preliminary data strongly suggests that the two genes might interact with each other. Studies on the organization, regulation, biochemical properties and genetics of the members of this gene pair are in progress.

Concluding Remarks: As yet there is no solid model describing the molar and biochemical pathway by which ras oncogenes initiate cellular transformation and tumor progression. The biochemical properties of the human Ras proteins, and many other Ras and Ras-related proteins in other organisms, predict that they are probably part of the signal transduction systems which control proliferative programs at the cellular and tissue levels. However, despite lengthy efforts the nature of the putative signal transmitted by the Ras proteins is not known, and the identity of the genes which are upstream or downstream to ras in the signal transduction pathway is still obscure. Furthermore, the roles of ras genes in developmental processes such as morphogenesis and tissue differentiation have not been studied yet.

Genetic methods are very efficient in several invertebrates, particularly nematodes and fruit flies. For example, the first solid indication for any role of a ras gene in a developmental pathway has been recently discovered in the nematode (Han and Sternberg, 1990; Beitel et al, 1990). In the long run genetic analysis of mutations in ras genes may lead to the identification of genes which interact with ras. This is a very difficult task to accomplish when only biochemical approaches are taken, due to the fact that these signal transduction pathways are complex and interrelated. Furthermore, substrates or other target molecules can be extremely rare. Even in cases like insulin and EGF receptors, where the ligands have been well studied, it has been difficult to identify other steps of the pathway. The use of genetic tools in Drosophila has made it possible to identify several components associated with signals transduced by the abl tyrosine kinase (Gertler et al, 1989, 1990) and the sevenless receptor tyrosine kinase (Rogge et al, 1991). Thus a genetic approach in the fruit fly may also contribute invaluable data concerning the functions of the ras genes and details of the signal transduction pathway in which they participate. Mutations induced in the Drosophila Ras genes should lead to a better understanding of their role in cellular proliferation and in developmental processes.

REFERENCES

Barbacid, M. (1987). ras genes. Annu. Rev. Biochem. 56, 779-827.

Beitel, G.J., Clark, S.G., and Horvitz, H.R. (1990). Caenorhabditis elegans ras gene let-60 acts as a switch in the pathway of vulval induction. Nature 348, 503-509.

Bishop, J.G.I.I.I. and Corces, V.G. (1988). Expression of an activated ras gene causes developmental abnormalities in transgenic Drosophila melanogaster. Genes Dev. 2, 567-577.

Brock, H.W. (1987). Sequence and genomic structure of ras homologues Dmras85D and Dmras64B of Drosophila melanogaster. Gene 51, 129-137.

Chardin, P., Touchot, N., Zahraoui, A., Pizon, V., Lerosey, I., Olofsson, B., and Tavitian, A. (1989). Structure and organization of the ras gene family, in human. In ras oncogenes. D. Spandidos, ed. (New York: Plenum), pp. 1-10.

Cohen, N., Salzberg, A., and Lev, Z. (1988). A bidirectional promoter is regulating the Drosophila ras2 gene. Oncogene 3, 137-142.

Gertler, F.B., Bennett, R.L., Clark, M.J., and Hoffmann, F.M. (1989). Drosophila abl tyrosine kinase in embryonic CNS axons: A role in axonogenesis is revealed through dosage-sensitive interactions with disabled. Cell 58, 103-113.

Gertler, F.B., Doctor, J.S., and Hoffmann, F.M. (1990). Genetic suppression of mutations in the Drosophila abl proto-oncogene homolog. Science 248, 857-860.

Han, M. and Sternberg, P.W. (1990). let-60, a gene that specifies cell fates during C. elegans vulval induction, encodes a ras protein. Cell 63, 921-931.

Lennard, A.C. and Fried, M. (1991). The bidirectional promoter of the divergently transcribed mouse Surf-1 and Surf-2 genes. Mol. Cell. Biol. 11, 1281-1294.

Lev, Z., Kimchie, Z., Hessel, R., and Segev, O. (1985). Expression of ras cellular oncogenes during development of Drosophila melanogaster. Mol. Cell Biol. 5, 1540-1542.

Lev, Z., Segev, O., Cohen, N., Salzberg, A., and Shemer, R. (1989). Structure of the Drosophila ras2 bidirectional promoter. In ras oncogenes. D. Spandidos, ed. (New York: Plenum), pp. 75-81.

Lindsley, D.L. and Zimm, G. (1990). The genome of Drosophila melanogaster. Dros. Info. Service 68, 150.

Mozer, B., Marlor, R., Parkhurst, S., and Corces, V. (1985). Characterization and developmental expression of a Drosophila ras oncogene. Mol. Cell Biol. 5, 885-889.

Neuman Silberberg, F.S., Schejter, E., Hoffmann, F.M., and Shilo, B.Z. (1984). The Drosophila ras oncogenes: structure and nucleotide sequence. Cell 37, 1027-1033.

Padgett, R.W., St.Johnston, R.D., and Gelbart, W.M. (1987). A transcript from a Drosophila pattern gene predicts a protein homologous to the transforming growth factor-beta family. Nature 325, 81-84.

Perkins, K.K., Dailey, G.M., and Tjian, R. (1988). Novel Jun- and Fos-related proteins in Drosophila are functionally homologous to enhancer factor AP-1. EMBO J. 7, 4265-4273.

Rijsewijk, F., Schuermann, M., Wagenaar, E., Parren, P., Weigel, D., and Nusse, R. (1987). The Drosophila homolog of the mouse mammary oncogene int-1 is identical to the segment polarity gene wingless. Cell 50, 649-657.

Rogge, R.D., Karlovich, C.A., and Banerjee, U. (1991). Genetic dissection of a neurodevelopmental pathway: Son of sevenless functions downstream of the sevenless and EGF receptor tyrosine kinases. Cell 64, 39-48.

Schejter, E.D. and Shilo, B.Z. (1985). Characterization of functional domains of p21 ras by use of chimeric genes. EMBO J. 4, 407-412.

Segal, D. and Shilo, B.Z. (1986). Tissue localization of Drosophila melanogaster ras transcripts during development. Mol. Cell Biol. 6, 2241-2248.

Shilo, B. (1986). Proto-oncogenes in Drosophila melanogaster. Trends. Genet. 3, 69-72.

Steward, R. (1987). Dorsal, an embryonic polarity gene in Drosophila, is homologous to the vertebrate proto-oncogene, c-rel. Science 238, 692-694.

TRANSCRIPTIONAL REGULATION OF C-HA-*RAS*1 GENE

A. Pintzas
A. Kotsinas [1]
D.A. Spandidos [1,2]

Cancer Research Campaign
Beatson Laboratories
The Beatson Institute for Cancer Research
Garscube Estate
Switchback Road
Bearsden
Glasgow G61 1BD

1.   Institute of Biological Research and Biotechnology, National Hellenic
        Research Foundation, 48 Vas Constantinou Avenue, Athens 116 35
        Greece

2.   Medical School, University of Crete, Heraklion, Greece

INTRODUCTION

Ras genes (H-*ras*, K-*ras* and N-*ras*) appear to be highly conserved throughout the animal kingdom and have been identified in lower eukaryotes (eg yeasts), Dictyostelium, Drosophila and chickens in addition to mammals[1]. In mammals the *ras* genes appear to be expressed in all cell types and at all developmental stages. The fact that the genes are highly conserved and constitutively expressed in different cell types has led to the suggestion that they have some essential cellular functions[2].

Members of the *ras* gene family have been implicated in a variety of naturally occuring tumours[3]. Activated *ras* oncogenes are found in a high percentage of certain tumour types - up to 40% of colerectal cancers[4,5] and over 90% of pancreatic cancers[6].

The human *ras* gene family encodes structurally related proteins of 21 Kd (p21) which are localized at the inner surface of the plasma membrane[7], are palmitoylated and polyisoprenylated in order to bind to it[10], they also bind the guanine nucleotides8, have weak GTPase activity[9] and homology to the G protein which regulates adenylate cyclase activity[11].

Activation of ras family members mainly occurs by point mutations in codons 12 or 61[4,5] which results in deregulation of the p21 protein activity and is considered to result in the appearance of the transformed phenotype. However, in several human tumours activation of the oncogenes is considered to occur by elevated expression of ras RNA transcripts[12, 13] and ras p21 protein[14-17]. The probability that over-expression of ras genes is a special factor in tumorigenesis makes it important to define the promoter elements of the genes and analyse their transcriptional control.

STRUCTURE OF RAS GENES

Although all three functional ras genes-H-ras 1, K-ras2 and N-ras code for a protein of MW 21000 Da (p21), their structure is different in terms of coding sequences. The Ha-ras 1 gene contains five exons, while the K-ras 2 gene consists of six and the N-ras gene has seven. The coding sequences of H-ras 1 and N-ras are spread over four exons. K-ras[2] has two alternative fourth coding exons, which code for two isomorphic ras p21 proteins carrying 188 and 189 amino acids, which differ in their carboxy terminus. The conservation observed within the coding domains does not extend to flanking regions and the ras genes have a size from 4.5 Kb (H-ras 1) to over 4.5 Kb (K-ras 2), due to large differences in the size of their nonhomologous introns. Nonetheless, mammalian ras genes carry a further 5' noncoding exon (exon 1a).

REGULATORY SEQUENCES OF H-RAS 1

Additionally to oncogenic mutations which usually lead to loss of stimulation of the GTPase activity of p21 by GAP[18], the level of gene expression is also important for the activity of both normal[19] and mutant[20] ras proteins. Despite the importance of the genes in cell proliferation and differentiation not much is known about the regulation of expression of ras genes either in normal or in malignant cells.

The promoters of H-ras 1,[21-23,37], K-ras[24] and N-ras[24] have been identified and lie immediately 5' to exon 1a, but they do not contain the characteristic TATA box, although a consensus CAAT box has been identified within the H-ras 1 promoter[26,34].

A number of transcriptionally regulatory elements have been described at both the 5' and 3' end[27] of the c-Ha-ras 1 gene (Fig 1). The promoter sequence is rich in G/C content and contains a number of potential binding sites for transcription factor sp1[28] and potential sites for glucocorticoid receptor[29] as well as for AP-1 transcription factor[30] binding. Regulatory sequences have also been found in the first[31] and fourth[32] intron of the H-ras 1 gene. In the first case, a six base pair insertion in the first intron of the T24 H-ras 1 gene (Fig 1) increases transcriptional activity from the H-ras 1 promoter[31] and has been recently characterised as an additional binding site of Sp1-like factors[33]. In the latter case a point mutation in the fourth intron is responsible for increased expression and transforming activity of the oncogene[32]. Moreover, Ha-ras 1 promoter seems to contain ras[35] - or insulin[36] - responsive elements, which are potential mediators of distinct signal transduction pathways from the cell membrane to the nucleus. Despite the fact that the human c-Ha-ras promoter region has been identified, there exists a discrepancy in the literature about the position and regulation of the multiple transcription initiation sites of it[21, 22, 38].

Recently, studies have shown that sequences further upstream of the already identified promoters may play a role in the regulation and transforming potential of Ha-*ras*[33] or N-*ras*[40] genes.

CONCLUDING REMARKS AND FUTURE PROSPECTS

It has been shown that quantitative and/or qualitative changes in *ras* gene expression are responsible for triggering various steps of cell transformation, ie immortalization, tumorigenicity and metastasis[41]. The regulation of H-*ras* 1 gene expression is not completely understood although transcriptional regulatory elements have been identified at the 5'and 3' end of the gene as well as in the intron sequences of murine, rat and human genes. It is likely that further work concerning

(1)   the way the various DNA-binding factors interact with each other and with the transcriptional machinery as well as how they respond to signals thus activating or inactivating the gene and

(2)   the position and regulation of the multiple transcription sites of the Ha-*ras* 1 gene

will provide a very important contribution in understanding the role of the oncogene in cell proliferation and differentiation.

STRUCTURE OF THE 5' END OF THE HUMAN Ha-ras 1 GENE

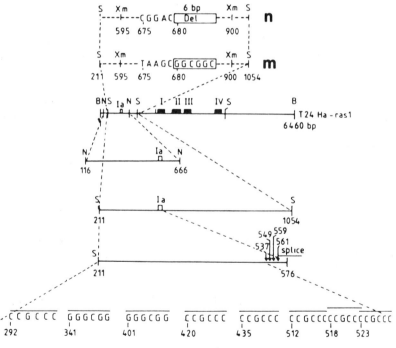

Figure 1. Organisation of the human H-*ras* gene and its promoter. The coding sequences are designated by black boxes and the non-coding 5' sequences by open boxes. The locations of the sequence GGGCGG and its complementary CCGCCC contained between nucleotides 220 and 576 of the map and situated at the 5' end of the H*ras* 1 gene are shown. The T24 H-*ras* insertion sequence GGGCGGC is shown. Transcriptional start sites are indicated by arrows. DEL, deletion; B, BamHI; S, Sst I; Xm, XmIII; N, Nae I; I, II, III and IV, coding exons; Ia, non-coding exon; n, normal; m, mutant T24.

REFERENCES

1. M. Barbacid. *Ras* genes. Annu. Rev. Biochem. 56:779 (1987).

2. D. Broek, N. Samily, O. Fasano, A. Fujiyama, F. Tomonoi, J. Northup and M. Wigler. Differential activation of yeast adenylate cyclase by wild-type and mutant *ras* proteins. Cell 46:477 (1985).

3. J. L. Bos. The *ras*-gene family and human carcinogenesis. Mutat. Res. 195:255 (1988).

4. J. L. Bos, E. R. Fearon, S. R. Hamilton, M. Verlaan-de Vries, J. H. Van Boom, A. J. Van der Eb and B. Vogelstein. Prevalence of *ras* gene mutations in human colorectal cancers. Nature 327:293 (1987).

5. K. Forrester, C. Almoguera, K. Han, W. E. Grizzle and M. Perucho. Detection of high incidence of K-*ras* oncogenes during human colon tumorigenesis. Nature 327:298 (1987).

6. C. Almoguera, D. Shibata, K. Forrester, J. Martin, N. Arnheim, and M. Perucho. Most human carcinomas of the exocrine pancreas contain mutant c-K-*ras* genes. Cell 53:549 (1988).

7. M. C. Willingham, I. Pastan, T. Y. Shih, E. M. Scolnick. Localisation of the *src* gene product of the Harvey Strain of MSV to plasma membrane of transformed cells by electron microscopic immuno-cytochemistry. Cell 19:1005 (1980).

8. T. Y. Shih, A. Papageorge, A. Stokes, M. O. Weeks and E. M. Scolnick. Guanine nucleotide-binding and autophosphorylating activities associated with the p21[src] protein of Harvey murine sarcoma virus. Nature 287:686 (1980).

9. J. P. McGrath, D. J. Capon, D. V. Goeddel and A. D. Levinson. Comparative biochemical properties of normal and activated human *ras* p21 protein. Nature 310:644 (1984).

10. J. F. Hancock, A. I. Magee, J. E Childs and C. J. Marshall. All *ras* proteins are polyisoprenylated but only some are palmitoylated. Cell 57:1167 (1989).

11. A. G. Gilman. G proteins and dual control of adenylate cyclase. Cell 36:577 (1984).

12. D. A. Spandidos and I. B. Kerr. Elevated expression of the human *ras* oncogene family in premalignant and malignant tumours of the colorectum Br. J. Cancer 49: 681 (1984).

13. D. J. Slamon, J. B. de Kernion, I. M. Verma and M. J. Cline. Expression of cellular oncogenes in human malignancies. Science 224:256 (1984).

14. A. R. W. Williams, J. Piris, D. A. Spandidos and A. H. Wyllie. Immunohistochemical detection of the *ras* oncogene p21 product in an experimental tumour and in human colorectal neoplasms. Br. J. Cancer 52:687 (1985).

15.   F. De Biasi, G. Del Sal and P. Horan Hand. Evidence of enhancement of the *ras* oncogene protein product (p21) in a spectrum of human tumours. Int. J. Cancer 43:431 (1989).

16.   T. Tanaka, D. J. Slamon, H. Battifare and M. J. Cline. Expression of p21 *ras* oncoproteins in human cancers. Cancer Res. 46:1465 (1986).

17.   N. J. Agnantis, A. Constantinidou, C. Pulios, A. Pintzas, A. Kakkanas and D. A Spandidos. Immunohistochemical study of the *ras* oncogene expression in human bladder endoscopy specimens. Eur. J. of Surg. Onc. 16:153 (1990).

18.   M. Trahey and F. McCormick. A cytoplasmic protein stimulates normal N-*ras* p21 GTPase, but does not affect oncogenic mutants. Science 238:542 (1987).

19.   S. Pulciani, E. Santos, L. K. Long, V. Sorrentino and M. Barbacid. *Ras* gene amplification and malignant transformation. Mol. Cell. Biol. 5:2836 (1985).

20.   S. A Hill, S. Wilson and A. F. Chambers. Clonal heterogeneity, experimental metastatic ability and p21 expression in H-*ras*-transformed NIH 3T3 cells J. Nat. Can. Inst. 80:484 (1988).

21.   H. Honkawa, W. Masahashi, S. Hashimoto and T. Hashimoto-Gotoh. Identification of the principal promoter sequence of the c-H-*ras* transforming oncogene. Deletion analysis of the 5' - flanking region by focus formation assay. Mol. Cell. Biol. 7:2933 (1987).

22.   S. Ishii, G. T. Merlino and I. Pastan. Promoter region of the human Harvey *ras* proto-oncogene: similarity to the EGF receptor, proto-oncogene promoter Science 230:1378 (1985).

23.   D. A. Spandidos and M. Riggio. Promoter and enhancer like activity at the 5'-end of normal and T24 H-*ras* 1 genes. FEBS Lett. 203:169 (1986).

24.   E. K. Hoffman, S. P. Trusko, N. Freeman and D. L. George. Structural and functional characterisation of the promoter region of the mouse C-K-*ras* gene Mol. Cell. Biol. 7:2592 (1987).

25.   A. Hall, R. Brown. Human N-*ras*: cDNA cloning and gene structure. Nuc. Ac. Res. 13:5255 (1985).

26.   W. S. Trimble and N. Hozumi. Deletion analysis of the c-Ha-*ras* oncogene promoter. FEBS Lett. 219:70 (1987).

27.   D. A. Spandidos and L. Holmes. Transcriptional enhancer activity in the variable tandem repeat DNA sequence downstream of the human Ha-*ras* 1 gene. FEBS Lett. 218:41 (1987).

28.   S. Ishii, J. T. Kadonaga, R. Tjian, J. N. Brady, G. T. Merlino and I. Pastan. Binding of the Sp1 transcription factor by the human Harvey *ras* 1 proto-oncogene promoter. Science 232:1410 (1986).

29. J. M. Strewhecker, N. A. Betz, R. Y. Neads, W. Houser and J. F. Pelling. Binding of the 97 kDa glucocorticoid receptor to the 5' upstream flanking region of the mouse c-Ha-*ras* oncogene. Oncogene 4, 1317 (1989).

30. D. A. Spandidos, R.A.B. Nichols, N. M. Wilkie and A. Pintzas. Phorbol ester-responsive H-*ras* 1 gene promoter contains multiple TPA-inducible/AP-1 binding consensus sequence elements. FEBS Lett. 240:191 (1988).

31. D. A. Spandidos and A. Pintzas. Differential potency and trans-activation of normal and mutant T24 human H-*ras* 1 gene promoters. FEBS LeH. 232:269 (1988).

32. J. B Cohen and A. D. Levinson. A point mutation in the last intron responsible for increased expression and transforming activity of the c-Ha-*ras* oncogene. Nature 334:119 (1988).

33. A. Pintzas and D. A Spandidos. Sp-1 specific binding sites within the human H-*ras* promoter: potential role of the 6bp insertion sequence in the T24 H-*ras* 1 gene. Submitted.

34. K. A. Jones, J. T. Kadonaga, P. J. Rosenfeld, T. J. Kelly and R. Tjian. A. Cellular DNA-binding protein that activates eukaryotic transcription and DNA replication. Cell 48:79 (1987).

35. A. Pintzas and D. A. Spandidos. *Ras* p21 oncoprotein is autoregulated and acts as a potential mediator of insulin action on the H-*ras* 1 promoter. Gene Analysis Techniques 6:125 (1989).

36. K. Lu, R. A. Levine and J. Campisi. *C-ras-* Ha gene expression is regulated by insulin or insulinlike growth factor and by epidermal growth factor in murine fibroblasts. Mol. Cell. Biol. 9:3411 (1989).

37. K. Brown, B. Bailleul, M. Ramsden, F. Fee, R. Krumlauf and A. Balmain. Isolation and characterisation of the 5' flanking region of the mouse c-Harvey-*ras* gene. Mol. Carc. 1:161 (1988).

38. N. F. Lowndes, J. Paul, J. Wu and M. Allan. C-Ha-*ras* gene bidirectional promoter expressed in vitro: location and regulation. Mol. Cell. Biol. 9: 3758 (1989).

39. A. K. Chakraborty, K. Cichutek and P.H. Duesberg. Transforming function of proto-*ras* genes depends on heterologous promoters and is enhanced by specific point mutations. PNAS 88:2217 (1991).

40. N. Nicoloiew, G. Triqueneaux and F. Dautry. Orgnaisation of the human N-*ras* locus: characterisation of a gene located immediately upstream of N-*ras*. Oncogene 6:721 (1991).

41. D.A. Spandidos. *Ras* oncogenes in cell transformation. ISI Atlas of Science. Immunology 1:1 (1988).

EFFECTS OF v-H-ras ON IMMORTALIZED NON-TUMORIGENIC HUMAN MAMMARY EPITHELIAL CELLS

Natasha Kyprianou, Jiri Bartek[1] and Joyce
Taylor-Papadimitriou

Imperial Cancer Research Fund, P O Box 123, Lincoln's Inn
Fields, London WC2A 3PX, U.K.
[1]Present Address: Institute of Medical Research, Research
Institute of Clinical and Experimental Oncology, 65601
Brno, Czechoslovakia

INTRODUCTION

Oncogenes are genes implicated in carcinogenesis which have been
identified by virtue of their association with oncogenic viruses, with
tumour-specific chromosomal abnormalities and with DNA sequences that
transform cultured cells to a tumorigenic state (Bishop, 1985).
Activated ras oncogenes have been detected in a wide variety of human
cancers (Bos, 1989) and expression of activated ras oncogenes or
overexpression of the normal ras proto-oncogenes stimulates
proliferation and induces transformation in a number of cell lines
(Barbacid, 1987). Recent studies however demonstrated that ras
oncogenes can promote either tumorigenic transformation or growth
inhibition depending on the phenotype of the cell in which they are
expressed. Thus, expression of v-H-ras oncogene leads to growth
arrest of mouse Schwann cells (Ridley et al., 1988), rat embryo
fibroblasts (Hirakawa and Ruley, 1988) and human thyroid epithelial
cells (Wynford-Thomas et al., 1990). These studies were performed
using cell lines expressing a temperature-sensitive mutant of large T
antigen at the restrictive temperature; it was found that in the
presence of a functional T antigen ras induced tumorigenicity, but in
its absence, cell growth was inhibited.

Carcinogenesis studies in animals have demonstrated a causal role
of activated v-H-ras oncogene in the initiation of mammary neoplasia
(Zarbl et al, 1985; Strange et al., 1989; Miyamoto et al, 1990; Kumar
et al., 1990). Moreover quantitative enhancement of c-H-ras
proto-oncogene and p21ras expression has been implicated in the
etiology and progression of human breast cancer (Spandidos and
Agnantis, 1984; Clair et al., 1987). In spite of an extensive
research however, qualitative changes in the c-H-ras oncogene by means
of mutational activation are very rarely associated with the
initiation or metastatic progression of human breast cancer (Liu et
al., 1988; Rochlitz et al., 1989). This suggests that the expression
of a mutated ras in the cell type from which breast cancer develops
does not usually confer a selective growth advantage. It was

The Superfamily of ras-Related Genes
Edited by D.A. Spandidos, Plenum Press, New York, 1991

therefore of interest to examine the effect of introduction of v-H-ras on the growth properties of normal human mammary epithelial cells.

The Malignant Phenotype and Normal Cell Lineages in the Human Breast

Although it is clear that breast cancers develop from epithelial cells lining the ductal-alveolar structures of the mammary tree, the detailed characterization of the malignant phenotype in terms of the normal cell lineages has been less well defined. There are two main epithelial phenotypes in the mammary gland, the basal or myoepithelial cells, and the luminal or secretory cells, which together form a double layer. The luminal cells which line the lumen of the ducts and lobules rest on the basal cells, but processes from them do contact the basement membrane on which the basal cells sit. It is the luminal cells which, under the influence of hormones produced at pregnancy and lactation, go on to become the milk producing cells.

It is possible to distinguish basal and luminal epithelial on the basis of the profile of intermediate filaments which they express (Taylor-Papadimitriou and Lane, 1987); all the luminal cells express keratins 8 and 18, and most also express keratin 19 (Bartek et al., 1985) while the basal cells express keratins 5 and 14, and vimentin (Taylor-Papadimitriou et al., 1989; Guelstein et al., 1988; Dairkee et al., 1985). Examination of primary and metastatic breast cancers revealed that the great majority (around 90%), express an intermediate filament pattern characteristic of luminal epithelial cells i.e. they express keratins 8, 18 and 19 (and usually 7) but not keratins 5, 14 or vimentin (Taylor-Papadimitriou and Lane, 1987; Bartek et al., 1985b; Dairkee et al., 1987a; Wetzels et al., 1989) A subset of primary breast cancers have been found to express vimentin or K14 but these constitute a small percentage (6-12%) of the tumours (Domagala et al., 1990; Gould et al., 1990; Dairkee et al., 1987b). The luminal origin of breast cancers is also indicated by the widespread expression of other luminal markers, namely a polymorphic epithelial mucin (more than 90% of cancers) (Taylor-Papadimitriou and Gendler, 1989; Gendler et al., 1989; Girling et al., 1989) and the estrogen receptor (in approximately 50% of tumours).

Immortalization of Luminal Epithelial Cells

Considering that the malignant phenotype in most breast cancers corresponds to that of the luminal epithelial cell, it is essential to culture those luminal cells for in vitro studies on human mammary carcinogenesis. Reduction mammoplasty tissue or human milk secretions provide a unique source of differentiated luminal epithelial cells(Taylor-Papadimitriou et al., 1989). Although relatively homogeneous cultures can be obtained from these sources, these cells have a very short in vitro life span which limits their use (Taylor-Papadimitriou et al,. 1977). Recent studies from our laboratory described the development of a series of non-tumorigenic immortal cell lines from milk-derived luminal cells, which have been immortalized by introducing SV40 large T antigen either by microinjecting or transfecting SV40 DNA, or by infecting cells with an amphotropic retrovirus carrying the sequence coding for SV40 T antigen

(Bartek et al. 1990; 1991). On the basis of their expression of
keratins 19 and 14, the immortalized lines, (none of which are
tumorigenic or grow in agar), have been classified into 4 groups
(Table 1); groups 1-3 represent luminal epithelial cells, some of
which (groups 2 and 3) are induced to express keratin 14 in culture.
Cell lines in group 1 represent an important phenotype since this
represents the phenotype of the majority of breast cancers.

**Table 1. Phenotypes of immortalized lines as defined by
immunohistochemical analysis**

		Keratins		Vimentin	PEM[a]
	7,8,18,	19	14		
Group 1. (7)[b]	+	+	−	−	+
Group 2. (4)	+/−	+/−		−	+
Group 3. (4)	+	−	+/−	−	+
Group 4. (3)	+	−	−	+	±

The expression of the antigens was determined by immunohistochemical
staining with monoclonal antibodies (Bartek et al., 1991).
[a] A polymorphic epithelial mucin expressed by breast luminal epithelial
cells.
[b] Number of cell lines developed.

Effect of Introducing v-H-ras into Immortalized Luminal Epithelial Cells

Using a recombinant amphotropic retrovirus based on Zipneo SV(X)
plasmid (Cepko et al,. 1984), an activated H-ras oncogene was
introduced into three SV40 immortalized cell lines, MTSV1-7, MMSV-1
(group 1) and MRSV3-1 (group 2). The results of analysis of the
hygromycin resistant clones developed from the infected lines are
summarized in Table 2. In each case most of the selected clones
senesced within days and only one clone continued to proliferate for
several passages. In each case the clones entered crisis and although
the MMSV-1 and MRSV-1-derived clones did not emerge, one MTSV1-7 clone
emerged after crisis as a single transformed cell line [MTSV1-7 (hygro
ras)].

Induction of Senescence of Luminal Epithelial Cells by v-Ha-ras

The above experiments support the notion that in the luminal
mammary epithelial cells, expression of the v-H-ras gene, even in the
presence of T antigen, induces growth inhibition leading to
senescence. However since this happens quickly it is difficult to
analyze the phenomenon. To overcome this difficulty we have
introduced into MTSV1-7 cells a mutated ras gene (EJ ras) under the
control of the inducible metallothionine promoter, using an
amphotropic retrovirus (Wynford-Thomas et al., 1990). Most of the

Table 2

v-H-ras effect on human mammary epithelial cell lines

Recipient cell line	Number of hygromycin resistant clones	Number of clones which senesced immediately	Crisis after passage	Long-term Growth
MMSV-1 (Group 1)	10	9	1 (p3)	–
MTSV1-7 (Group 1)	7	6	1 (p7)	Grew after 3 months. Cell line grows in agar and in the nude mouse
MRSV3-1 (Group 2)	5	4	1 (p7)	–

The indicated cell lines were infected with the amphotropic retrovirus pZiprashygro (Bartek et al., 1991) and clones selected in hygromycin.
Cells infected with pZiphygro (without ras) were not inhibited.

clones selected for neomycin resistance after infection with the ras-containing retrovirus in this case did not senesce although a few did go though a crisis period. After addition of zinc however the clones which showed a normal growth pattern in the absence of inducer, senesced after 1-2 weeks. MTSV1-7 clones selected with neomycin after infection with an amphotropic retrovirus containing only the neomycin gene, were not however affected by Zn. The ability to induce growth inhibition and senescence by inducing ras expression will now allow an analysis of the mechanisms involved.

Ras-transformed Tumorigenic Cell Lines

Although in the original experiments introduction of v-H-ras resulted in senescence of most of the selected clones, the MTSV1-7 (hygro ras) cell line was isolated. This cell line formed colonies in soft agar and gave rise to tumours in nude mice (Bartek et al., 1991). One could speculate that the growth inhibitory effect of v-H-ras, evident in the senescing clones was overcome by some unknown factor. In the case of the MTSV1-7 (hygro ras) line, it should be noted that the cell line emerged only after a crisis period of 3 months during which time the cells did not die but neither did they proliferate. We have now isolated revertants from the v-H-ras transformed MTSV1-7 cell line, which are not tumorigenic, and which can be used for studying the mechanisms involved in suppression of ras induced transformation.

Azatyrosine Induced Reversion of the Transformed Phenotype

Azatyrosine (Figure 1) is an antibiotic which was isolated from Streptomyces chibanesis and which has been shown to induce reversion of ras-transformed 3T3 cells to a non-transformed phenotype with very high efficiency (Shindo-Okada et al., 1989). Although the mechanism of action is obscure, the effect is dramatic. Not only did azatyrosine selectively inhibit the growth of the ras-3T3 cells but more than 70% of clones isolated after 7 days of treatment with the antibiotic showed a permanent reversion of the transformed phenotype. Moreover the suppression of the tumorigenic properties in the revertants occurred in the presence of high levels of expression of the p21 ras protein. We have now repeated these studies with MTSV1-7 (hygro ras) using azatyrosine (kindly provided by Dr Nishimura). Azatyrosine significantly inhibited the growth of the ras-transformed cells, but not of the normal MTSV1-7 cells. Although the growth inhibition of ras-transformed mammary epithelial cells was not as dramatic as it was

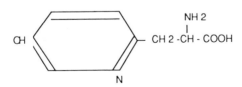

Figure 1. Structure of the antibiotic azatyrosine [L-β(5-hydroxy-2-pyridyl)-alanine].

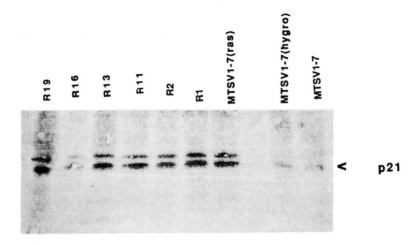

Figure 2. Expression of p21ras protein in azatyrosine-induced revertants. The p21 protein encoded by the H-ras oncogene was identified in cell lysates of parental, (hygro) control and ras-transformed MTSV1-7 and individual revertants, by Western blotting. The antibody used (NIE 704; Du Pont) detects both normal (lower band) and activated (upper band) forms of p21ras protein. Activated p21ras protein is detected in all the revertants analyzed, as well as the MTSV1-7 ras transformed cells. Each lane contained approximately 100 μg of protein.

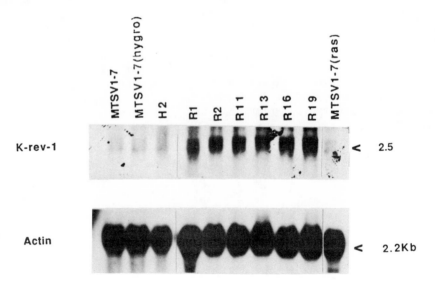

Figure 3. Northern blot analysis of K-rev-1 mRNA expression in azatyrosine-induced revertants. Hybridizations of Northern blots originally to ^{32}P-labelled K-rev-1 cDNA probe and subsequently to ^{32}P-labelled β-actin cDNA probe. Each lane contains 20 μg of total cellular RNA from the indicated cell line.

reported for the ras-transformed NIH/3T3 cells (Shindo-Okada et al., 1989), the reversion efficiency was similarly high (Kyprianou and Taylor-Papadimitriou, 1991). From the revertant clones selected, six have been analyzed in detail and all showed loss of ability to form colonies in soft agar and induce tumours in nude mice, while still retaining the integrated v-H-ras gene and expressing high levels of p21ras protein (Fig. 2). Thus the revertant phenotype induced by azatyrosine does not result from inactivation of v-H-ras oncogene or inhibition of its expression.

K-rev-1 Expression in the Azatyrosine-induced Revertants

In the search for genes whose expression is involved in suppression of tumorigenicity of ras-transformed cells, Noda and colleagues isolated a gene (Krev-1) from flat revertants of ras-transformed 3T3 cells transfected with a cDNA expression library from normal cells (Noda et al., 1989). K-rev-1 shows a high homology with the ras genes (Kitayama et al., 1989) and is equivalent to RAPIA (Kitayama et al., 1990). It has been shown to bind with high affinity to ras GAP (Frech et al., 1990) and may influence ras action by competing for this protein. As an initial step in looking for genes whose expression is related to the reversion of the ras-transformed phenotype of mammary epithelial cells, we examined the revertants for the expression of Krev-1. We found that all of the revertants expressed markedly increased levels of Krev1 mRNA as compared to the original tumorigenic cell line (Fig. 3). Whether this increased expression of K-rev-1 is causative or correlative in reversing the transformed phenotype in the mammary cells has not yet been determined. Introduction of K-rev-1 into ras-transformed 3T3 cells has proved to be very inefficient in causing reversion (only 2% revert). Probably the most conclusive demonstration of a causative action of Krev-1 would be to use antisense RNA or ribozymes to inhibit mRNA expression.

SUMMARY

Most breast cancers show the phenotype of the differentiated luminal epithelial cell, and only rarely have activated ras genes been found in breast cancers. Normal mammary epithelial cells cultured from milk have been immortalized using SV40 T antigen to give non-tumorigenic cell lines which exhibit many of the features of the luminal epithelial cell. When a mutated ras gene is introduced into these lines, most of the selected clones senesce, although a rare clone can be isolated which expresses p21 ras and shows a transformed phenotype. This observation suggests a reason why mutated ras genes are seldom found in breast cancers. In order to study the phenomenon of ras-induced senescence, the oncogene has been introduced into a mammary epithelial cell line under the control of the inducible metallothionine promoter.

By treating a ras-transformed tumorigenic mammary epithelial cell line [MTSV1-7(hygro ras)] with azatyrosine, non-tumorigenic revertants can be induced with high efficiency. The revertants express high levels of p21ras, but in addition, increased levels of Krev-1 mRNA are expressed suggesting that the expression of this gene may be involved in the suppression of the tumorigenic phenotype in the ras-transformed mammary epithelial cells.

REFERENCES

Barbacid, M., 1987, ras genes Ann. Rev. Biochem, 56:779.

Bartek, J., Durban, E. M., Hallowes, R. C. and Taylor-Papadimitriou, J., 1985a, A subclass of luminal epithelial cells in the human mammary gland, defined by antibodies to cytokeratins. J. Cell Sci., 75:17.

Bartek, J., Taylor-Papadimitriou, J., Miller, N. and Millis, R. 1985b, Patterns of expression of keratin 19 as detected with monoclonal antibodies in human breast tissues and tumours. Int. J. Cancer, 36:299.

Bartek, J., Bartkova, J., Lalani, E-N., Brezina, V. and Taylor-Papadimitriou, J. 1990, Selective immortalization of a phenotypically distinct epithelial cell type by microinjection of SV40 DNA into cultured human milk cells Int. J. Cancer, 45:1105.

Bartek, J., Bartkova, J. Kyprianou, N., Lalani, E.-L., Staskova, L., Shearer, M., Chang, S. and Taylor-Papadimitriou, J., 1991, Efficient immortalisation of luminal epithelial cells from the human mammary gland by introduction of SV40 large T antigen using a recombinant retrovirus. Proc. Natl. Acad. Sci, USA, 88:3520.

Bishop, M., 1985, Viral oncogenes. Cell, 42:23.

Bos, J., 1989, ras oncogenes in human cancer, A Review. Cancer Res, 49:4682.

Clair, T., Miller, W. R. and Cho-Chung, Y. S., 1987, Prognostic significance of the expression of a ras protein with molecular weight of 21,000 by human breast cancer. Cancer Res, 47:5290.

Cepko, C. L., Roberts, B.E. and Mulligan, R. C., 1984, Construction and applications of a highly transmissible murine retrouirus shuttler vector. Cell, 37:1053.

Dairkee, S. H., Blayney, C., Smith, H. S. and Hackett, A. J., 1985, Monoclonal antibody that defines human myoepithelium. Proc. Natl. Acad. Sci, USA, 82:7409.

Dairkee, S. H., Ljung, B. M., Smith, H. and Hackett, A., 1987a, Immunolocalization of a human basal epithelium specific keratin in benign and malignant breast disease. Breast Cancer Res. Treatment, 10:11.

Dairkee, S. H., Mayall, BH, and Smith H.S. 1987b, Monoclonal marker that predicts early recurrence of breast cancer. Lancet, 1:514.

Domagala, W., Lasota, J., Dukowicz, A., Markiewski, M., Striker, G., Weber, K. and Osborn, M., 1990, Vimentin expression appears to be associated with poor prognosis in node-negative ductal breast carcinomas. Am. J. Pathol, 137:1299.

Frech, M., John, J., Pizon, V., Chardin, P., Tavitian, A., Clark, R., McCormick, F. and Wittinghofer, A., 1990, Inhibition of GTPase activating protein stimulation of ras p21 GTPase by the K Rev-1, gene product Science, 249:169

Gould, V. E., Koukoulis, G. K., Janson, D. S., Nagle, R. B., Frankle, W. W. and Moll, R., 1990, Coexpression patterns of vimentin and glial filament protein with cytokeratins in the normal, hyperplastic and neoplastic breast. Am. J. Pathol, 137:1143.

Gendler, S., Taylor-Papadimitriou, J., Burchell, J. and Duhig, T., 1989, A polymorphic epithelial mucin expressed by breast and other carcinomas: Immunological and molecular studies. in:

"Human Tumor Antigens and Specific Tumor Therapy", R. Metzgar and M. Mitchell, eds., Alan R. Riss, Inc., New York, pp 11.

Girling A., Bartkova, J., Burchell, J., Gendler, S., Gillet, C. and Taylor-Papadimitriou, J., 1989, A core protein epitope of the polymorphic epithelial mucin detected by the monoclonal antibody SM-3 is selectively exposed in a range of primary carcinomas. Int. J. Cancer, 43:1072.

Guelstein, V. I., Tchypysheva, T. A., Ermilova, V. D., Litviyova, L. V., Troyanovsky, S. M. and Bannikov, G. A. 1988,Monoclonal antibody mapping of keratins 8 and 17 and of vimentin in normal human mammary gland, benign tumours, dysplasias and breast cancers. Int.J. Cancer, 42:147.

Hirakawa, T., and Ruley, H.E. 1988, Rescue of cells from ras oncogene – induced growth arrest by a second, complementing oncogene. Proc. Natl. Acad. Sci, USA, 85:1519.

Kumar, R., Sukumar, S., Barbacid, M., 1990, Activation of ras oncogenes preceding the onset of neoplasia. Science, 248:1101.

Kitayama, H., Sugimoto, Y., Matsuzaki, T., Ikama, Y. and Noda, N., 1989, A ras – related gene with transformation suppressor activity. Cell, 56:77.

Kitayama, H., Matzukaki, T., Ikawa, J. and Noda, M., 1990, Genetic analysis of the Kirsten-ras – revertant 1 gene. Potential of its tumor suppressor activity by specific point mutations. Proc. Natl. Acad. Sci, USA, 87:4284.

Kyprianou, N. and Taylor-Papadimitriou, J., 1991, Isolation of azatyrosine-induced revertants from ras-transformed human mammary epithelial cells. Manuscript submitted for publication.

Liu, E., Dollbaum, C., Scott, G., Rochlitz, C., Benz, C. C. and Smith, H. S., 1988, Molecular lesions involved in the progression of a human breast cancer. Oncogene, 3:323.

Miyamoto, S., Guzman, R. C., Shiurba, R. A., Firestone, G. L. and Nandi, S., 1990, Transfection of activated H-ras Protooncogenes causes mouse mammary hperplasia. Cancer Res., 50:6010.

Noda, M., Kitayama, H., Matsuzaki, T., Sugimotao, Y., Okyama, H., Bassin, R. H. and Ikawa, Y., 1989, Deletion of genes with a potential for suppressing the transformed phenotype associated with activated ras genes. Proc. Natl. Acad. Sci, USA, 86:162.

Ridley, A. J., Paterson, H. F., Noble, M. and Land, H., 1988, ras-mediated cell cycle arrest is altered by nuclear oncogenes to induce Schwann cell transformation. EMBO J., 7:1635.

Rochlitz, C. F., Scott, G. K., Dodson, J. M., Liu, E., Dollbaum, C., Smith, H. S. and Benz, C. C., 1989, Incidence of activating ras oncogene mutations associated with primary and metastatic human breast cancer. Cancer Res., 49:357.

Strange, R., Anguillar, C. E., Young, L. J. T., Billy, H. T., Dandeker, S. and Cardiff, R. D., 1989, Harvey – ras mediated neoplastic development in the mouse mammary gland. Oncogene, 4:309.

Spandidos, D. A. and Agnantis, N. J., 1984, Human malignant tumors of the breast, as compared to their respective normal tissue, have elevated expression of the Harvey-ras oncogene. Anticancer Res., 4:269.

Shindo-Okada, N., Makase, O., Nagahara, H., Nishimura, S., 1989,
        Permanent conversion of mouse and human cells transformed by
        activated ras or rat genes to apparently normal cells by
        treatment with the antibiotic azatyrosine. Mol.
        Carcinogenesis, 2:159.
Taylor-Papadimitriou, J. and Lane, E. B., 1987, Keratin expression in
        the mammary gland, in: "The Mammary Gland", M. C. Neville and
        C. W. Daniel, eds, Plenum Publishing Corporation, pp 181.
Taylor-Papadimitriou, J., Stampfer, M., Bartek, J., Lewis, A.,
        Boshell, M., Lane, E. B. and Leigh, I. M., 1989, Keratin
        expression in human mammary epithelial cells cultured from
        normal and malignant tissue: relation to in vivo phenotypes
        and influence of medium. J. Cell Sci., 94:403.
Taylor-Papadimitriou, J. and Gendler, S. J., 1989, Molecular aspects
        of mucins, in: "Cancer Reviews", J. Hilgers, and S. Zotter,
        eds, Munksgaard, pp 11.
Taylor-Papadimitriou, J., Shearer, M. and Tilly, R., 1977, Some
        properties of cells cultured from early lactation and human
        milk. J. Natl. Cancer Inst., 58:1563.

Wynford-Thomas, D., Bond, J. A., Wyllie, F. S. Burns, J. S., Williams,
        E. D., Jones, T., Sheer, D., Lemoine, N. R., 1990,
        Conditional immortalization of human thyroid epithelial
        cells: A tool for analysis of oncogene action. Mol. Cell.
        Biol, 10:5365.
Wetzels, H. W., Holland, R., Van Haelst, U. J. G. M., Lane, E. B.,
        Leigh, I. M. and Ramaekers, F. C. S., 1989, Detection of
        basement membrane components and basal cell keratin 14 in
        noninvasive and invasive carcinomas of the breast. Am. J.
        Pathol., 134:571.
Zarbl, H., Sukumar, S., Arthur, A. V., Martin-Zanca, D. and Barbacid,
        M., 1985, direct mutagenesis of Ha-ras-1 oncogenes by
        N-nitroso-N-methylurea during initiation of mammary
        carcinogenesis in rats. Nature, 315:382.

SDC25, A NEW GENE OF SACCHAROMYCES CEREVISIAE, HOMOLOGOUS TO CDC25: THE 3'-PART OF SDC25 ENCODES AN EXCHANGE FACTOR ABLE TO ACT ON RAS PROTEINS

Faten Damak, Emmanuelle Boy-Marcotte, Pranvera Ikonomi, and Michel Jacquet

Groupe Information Génétique et Développement. Université Paris-Sud Bât. 400 - 91405 Orsay Cédex, France

INTRODUCTION

R*as* genes were first described as the transforming genes of the Kirsten and Harvey oncogenic retroviruses. They belong to a family of ubiquitous eucaryotic genes involved in growth control and in cell differentiation. *ras* gene products are GTP-binding proteins with an intrinsic GTPase activity. They activate their effector when bound to GTP but not when bound to GDP (Barbacid 1987). In *Saccharomyces cerevisiae*, the *RAS1* and *RAS2* genes are closely related to the *ras* genes of higher eucaryotic organisms (Dhar, Nieto et al. 1984; Powers, Kataoka et al. 1984) and their products activate adenylate cyclase encoded by the *CYR1* gene (allelic to *CDC35*) (Broek, Samiy et al. 1985). As a result of this activation, cyclic AMP stimulates the cAMP-dependent protein kinase, whose regulatory subunit is encoded by the *BCY1* gene (Toda, Cameron et al. 1987a). Three genes *TPK1*, *TPK2* and *TPK3* encode interchangeable catalytic subunits (Toda, Cameron et al. 1987b). The cAMP-dependent protein kinase pathway is essential for the cell division cycle since cAMP is required for the G1/S transition (Matsumoto, Uno et al. 1983a). This pathway relays also nutritional information which controls different processes such as carbohydrate storage (Martegani, Vanoni et al. 1986), sporulation (Iida and Yahara 1984) and a G1/G0 switch (Boy-Marcotte, Garreau et al. 1987).

The product of the *CDC25* gene is required for cAMP production (Camonis, M. Kalekine et al. 1986). Its effect is mediated by RAS proteins, as deduced from the existence of different mutations in the *RAS2* gene that suppress the *cdc25* mutations. One is the Gly->Val19 substitution, equivalent to the oncogenic substitution Gly->Val12 in the mammalian c-H-ras gene which leads to a lower GTPase activity (Kataoka, Powers et al. 1984). The second is a spontaneous mutation selected as a suppressor of a *cdc25-5* thermosensitive mutation which has been shown to be a Thr->Ile152 substitution (Camonis and Jacquet 1988). This mutation leads to spontaneous GDP/GTP exchange by increasing the guanyl nucleotide dissociation of the RAS2·GDP and RAS2·GTP complexes (Créchet, Poullet et al. 1990a). This finding suggests that the *CDC25* gene product is a positive regulator which activates RAS proteins, most likely as a GDP -> GTP exchange factor. Negative regulators of RAS proteins have also been characterized in yeast. They are encoded by *IRA1* and *IRA2*. The products of these genes are partially homologous to mammalian GAP and NF1 proteins. They can be functionally replaced by the mammalian GAP and NF1 in yeast suggesting that IRA proteins play the same role as GAP and NF1 (Tanaka, Nakafuku et al. 1990).

The *CDC25* gene has been cloned and contains an open reading frame (ORF) of 1,589 amino acids (Camonis, M. Kalekine et al. 1986; Daniel 1986; Martegani, Baroni et al. 1986;

Broek, T. Toda et al. 1987). The essential portion of the gene product, which is required to activate the RAS-dependent adenylate cyclase, corresponds to the C-terminal third of the polypeptide product (Camonis, M. Kalekine et al. 1986). The product of the *CDC25* gene has been detected immunologically as associated with the membrane. It is a polypeptide of 180 kDa (Garreau, Camonis et al. 1990; Vanoni, Vavassori et al. 1990; Jones, Vignais et al. 1991).

To extend our knowledge of the elements involved in the control of the RAS-dependent adenylate cyclase activation, we looked for extragenic suppressors of the *cdc25-5* thermosensitive mutation which do not suppress *cdc35* mutations. We have isolated a yeast DNA fragment which is able to restore growth of the *cdc25-5* mutant at restrictive temperature. Nucleotide sequence analysis of this DNA fragment shows a 5'-truncated ORF that we have named *SDC25*. This shares 47% identity at the amino acid level with the C-terminal part of the *CDC25* gene product (Boy-Marcotte, Damak et al. 1989). The product of this fragment is able of activating GDP to GTP exchange on yeast pRAS2 and on human p21 H-ras *in vitro* (Créchet, Poullet et al. 1990b). It is also capable to act in mammalian cells by activating ras-dependant transactivator (Rey, Schweighoffer et al. 1990). The DNA fragment corresponding to the N-terminal part of the ORF was also cloned and sequenced. Homology with the *CDC25* gene is also present in the N-terminal part, although this homology is weaker than that found in the C-terminal part. In contrast with the C-terminal part, the complete *SDC25* gene on a multicopy plasmid did not suppress the *CDC25* gene defect although it is transcribed and translated. The ability to promote suppresion was restored by a deletion in the N-terminal portion of the protein (Damak, Boy-Marcotte et al. 1991). The physiological role of the *SDC25* gene product is discussed.

RESULTS AND DISCUSSION

Cloning and characterization of a DNA fragment which suppresses the thermosensitive
cdc25-5 mutation

Yeast genomic DNA was prepared from a strain which contains the thermosensitive mutation *cdc35-10* in the structural gene of adenylate cyclase and the thermosensitive *cdc25-5* mutation. This avoids the cloning of the wild type allele of these genes in a *cdc25-5* mutant strain. This genomic DNA was cloned in a multicopy plasmid (YRp7) and used directly to transform the *cdc25-5, trp1* strain (OL86). Transformants were selected at restrictive temperature in the absence of tryptophan. A recombinant plasmid containing a 3.4-kb DNA fragment was isolated from one transformant (figure 1). This plasmid also restored the normal level of cAMP in tranformed cells, but was unable to suppress the *cdc35-10* and the *ras1ras2*[ts] mutations (Boy-Marcotte, Damak et al. 1989).

The sequencing of 2940 bp of this DNA segment was performed and two ORFs, separated by 411 nucleotides, were found. The left hand ORF, included in the segment required for suppression, corresponds to 584 C-terminal codons of a larger ORF. This was named ORF-SDC25 for suppression of *cdc25*. The stop codon is followed by putative transcription termination consensus sequenses. The second ORF was named ORF-X and corresponds to the 255 N-terminal codons of a larger ORF (Figure 1). The partial ORF-SDC25 was compared to the amino acid sequence of the C-terminal part of the ORF-CDC25. In the alignment proposed in figure 2, the two sequences share 47% identical residues. The similarities increase up to 66% if conservative substitutions are taken into account (Boy-Marcotte, Damak et al. 1989). The capacity of the *SDC25* fragment to suppress the *cdc25-5* mutation and its homology to the C-terminal part of the *CDC25* led us to propose that the 3'-part of *SDC25* and *CDC25* genes each encode a protein domain capable of carring out a similar biochemical function.

The product of the 3'-part (1,9-kb *Xba*I-*Nru*I fragment; see figure 1) of the *SDC25* gene has been expressed in *E. coli*. The partially purified protein *in vitro* strongly enhances the release of GDP from the *S. cerevisiae* RAS2·GDP and from the human p21 H-ras·GDP complexe and promotes faster GDP/GTP exchange (Créchet, Poullet et al. 1990b). Thus, the SDC25 C-terminal domain is the first known element which acts directly on RAS proteins from both yeast and mammals and behaves as an upstream regulator. This finding provides the

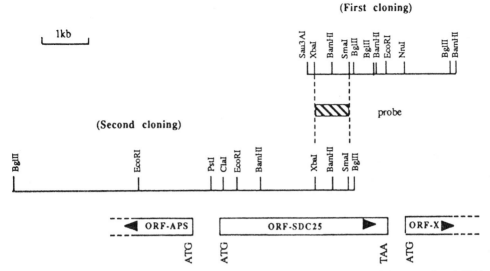

Fig. 1. Strategy for cloning of the *SDC25* gene. The restriction map of the two cloned DNA fragments is presented. The 3.4-kb DNA fragment containing the 3'-part of the *SDC25* gene has been cloned in the YRp7 vector (Boy-Marcotte, Damak et al. 1989). To identify the 7-kb-DNA fragment containing the 5'-part inserted in YRp7, the *Xba*I-*Sma*I fragment was used as a probe (Damak, Boy-Marcotte et al. 1991). The ORF-SDC25 flanked by ORF-X and ORF-APS is indicated.

```
CDC25   998  NLEFLNNSDDFKDACEKYVEISNLACIIVDQLIEERENLLNYAARMMKNNLTAELLKGEQEKWFDIYSEDYSDD-DSENDEAIIDDELGS
             .: :.:     ::  :. : ..  ::  :.::: ::::::::::::::::.:  ::.::. .:     :  :. :
SDC25     1  DLKFFNLAHVFKKSCDDYFDVLKLAIEFVNQLILERENLLNYAARMMKNNITELLLRGEEG--YGSYDG--GETAEKSDTNAVYADS-DT

CDC25  1087  EDYIERKAANIEKNLPWFLTSDYETSLVYDSRGKIRGGTKEALIEHLTSHELVDAAFNVTMLITFRSILTTREFFYALIYRYNLYPPEGL
             : :..    :::..: .  :  .:.:: : ::: ::: :: :.:  : ::::::::.:: ::. ::: :::  :
SDC25    86  KDNDEWRDSQV--KLPRYLQREYDSELIWGSNNRIKGGSKHALISYLTDNEKKDLFFNITFLITFRSIFTTTEFLSYLISQYNLDPPEDL

CDC25  1176  SYDDYNIWIEKKSNPIKCRVVNIMRTFLTQYWTRNYYEPGIP-LILN-FAKMVVSEKIPGAEDLLQKINEKLINENEKEPVDP-KQ-QDS
             ....: :. ::  :.:::::: ::: ::: ::: ::::  :  ::::..  ::: .:.  ..:.  : :  :
SDC25   174  CFEEYNEWVTKKLIPVKCRVVEIMTTFFKQYWFPGYDEPDLATLNLDYFAQVAIKENITGSVELLKEVNQKFKLGNIQEATAPMKTLDQQ

CDC25  1263  VSAVV-QTTKRDNKSPIHMSSSSLPSSASSAFFRLKKLKLLDIDPYTYATQLTVLEHDLYLRITMFECLDRAWGTKYC-NMGGSPNITKF
             .        :          :       : .:: .:::::::.::..: ::.::.., : :: :  : :  :
SDC25   264  ICQDHYSGTLYSTTESI----------------------LAVDPVLFATQLTILEHEIYCEITIFDCLQKIWKNKYTKSYGASPGLNEF

CDC25  1351  IANANTLTNFVSHTIVKQADVKTRSKLTQYFVTVAQHCKELNNFSSMTAIVSALYSSPIYRLKKTWDLVSTESKDLLKNLNNLMDSKRNF
             : :: :::::. .::.::  : :  :. :.. .:::::::::.::::::::: ::: .  ..: :: ::: :.:
SDC25   331  ISFANKLTNFISYSVVKEADKSKRAKLLSHFIFIAEYCRKFNNFSSMTAIISALYSSPIYRLEKTWQAVIPQTRDLLQSLNKLMDPKKNF

CDC25  1441  VKYRELLRSVTDVACVPFFGVYLSDLTFTFVGNPDFL---------HNSTNIINFSKRTKIANIVEEIISFKRFHYKLKRLDDIQTVIEA
             . :: .:.  :::::::::::::::::::: ::  : . :   : : .  :   :   : .  :  :
SDC25   421  INYRNELKSLHSAPCVPFFGVYLSDLTFTDSGNPDYLVLEHGLKGVHDEKKYINFNKRSRLVDILQEIIYFKKTHYDFTKDRTVIECISN

CDC25  1522  SLENVPHIEKQYQLSLQVEPRSGNTKGSTHASS-ASGTKTAKFLSEFTDDKNGNF------LKLGKKKPPSRLFR
             :::::.::::::::::: .::..  :      :  : : :  : .  .       : : :
SDC25   511  SLENIPHIEKQYQLSLIIEPKPRKKVVPNSNSNNKSQEKSRDDQTDEGKTSTKKDRFPKFQLHKTKKKAP-KVSK
```

Fig. 2. Alignment of *CDC25* and *SDC25* ORFs. The optimal alignment of the amino acid sequences deduced from the C-terminal part of the CDC25 ORF (aa 998-1589; (Camonis, M. Kalekine et al. 1986; Broek, T. Toda et al. 1987) and the SDC25 ORF (aa 1-584, number 1 corresponds to the first codon of the truncated ORF; Boy-Marcotte, F. Damak et al. 1989) is shown. Double dots indicate identity; single dots indicate a conservative change.

biochemical explanation for the suppression of the *cdc25-5* mutation by the 3'-portion of *SDC25*. The sequence homology between *SDC25* and *CDC25* strongly suggests that the latter should act also as an exchange factor. However, when similar GDP release experiments were performed *in vitro* with the partially purified preparations of the C-terminal portion of CDC25, the results were negative. This could be due to technical problems, such as defective expression of CDC25 in *E. coli*, or CDC25 may require another element for its activation which is absent in *E. coli*.

These results suggest the existence of a family of proteins mediating the GDP/GTP exchange of ras proteins in different organisms. Indeed, the presence of analogous factors in mammalian cells is suggested by the capacity of the SDC25 C-domain to promote this reaction both with *S. cerevisiae* pRAS2 and human p21 ras proteins. In fact, two reports have described the identification of an H-ras-guanyl nucleotide exchange factor activity in bovine brain tissue (Wolfman and Macara 1990; West, Kung et al. 1990). Morever, in *Schizosaccharomyces pombe*, a gene *ste6* with similarities to the 3'-part of both *SDC25* and *CDC25* genes has been identified and could also be involved in the activation of the *S. pombe ras1* gene product (Hughes, Fukui et al. 1990).

Cloning of the N-terminal portion and characterization of the complete *SDC25* gene

The DNA region corresponding to the N-terminal part of the ORF-SDC25 was cloned as an overlapping segment inserted into a plasmid vector. This DNA was detected by cross hybridization with the previously isolated DNA fragment. Sequencing of this new 7-kb-DNA fragment was performed and revealed that the complete SDC25 ORF contains 1,251 codons and is flanked by two other ORFs: ORF-X and ORF-Y (Figure 1). A DNA fragment containing ORF-Y has also been cloned and sequenced by Erbs *et al.* (P. Erbs and R. Jund. personal communication). They demonstrated that ORF-Y corresponds to the 5'-part of the *APS* gene that encodes the aspartyl-tRNA synthetase. *SDC25* and *APS* genes are separated by 653 nucleotides. This small intergenic region could contain some transcriptional regulatory sequences in common for these two genes (Damak, Boy-Marcotte et al. 1991).

The SDC25 ORF is slightly smaller than the CDC25 ORF (1,251 residues instead of 1,589). The relative amino acid composition of the *SDC25* and *CDC25* gene products is very similar. The amino acid sequence of the SDC25 and CDC25 ORFs were compared by dot matrix analysis (figure 3). The SDC25 amino acid sequence presents similarities throughout its length with that of the CDC25 ORF. In the N-terminal part of both ORFs (SDC25: positions 1 to 652 and CDC25: positions 1 to 982), 5 segments are partially homologous and are separated by non homologous sequences. These sequences contain most of the 338 additional amino acids of the CDC25 ORF. The SDC25 ORF from position 22 to position 101 and the CDC25 ORF from position 54 to 130 correspond to the first homologous segment, each of them contains a SH3 (src homology region 3) consensus sequence. This SH3 sequence has been identified in proteins that associate with the membrane cytoskeleton such as products of proto-oncogenes c-src (Kato, Takeya et al. 1986), and c-abl (Jackson and Baltimore 1989), myosin I (Jung, Saxe III et al. 1989) and the *S. cerevisiae* ABP1 protein (Drubin, Mulholland et al. 1990). It has been proposed that the SH3 domain is involved in actin binding and could serve to bring together signal transduction proteins and their targets or regulators. The SDC25 ORF contains a putative phosphorylation site for the cAMP protein kinase (Cohen 1985) at position 439 which is not conserved in the CDC25 ORF. The C-terminal part of the two ORFs (SDC25: positions 650 to 1,200 and CDC25: positions 980 to 1,545) are more strongly related as previously described. These segments cover the minimal region sufficient for the suppression of the *CDC25* defect. A hydrophobic domain which has been postulated to be a transmembrane domain in CDC25 (position 1,455 to 1,469) (Daniel and Simchen 1986) is conserved in the C-terminal part of the SDC25 ORF (position 1,102 to 1,116). The 50 last C-terminal amino acids of SDC25 ORF are not related to the 45 last C-terminal amino acids of the CDC25 ORF, although there are both very polar.

The *SDC25* gene is transcribed into a 4-kb long mRNA during exponential growth. Its transcription product is at least ten fold more abundant than the *CDC25* mRNA (Damak, Boy-Marcotte et al. 1991). At the 5'-end of the gene, three major transcriptional start points corresponding to positions -42, -27 and -15 have been identified. RNA starting at position -42

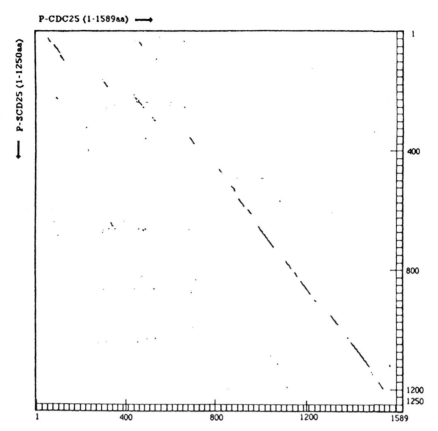

Fig. 3. Dot matrix comparison of the SDC25 and CDC25 amino acid sequences. The Mac Gene program was used with a window of 8 amino acids and an amino acid homology score of 72%.

is the most abundant. The *SDC25* gene has been located on chromosome XII by hybridization to electrophoretically separated chromosomes. Genetic studies confirmed this localization and showed it to be 5.5 centimorgans (cM) from the centromere and 8.7 cM from the *PPR1* gene located on the other arm of the same chromosome (Damak, Boy-Marcotte et al. 1991).

The complete *SDC25 and CDC25* genes with their own promoter as well as a truncated version encoding the SDC25 C-terminal domain were reconstructed in the multicopy plasmid YEp352, yielding the recombinant plasmids pRG3, pPI1 and pRG1, respectively. The ability of these three plasmids to suppress the *cdc25-5* mutation was tested. As shown in figure 4, in contrast to plasmids pPI1 and pRG1, plasmid pRG3 expressing the complete *SDC25* gene product did not suppress the thermosensitivity of the *cdc25-5* strain. However, the amount of *SDC25* mRNA transcribed from plasmid pRG3 was large and similar to that of the *SDC25* mRNA transcribed from plasmid pRG1. Efficient translation of the SDC25 ORF was also demonstrated by production of active beta-Gal in the in-frame SDC25-beta-galactosidase fusions promoted by transposition of the mini-Mu phage (Damak, Boy-Marcotte et al. 1991).

To investigate the role of the region within the N-terminal part of the *SDC25* gene product which prevents its suppressing the *cdc25-5* mutation, a set of deletions obtained by BAL31 digestion starting from the *Eco*NI site, was screened for the ability to suppress the *cdc25-5* thermosensitive mutation. Plasmid pRG3-9 contains a deletion of 1,032 bp (figure 4) resulting in an in-frame junction between amino acids 263 and 608. To ensure that activation of the suppressing activity of *SDC25* was due to this deletion, the 2-kb *Cla*I-*Xba*I fragment was replaced by the deleted 1-kb *Cla*I-*Xba*I fragment to create plasmid pRG3-9*. Both pRG3-9*

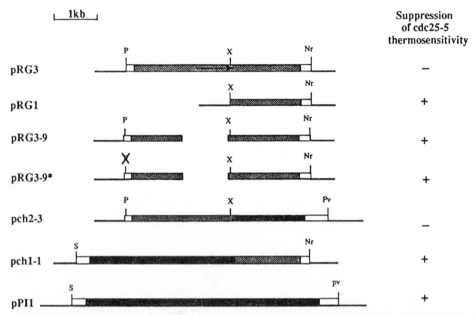

Fig. 4. Suppressing ability of the truncated and deleted *SDC25* gene, of the complete *SDC25* and *CDC25* genes and of the chimeric genes *SDC25/CDC25*. To construct the pRG1 plasmid, the 2-kb *XbaI-NruI* fragment, containing the 3'-part of the *SDC25* gene, was inserted into the YEp352 vector. The pRG3 plasmid, which contains the complete *SDC25* gene, is derived from pRG1 by insertion, between the *PstI* and the *XbaI* sites, of the 2.4-kb *PstI-XbaI* fragment which contains the 5'-part of the *SDC25* gene. The pPI1 plasmid was obtained by cloning the 5.5-kb *SalI-PvuII* fragment, containing the complete *CDC25* gene, into the YEp352 vector. Open boxes indicate yeast genomic DNA; Stippled boxes, *SDC25* sequences; blackened boxes, *CDC25* sequences. Plasmid pRG3-9 was obtained in the following way. Plasmid pRG3 was linearized at its unique *Eco*NI site and incubated with nuclease BAL31. After treatment with the Klenow fragment of DNA polymerase I and T4 DNA ligase, the ligation product was used to transform the yeast strain, OL971.11B (*cdc25-5 ura3*). Prototrophic Ura+ transformants were isolated at the permissive temperature, replica plated, and incubated at the restrictive temperature of 36°C. Plasmid pRG3-9 was recovered from one thermoresistant transformant strain, and the extent of the deletion in this plasmid was determined by restriction mapping and sequencing. Plasmid pRG3-9* was constructed in the following way. First, the unique *PstI* site in pRG3, situated at the junction between the YEp352 vector and the insert, was destroyed by digesting this plasmid with *PstI*, removing the resultant protruding nucleotides with T4 DNA polymerase, and recircularizing by using T4 DNA ligase. Second, the 2-kb *ClaI-XbaI* fragment of this plasmid was replaced by the deleted 1-kb *ClaI-XbaI* fragment of pRG3-9, generating plasmid pRG3-9*. Chimeric sequences in the pch1-1 and pch2-3 plasmids have been obtained by *in vivo* recombination of the partially homologous sequences *SDC25* and *CDC25* using the method based on gap repair plasmid described by Pompon and Nicolas (1989). The structures of the chimeric genes were deduced from the restriction pattern and their nucleotide sequence. Strain OL971.11B (*cdc25-5 ura3 leu2 his3 His7*) was transformed with these plasmids and the suppression of the *cdc25-5* mutation was assessed by the recovery of growth of the Ura+ transformants at 36°C. Symbols + and - indicate the growth and the absence of growth, respectively, of the transformants at 36°C.

and pRG3-9 suppressed the thermosensitivity of the *cdc25-5* mutation as well as plasmid pRG1 encoding the original C-terminal domain (Figure 4). Thus, a deletion within the *ClaI-XbaI* fragment in the N-terminal part of the SDC25 ORF is able to activate the *SDC25* gene product to suppress the *CDC25* gene defect. This result suggests that the 3'-part of the *SDC25* gene contains elements involved in the regulation of the *SDC25* gene product. Two plasmids containing chimeric genes between *SDC25* and *CDC25* have also been constructed. The first one, named pch1-1, encodes the 1,033 N-terminal amino acids of CDC25 and the 541 C-terminal amino acids of SDC25. The second one, named pch2-3, encodes the 714 N-terminal amino acids of SDC25 and the 545 C-terminal amino acids of CDC25. As shown in figure 4, pch1-1 is able to suppress the thermosensitivity of the *cdc25-5* mutation but not pch2-3. Thus, the N-terminal part of the *SDC25* gene product acts in the same way on the C-terminal parts of both CDC25 and SDC25 proteins.

In our search for a function, two disruptions of the *SDC25* gene have been constructed: the first consists of the replacement of the majority of the gene (2.3-kb-*Bam*H1 fragment) by the *HIS3* gene, the second consists of the replacement of the 0.4-kb-*Bgl*II fragment by the *HIS3* gene (Damak, Boy-Marcotte et al. 1991). Disrupted strains were viable. Thus, the *SDC25* gene is not essential for growth. None of the phenotypic modifications that have been described as associated with *cdc25*, *ras*, and *cdc35* mutations were observed in the *sdc25::HIS3* disrupted strain: the cAMP level was the same as in the wild type strain; no significant difference in glycogen accumulation was observed between disrupted and wild type strains; and growth on glycerol medium was not affected. Other phenotypes, such as generation time on fermentable and non fermentable carbon sources, cellular density in stationary phase, efficiency of sporulation, efficiency of conjugation, cryosensitivity and thermosensitivity, and secretion by measurement of the sectreted invertase, were tested: no significant differences from the wild type strain were detected. To test the possibility that the wild type *CDC25* gene suppresses the defect of the *SDC25* disruption in a pathway different from that for cAMP production, we constructed the double disruptant, *sdc25::HIS3 cdc25::HIS3*. The strain is viable in the presence of the allele *RAS2Ile-152*, which rescues the cAMP defect. Thus, a possible essential role of *SDC25*, which can be rescued by *CDC25*, seems to be excluded. However, we cannot exclude the possibility that *SDC25* is an activator of RAS in a function other than the activation of adenylate cyclase, with the use of the activated allele *RAS2Ile-152*, overcoming the requirement for *SDC25* or *CDC25*.

Thus, the *SDC25* gene appears to be dispensable for cell growth under a variety of conditions and no phenotype was detected after either disruption or overexpression of the *SDC25* gene. The lack of a phenotype for a deleted gene is often explained by the existence of redundant gene. Several examples of such redundancy already exist in the cAMP pathway: *RAS1* and *RAS2* genes are functionally interchangeable (Kataoka, Powers et al. 1984)(Tatchell, Chaleff et al. 1984), and the same is true for *TPK1*, *TPK2* and *TPK3* (Toda, Cameron et al. 1987b), for *PDE1* and *PDE2* (Nikawa, Sass et al. 1987), and most likely for *IRA1* and *IRA2* (Tanaka, Nakafuku et al. 1990). Despite these homologies, a strict redundancy between *CDC25* and *SDC25* for the activation of adenylate cyclase through *RAS* can be eliminated because (i) the lack of a functional *CDC25* gene is lethal and (ii) no difference in growth capabilities was detected in cells harboring an *SDC25* disruption in a *CDC25* disrupted background. Thus, if the *SDC25* gene product interferes with the RAS-cAMP pathway *in vivo*, it must do it under conditions that have not yet been identified. It seems more likely that *SDC25* is involved in another system as a dispensable regulator. Many small GTP binding proteins have been described in *S. cerevisiae* and these proteins might also require an exchange factor. Utilising the activity of the C-domain, the *SDC25* gene product could promote such an exchange on a ras-related GTP-binding protein, with the N-terminal part of the molecule being involved in regulation or targeting.

REFERENCES

Barbacid, M. (1987). "*ras* genes." Ann. Rev. Biochem.,. **56**: 779-827.

Boy-Marcotte, E., F. Damak, J. Camonis, H. Garreau and M. Jacquet. (1989). "The C-terminal part of a gene partially homologous to *CDC25* gene suppresses the *cdc25-5* mutation in *Saccharomyces cerevisiae*." Gene. **77**: 21-30.

Boy-Marcotte, E., H. Garreau and M. Jacquet. (1987). "Cyclic AMP controls the switch between division cycle and resting state programs in response to ammonium availability in *Saccharomyces cerevisiae*." Yeast. **3**: 85-93.

Broek, D., N. Samiy, O. Fasano, A. Fujiyama, F. Tamanoi, J. Northup and M. Wigler. (1985). "Differential activation of yeast adenylate by wild-type and mutant RAS proteins." Cell. **41**: 763-769.

Broek, D., T. Toda, T. Michaeli, L. Levin, C. Birchmeier, M. Zoller, S. Powers and M. Wigler. (1987). "The *S. cerevisiae CDC25* gene product regulates the RAS/adenylate cyclase pathway." Cell. **48**: 789-799.

Camonis, J. H. and M. Jacquet. (1988). "A new RAS mutation that suppresses the *CDC25* gene requirement for growth of *Saccharomyces cerevisiae*." Mol. Cell. Biol. **8**: 2980-2983.

Camonis, J. H., M. Kalekine, B. Gondré, H. Garreau, E. Boy-Marcotte and M.Jacquet. (1986). "Characterization , cloning and sequence analysis of the *CDC25* gene which controls the cyclic AMP level of *Saccharomyces cerevisiae*." EMBO J. **5**: 375-380.

Cohen, P. (1985). "The role of protein phosphorylation in the hormonal control of enzyme activity." Eur. J. Biochem. **151**: 439-448.

Créchet, J. B., P. Poullet, J. Camonis, M. Jacquet and A. Parmeggiani. (1990a). "Different kinetic properties of the two mutants *RAS2ile152* and *RAS2val19* that suppress the *CDC25* requirement in RAS/adenylatecyclase pathway in *S. cerevisiae*." J. Biol. Chem. **265**: 1563-1568.

Créchet, J. B., P. Poullet, M. Mistou, A. Parmeggiani, J. Camonis, E. Boy-Marcotte, F. Damak and M. Jacquet. (1990b). "Enhancement of the GDP-GTP exchange of ras proteins by the carboxyl-terminal domain of SCD25." Science. **248**: 866-868.

Damak, F., E. Boy-Marcotte, D. Le-Roscouet, R. Guilbaud and M. Jacquet. (1991). "SDC25, a CDC25-like gene which contains a Ras-activating domain and is a dispensable gene of *Saccharomyces cerevisiae*." Mol. Cell. Biol. **11**: 202-212.

Daniel, J. (1986). "The *CDC25* "Start" gene of *Saccharomyces cerevisiae*: sequencing of the active C-terminal fragment and regional homologies with rhodopsin and cytochrome P450." Curr.Genet. **10**: 879-885.

Daniel, J. and G. Simchen. (1986). "Clones from two different genomic regions complement the *cdc25* start mutation of *Saccharomyces cerevisiae*." Curr.Genet. **10**: 643-646.

Dhar, R., A. Nieto, R. Koller, D. Defeo-Jones and E. Scolnick. (1984). "Nucleotide sequence of two ras-related genes isolated from the yeast *Saccharomyces cerevisiae*." Nucl. Acids Res. **12**: 3611-3618.

Drubin, D. G., J. Mulholland, Z. Zhu and D. Botstein. (1990). "Homology of yeast actin binding protein to signal transduction proteins and myosin-I." Nature (London). **343**: 288-290.

Garreau, H., J. H. Camonis, C. Guitton and M. Jacquet. (1990). "The *Saccharomyces cerevisiae CDC25* gene product is a 180 kDa polypeptide and is associated with a membrane fraction." FEBS. **269**: 53-59.

Hughes, D. A., Y. Fukui and M. Yamamoto. (1990). "Homologous activators of ras in fission and budding yeast." Nature. **344**: 355-357.

Iida, H. and I. Yahara. (1984). "Specific early-G1 blocks accompanied with stringent response in *Saccharomyces cerevisiae* lead to growth arrest in resting state similar to the G0 of higher eucaryotes." J. Cell Biol. **98**: 1185-1193.

Jackson, P. and D. Baltimore. (1989). "N-terminal mutations activate the leukemogenic potential of the myristoylated form of c-abl." EMBO J. **8**: 449-456.

Jones, S., M. Vignais and J. R. Broach. (1991). "The CDC25 protein of *Saccharomyces cerevisiae* promotes exchange of guanine nucleotides bound to ras." Mol. Cell. Biol. **11**: 2641-2646.

Jung, G., C. L. Saxe III, A. R. Kimmel and J. A. Hammer III. (1989). "*Dictyostelium discoideum* contains a gene encoding a myosin I heavy chain." Proc. Natl. Acad. Sci. USA. **86**: 6186-6190.

Kataoka, T., S. Powers, C. McGill, O. Fasano, J. Strathern, J. Broach and M. Wigler. (1984). "Genetic analysis of yeast *Saccharomyces cerevisiae RAS1* and *RAS2* genes." Cell. **37**: 437-446.

Kato, J. Y., T. Takeya, C. Grandori, H. Iba, J. B. Levy and H. Hanafusa. (1986). "Amino acid substitutions sufficient to convert the non transforming p60c-src protein to a transforming protein." Mol. Cell. Biol. **6**: 4155-4160.

Martegani, E., M. D. Baroni, G. Frascotti and L. Alberghina. (1986). "Molecular cloning and transcriptional analysis of the start gene *CDC25* of *Saccharomyces cerevisiae*." EMBO J. **5**: 2363-2369.

Martegani, E., M. Vanoni and M. Baroni. (1986). "Macromolecular syntheses in the cell cycle mutant *cdc25* of budding yeast." Eur. J. Biochem. **144**: 205-210.

Matsumoto, K., I. Uno and T. Ishikawa. (1983a). "Control of the cell division in *Saccharomyces cerevisiae* defective in adenylate cyclase and cAMP-dependent protein kinase." Exp. Cell Res. **146**: 151-161.

Nikawa, J.-I., P. Sass and M. Wigler. (1987). "Cloning and characterization of the low-affinity cyclic AMP phosphodiesterase gene of *Saccharomyces cerevisiae*." Mol. Cell. Biol. **7**: 2629-2636.

Pompon, D. and A. Nicolas. (1989). "Protein engineering by cDNA recombination in yeasts: shuffling of mammalian cytochrome P-450 functions." Gene. **83**: 15-24.

Powers, S., T. Kataoka, O. Fasano, M. Goldfarb, J. B. Strathern, J. Broach and M. Wigler. (1984). "Genes in *S. cerevisiae* encoding proteins with domains homologous to the mammalian ras proteins." Cell. **36**: 36: 607-612.

Rey, I., F. Schweighoffer, I. Barlat, J. Camonis, E. Boy-Marcotte, R. Guilbaud, M. Jacquet and B. Tocqué. (1990). "The COOH-domain of the product of the *Saccharomyces cerevisiae SDC25* gene elicits activation of p21 ras proteins in mammalian cells." Oncogene.

Tanaka, K., M. Nakafuku, T. Satoh, M. S. Marshall, J. B. Gibbs, K. Matsumoto, Y. Kaziro and A. Toh-e. (1990). "*S. cerevisiae* genes *IRA1* et *IRA2* encode proteins that may be functionally equivalent to mammalian ras GTPase activating protein." Cell. **60**: 803-807.

Tatchell, K., D. T. Chaleff, D. DeFeo-Jones and E. M. Scolnick. (1984). "Requirement of either of a pair of ras related genes of *Saccharomyces cerevisiae* for spore viability." Nature. **309**: 523-527.

Toda, T., S. Cameron, P. Sass, M. Zoller, J. D. Scott, B. McMullen, M. Hurwitz, E. B. Krebs and M. Wigler. (1987a). "Cloning and characterization of *BCY1*, a locus encoding a regulatory subunit of cyclic AMP-dependent protein in *Saccharomyces cerevisiae*." Mol. Cell. Biol. **7**: 1371-1377.

Toda, T., S. Cameron, P. Sass, M. Zoller and M. Wigler. (1987b). "Three different genes in *S. cerevisiae* encode the catalytic subunit of the cyclic AMP-dependent protein kinase." Cell. **50**: 277-287.

Vanoni, M., M. Vavassori, G. Frascotti, E. Martegani and L. Albefghina. (1990). "Overexpression of the *CDC25* gene, an upstream element of the RAS/Adenylyl cyclase pathway in *Saccharomyces cerevisiae*, allows immunological identification and characterization of its gene product." Boichem. Biophys. Res. Commun. **172**: 61-69.

West, M., H. Kung and T. Kamata. (1990). "A novel membrane factor stimulates guanine nucleotide exchange reaction of ras proteins." FEBS Lett. **259**: 245-248.

Wolfman, A. and I. G. Macara. (1990). "A cytosolic protein catalyzes the release of GDP from p21ras." Science. **248**: 67-69.

INDEX

DATE DUE

DATE DUE			
MAY 1 5 2001			
DEC 1 6 2004			